Frontiers in Biotransformation
Vol. 2

Frontiers in Biotransformation   Volume 2

# Principles, Mechanisms and Biological Consequences of Induction

Edited by
Klaus Ruckpaul and Horst Rein

Academy of Sciences
GDR

Taylor & Francis
London, New York and Philadelphia 1990

UK      Taylor & Francis Ltd., 4 John St., London WC1N 2ET

USA     Taylor & Francis Inc., 242 Cherry St., Philadelphia, PA 19106—1906

**British Library Cataloguing in Publication Data**

Principles, Mechanisms and Biological Consequences of Induction
(Frontiers in Biotransformation; v. 2).
    1. Organisms. Cytochrome P-450
    I. Ruckpaul, Klaus II. Rein, Horst III. Series
    574.19

    ISBN 0-85066-799-2


**Library of Congress Cataloging-in-Publication Data is available**

Printed in the GDR

# Contents

# Introduction

H. REIN and K. RUCKPAUL

Volume 2 of the series "Frontiers in Biotransformation" is focussed on the induction of enzymes involved in biotransformation. Induction as an effector mediated stimulation of the biosynthesis of distinct enzymes represents a regulation mechanism of enzymatic activity which is controlled on the cellular level. Thus the content of volume 2 continues and extends the scope of volume 1, which dealt with problems of regulation of cytochrome P-450 catalysed reactions on the molecular and membrane level of integration.

Biological and medical consequences of induction resulting in changed biotransformation activities are manifold and important for toxification and detoxification as well. In addition, inducing agents have become to be seen as promotors and may thus initiate malignant transformation. Therapeutic effects can be altered by inducing drugs and in this way, e.g., the protective effect of contraceptive steroids may vanish. Induction modifies the isozymic pattern, which originates from the genetically determined polymorphism. The changed isozymic pattern determines their individual capability of metabolizing endogenous and exogenous compounds alike.

A huge number of articles dealing with the induction of enzymes involved in biotransformation have generated a real need for reviews. Therefore the interest of the editors in selecting the contributions for this volume was not to look for outstanding casuistic articles but rather to put together such papers which would comprehend and condense problems more general in character. Volume 2 is introduced by an article by PARKE, summarizing the general principles and biological consequences of induction. More detailed studies of molecular orbitals of inducers have revealed that the type of cytochrome P-450 induced is governed by the spatial dimensions and electronic structures of the inducing agents. Corresponding to the different types of inducers there exist different mechanisms of induction. An increase of the monooxygenase concentration but also of the NADPH dependent cytochrome P-450 reductase and further enzymes is observed at induction by the phenobarbital type inducers. Moreover, proliferation of the endoplasmic reticulum and cellular hypertrophy of liver is observed. Obviously, phenobarbital increases both the transcriptional and translational activity.

The amino acid sequences of about 100 cytochromes P-450 led to a proposal for a unified nomenclature of the P-450 gene superfamily described by NEBERT and GONZALEZ in this volume. Most members of these families are inducible, with the assignment to a distinct family being mainly determined by the degree of homology but also by the specific inducer. The considerable increase in molecular biological studies on cytochrome P-450 in the last few years has led to a better understanding of constitutive and inducible gene expression of these key enzymes of biotransformation. The identification and characterization of cis-acting DNA regulatory elements, and the cloning of genes that encode trans-acting regulatory proteins interacting with these elements, are important steps in our knowledge of gene expression as well as of the developmental, sex and tissue specificities.

In the review of BRESNICK and HAUSER the polycyclic hydrocarbon specific cytochrome P-450 is described in detail. The gene of cytochrome P-450c was isolated and its DNA sequence determined. By genetic construction of specific gene segments their importance for interaction with the 4S polycyclic hydrocarbon-binding protein could be shown. The authors suggest that this protein could be the transcriptional factor which shuttles between the cytosolic compartment and the nucleus.

Already very early a sex-specific effect of inducers was observed, especially concerning the sex-related metabolism, such as hexobarbital hydroxylation and aminopyrine N-demethylation. The molecular basis of these differences is now understood in terms of a sex-specific induction of cytochrome P-450 isozymes in female and male animals. In the review of KATO and YAMAZOE not only this aspect of sex-dependent regulation of cytochrome P-450 expression is described but also a detailed analysis of sex-related cytochrome P-450 forms is given. The authors developed an immunoquantitative method for quantitation of cytochrome P-450-male and cytochrome P-450-female. These sex-specific forms clearly differ from each other in catalytic activity, molecular weight, physico-chemical properties and amino acid sequence too. Their specificity in drug oxidation was well documented by use of highly specific antibodies. Such inhibition studies offer an excellent method for classification of isozyme specific contributions in drug metabolism.

Doubtless the ontogenetic development of biotransformation activity is regulated and, or moreover, determined by the induction since it is known that the administration of phenobarbital to neonates has been shown to "imprint" or "programme" the microsomal enzymes of the adult. Distinct forms of cytochrome P-450 are induced in the foetal liver after administration of selected inducers to pregnant rats and treatment of neonates with inducers or their uptake via the maternal milk increases distinct cytochrome P-450 dependent enzyme activities. These facts are considered in the review of

KLINGER describing the biotransformation during the ontogenetic development. Although the total cytochrome P-450 concentration is a function of age, evidently, this enzyme concentration is responsible neither for the age dependence of drug oxidation, nor for the age dependent induction by the inducers of the phenobarbital and 3-methylcholanthrene group. Several reasons determine this behaviour. Different developmental patterns are observed for the multiple forms of cytochrome P-450. For the activity of a relevant isozyme, however, developmental changes in the microenvironment of the enzyme are of importance, i.e. phospholipids and their fatty acid composition which change with age, thereby changing structural and functional properties of the membrane and the cytochrome P-450 reducibility likewise. This knowledge is of importance to understand the prolonged half-life times in blood for most drugs in human neonates and to avoid therapeutic accidents as observed after chloramphenicol administration in newborns.

From the toxicological point of view, postoxidation reactions of xenobiotics are of importance because by these reactions, not in all cases highly water soluble metabolite of decreased pharmacological activity but also more toxic products are formed. Examples of such activation pathways are given in the chapter of LANGNER, BORCHERT and PFEIFER who describe the sulfo-, methyl-, and acetyltransferases in detail, including a lot of substrates which are conjugated by these enzymes. The fact that postoxidation enzymes are also inducible by different inducers is of interest in medicinal practice because enhanced bilirubin blood levels can be normalized by induction of glucuronosyltransferase which conjugates bilirubin. Instead of phenobarbital with many undesired side effects, the development of specific inducers for glucuronosyltransferase is necessary which requires deep insight into its structure and regulation. Definite data currently available for rat cDNAc are the basis of the review of MACKENZIE dealing with the analysis of structural features, substrate preferences and regulation of glucuronosyltransferases predominantly of this species.

One of the main conjugation pathways is catalysed by glutathione transferases. These multifunctional enzymes are associated with the detoxification of drugs and carcinogens, the reduction of organic hydroperoxides and nitrate, the binding and intracellular transport of lipophiles and the biosynthesis of the local hormone leucotriene C. In the review of KETTERER and TAYLOR the reader will not only find a broad account of reactions with GSH which are catalysed by GSH transferases but the current situation regarding the distribution of these enzymes in the animated world, their structure and, finally, what is known at present of their genetic regulation.

In a last chapter THOMAS and OESCH deal with the most recent findings about the enzyme epoxide hydrolase which catalyses the hydration of epoxides. Normally this reaction is considered a step of detoxification. However, in the special case when dihydrodiol is formed from benzo(a)pyrene 7,8-oxide, a further activation by a monooxygenase takes place resulting in the ultimate carcinogen, i.e. benzo(a)pyrene 7,8-diol-9,10-oxide. Biotransformation of foreign compounds is a complex process including steps of both toxification and detoxification. Clearly, the balance between the toxification and detoxification pathways depends on several factors; most importantly the genetically determined concentration of individual enzymes involved in biotransformation steps and their inducibility, too.

# Chapter 1
# Induction of Cytochromes P-450 –
# General Principles and Biological Consequences

D. V. PARKE

# 1. Introduction

The cytochromes P-450, a ubiquitous family of enzymes which have been detected in all living systems examined, function as mixed-function oxygenases involved in the biogenesis of sterols, steroid hormones and prostanoids, the oxidation of fatty acids and steroids, and the metabolic oxygenation of a multitude of diverse exogenous xenobiotic chemicals. These chemicals range from the polycyclic aromatic hydrocarbons from combustion of fossil fuels to the products of human ingenuity for creating chemicals of high social value, such as the polychlorinated biphenyls (electrical insulators) or the benzodiazepine drugs (hypnotics). The cytochromes P-450 were among the first mammalian enzymes discovered to exhibit substrate-induced genomal regulation of their enzyme activity (enzyme induction), which until 30 years ago had been considered to be a phenomenon confined to microorganisms. In 1956 CONNEY et al. showed that treatment of animals with the polycyclic aromatic hydrocarbon, 3-methylcholanthrene, increased their ability to metabolize methylated azo dyes, and REMMER and ALSLEBEN (1958) found that tolerance to the barbiturate drugs was the result of these drugs enhancing their own metabolism, by induction of the cytochromes P-450.

These two distinctly different examples of induction of the cytochromes P-450 were associated with two different forms of this enzyme, which at that time were termed 'cytochrome P-450' (induced by phenobarbitone) and 'cytochrome P-448' (induced by carcinogenic polycyclic hydrocarbons), the names being derived from the wavelength (450 and 448 nm) of their reduced CO-ligand absorption spectra. Although initially it was considered that cytochrome P-448 was merely a derivative of cytochrome P-450 in which the polycyclic hydrocarbon inducing agent had become irreversibly bound to the enzyme molecule, later studies in which the liver microsomal enzyme was solubilized and rigorously purified revealed that these were two distinctly different enzyme proteins, under separate genetic control. Today, the cytochromes P-450 are considered to comprise a superfamily of enzymes, many of which have been shown to exhibit induction following exposure of the intact animal to a typical substrate or certain chemical agents. Studies of the amino acid structures and gene regulation of the many forms of cytochrome P-450 have revealed valuable information concerning the structural and functional relationships of the members of this superfamily of enzymes, but unfortunately have also resulted in a confused variety of different nomenclatures. This has led to some obfuscation of the functional differences of these isozymes and of the differences in their regulation. Recent endeavours to clarify the problem has resulted in the publication of new recommendations for cytochrome P-450 nomenclature, based on the evolution of these enzymes (NEBERT and GONZALEZ, 1987; NEBERT et al., 1987). This new nomenclature will be used in the present chapter, in addition to the more distinctive, familiar, original terminology (see Table 1).

Evidence is now available for the existence of eight mammalian P-450 gene families (I, II, III, IV and the steroid hydroxylases XI, XVII, XIX and XXI), and of at least five sub-families within the P-450 II family (NEBERT and GONZALEZ, 1987; NEBERT et al., 1987). The P-450 I gene family of polycyclic aromatic hydrocarbon-inducible cytochromes comprises only one sub-family with only two genes, A1 and A2. The P-450 II gene family comprises P-450 II A, inducible by 3-methylcholanthrene (MC) and possibly also by phenobarbital (PB), a number of PB-inducible forms (P-450 II B and C — some non-inducible), the ethanol-inducible form, II E, and form II D. The P-450 III family comprises one sub-family (A1 and A2) of steroid-inducible genes which are also induced by phenobarbital and by macrolide antibiotic

**Table 1.** Cytochromes P-450 and their induction

| Isozyme | | Typical inducing agent | Specific enzymic activity induced |
|---|---|---|---|
| **New nomenclature** | **Original name** | | |
| P-450 I A1 | P-448 (rat P-450c, rabbit LM6, mouse $P_1$-450) | 3-methyl-cholanthrene | 7-ethoxyresorufin-O-deethylase |
| P-450 I A2 | P-448 (rat P-450d, rabbit LM4, mouse $P_3$-450) | 3-methyl-cholanthrene, isosafrole | 7-ethoxyresorufin-O-deethylase, phenacetin O-deethylase |
| P-450 II A1 | P-450 (rat P-450a) | 3-methyl-cholanthrene | testosterone-7-hydroxylase |
| P-450 II B1 | P-450 (rat P-450b) | phenobarbital | 7-pentoxyresorufin-O-deethylase |
| P-450 II B2 | P-450 (rat P-450e) | phenobarbital | 7-pentoxyresorufin-O-deethylase |
| P-450 II C1—10 | (rat P-450, PB1 and P-450f) | phenobarbital | — |
| P-450 II D1 & 2 | P-450 | — | debrisoquine-4-hydroxylase |
| P-450 II E | (rat P-450j) | ethanol, isoniazid | aniline hydroxylase |

**Table 1.** (continued)

| Isozyme | | Typical inducing agent | Specific enzymic activity induced |
| --- | --- | --- | --- |
| New nomenclature | Original name | | |
| P-450 III A1 | P-450 (rat pcn 1) | pregnenolone-16α-carbonitrile | aminopyrine-N-demethylase, ethylmorphine-N-demethylase |
| P-450 III A2 | P-450 (rat pcn 2) | pregnenolone-16α-carbonitrile | aminopyrine-N-demethylase, ethylmorphine-N-demethylase |
| P-450 IV | P-452 | clofibrate | lauric acid $\omega$ and $\omega-1$ hydroxylase |
| P-450 XI A1 | bovine and human scc | | |
| P-450 XI B1 | bovine and human 11$\beta$ | | |
| P-450 XVII A1 | bovine and human 17α | | |
| P-450 XIX A1 | human aromatase | | |
| P-450 XXI A1 | bovine, murine and human | | |
| P-450 XXI A2 | steroid 21-hydroxylase | | |
| P-450 LI A1 | yeast 1 an | | |
| P-450 CI A1 | *Pseudomonas putida* cam | | |

and the clofibrate-inducible P-450 IV comprises yet another family, with probably two or three genes (NEBERT et al., 1987). The P-450 XI, XVII, XIX and XXI familes are the proteins involved in steroidogenesis, and the P-450 LI and CI families comprise the yeast and *Pseudomonas* cytochromes respectively. Several other unique P-450 gene families, including those comprising the microsomal cholesterol $7\alpha$-hydroxylase and the renal mitochondrial 25-hydroxyvitamin $D_3$-$1\alpha$-hydroxylase, remain to be cloned and characterized (NEBERT and GONZALEZ, 1987). From considerations of the structures of the P-450 I and P-450 II gene families, the amino acid residue numbers at each exon-intron junction, the location in the coding triplets in which the exons are split, and homologies among introns and exons, it has been deduced that these two gene families probably diverged from a common ancestor more than 200 million years ago and that P-450 I A1 and A2 genes split from each other about 65 million years ago (GONZALEZ et al., 1985).

The cytochrome P-450-mediated mixed-function oxidation of xenobiotics (drugs, pesticides, industrial chemicals, etc.) involves primarily the P-450 I − IV families, and is concerned with detoxication, namely, the oxidative metabolism of xenobiotics to more-polar, biologically-inactive, readily-excretable metabolites. However, the same microsomal enzymes can also catalyse the oxidative formation of reactive intermediates and ultimate carcinogens, leading to covalent binding, DNA damage, mutations and malignancy, and other pathological processes. This catalysis of the opposing pathways of detoxication and activation of chemical carcinogens by the same group of enzymes has been described by GELBOIN (1983) as the paradox of chemical carcinogenesis and cancer. However, chemical carcinogenesis has long been associated with the specific induction of the cytochromes P-448 (P-450 I) (CREAVEN and PARKE, 1966), and more recently it has been shown that the P-448 family of enzymes is specifically concerned with the activation of chemicals and carcinogens to reactive intermediates and the formation of mutagens and carcinogens (IOANNIDES et al., 1984).

The present review will be limited to considerations of the induction of the P-450 I − IV families of microsomal cytochromes P-450, and will not be concerned with induction of the mitochondrial cytochromes P-450 of steroidogenesis (XI, XVII, XIX and XXI), or the induction of the cytochromes P-450 of yeast (LI) or of *Pseudomonas putida* (CI).

## 2.  Induction of specific cytochromes P-450

### 2.1.  Cytochromes P-448 (P-450 I)

The cytochromes P-448 are induced by planar molecules, which are their preferred substrates, such as the polycyclic aromatic hydrocarbons, e.g. benzo(a)pyrene and 3-methylcholanthrene (MC) (THOMAS et al., 1983; IOANNIDES et al., 1984), and by planar polyhalogenated biphenyls (PARKINSON et al., 1983) such as 3,3′,4,4′,5-pentachlorobiphenyl (OZAWA et al., 1979) and 2,3,3′,4,4′,5-hexabromobiphenyl (ROBERTSON et al., 1981), by 3,3′,4,4′-tetrachloroazobenzene (HSIA and KREAMER, 1979), aminoazobenzenes (DEGAWA et al., 1985), the anticancer drug, ellipticine (CRESTEIL et al., 1982) and the muscle relaxant, dantrolene (JAYYOSI et al., 1987), and to a lesser extent by aromatic amines (IOANNIDES et al., 1984) and amides (ASTRÖM and DE PIERRE, 1985; IWASAKI et al., 1986).

The most potent inducer of cytochrome P-448 is 2,3,7,8-tetrachlorodibenzo-p-dioxin (TCDD) a contaminant of the herbicide 2,4,5-trichlorophenoxyacetic acid which, because of its resistance to biodegradation, is a persistent environmental contaminant (POLAND and GLOVER, 1974). TCDD is $10^4$ times more potent than MC in the induction of P-448. Of the analogues of TCDD, 2,3,7,8-tetrabromodibenzo-p-dioxin has 50% of the potency of TCDD, 1,2,3,4,7,8-hexachlorodibenzo-p-dioxin has 10%, and all other analogues had markedly lower activities ranging from 10% to $10^{-7}$ that of TCDD (BRADLOW et al., 1980). TCDD induces P-448 in human lymphocytes and in a number of different cell lines derived from a variety of animals (NIWA et al., 1975). Induction of P-448 by TCDD is tissue specific, and in rabbit liver it induces forms LM6 (P-450 I A1) and LM4 (P-450 I A2), but in lung and kidney only LM6 is induced (LIEM, 1980). TCDD markedly induces both isozymes of P-448 (P-450c and d) in rat liver, but only P-450c (P-450 I A1) in extrahepatic tissues; induction of P-450c is greatest in liver > kidney > lung > intestine > spleen > testes (GOLDSTEIN and LINKO, 1984).

β-Naphthoflavone, a potent non-carcinogenic inducer of cytochrome P-448, increases this enzyme in mouse liver by a mechanism similar to that evoked by MC (BOOBIS et al., 1977). Cigarette smoke is also a potent inducer of P-448 in lung and kidney, but not liver, of rats, hamsters and guinea pigs (BILIMORIA and ECOBICHON, 1980), and in human placenta (WELCH et al., 1968); placental P-448 has also been induced in humans by exposure to polychlorinated biphenyls (WONG et al., 1986). 1,1-Dichloroethylene, a plastics copolymer, also resulted in a tissue specific induction of P-448, increasing the levels in mouse kidney, especially in the male, but not in mouse liver (KRIJGSHELD and GRAM, 1984). A novel form of P-448, induced in hamster liver by MC, and distinct from the MC-induced rat liver P-448, has been purified and shown to have high specificity for aflatoxin $B_1$, which it activates to a mutagen 50 times more effectively than does rat liver P-448 (MIZOKAMI et al., 1986).

The cytochromes P-448 may be quantified specifically by the 7-ethoxyresorufin (EROD) assay (PHILLIPSON et al., 1984), and are induced in liver, kidney and lungs of rats, hamsters, guinea pigs and mice by a variety of carcinogens and non-carcinogenic inducing agents (IWASAKI et al., 1986). The MC-induced pulmonary cytochrome P-448 isolated from rats is structurally identical to the rat liver enzyme (ROBINSON et al., 1986). The mixed-function oxidases of the intestines are induced in rat by β-naphthoflavone (LINDESKOG et al., 1986), and by various chemicals of the diet, for example, indole-3-carbinol and ascorbigen, metabolites of glucobrassicin present in cabbage and other vegetables, which induce EROD of rat intestinal mucosa and liver

(McDanell et al., 1987). TCDD and $\beta$-naphthoflavone also induce EROD activity in the smooth muscle of rabbit aorta, which may be associated with the role of carcinogens and mutations in atherosclerosis (Serabjit-Singh et al., 1985; Paigen et al., 1986).

Carcinogenic primary aromatic amines (2-aminofluorene, aminobiphenyl), secondary aromatic amines (N-methyl-4-aminoazobenzene) (Kimura et al., 1985), and the arylamide, 2-acetylaminofluorene (2-AAF) (Lotlikar et al., 1984), are metabolically activated to mutagens and proximate carcinogens only by the high spin form of cytochrome P-448 (P-448 H, P-450d, or P-450 I A2). This isozyme of P-448 is selectively induced by the azo dyes, such as 3-methoxy-4-aminoazobenzene and several other methyl and methoxyl derivatives of 4-aminoazobenzene (Degawa et al., 1986), and by the 'mixed' P-448 and P-450 inducer, hexachlorobenzene (Linko et al., 1986). Isosafrole is also a selective inducer of P-448 H, but as the induced cytochrome forms a complex with isosafrole metabolites, the induced enzyme is largely inactivated.

**Table 2.** Selective induction of rat liver cytochromes P-450

| Inducing agent | Type of inducer | Cytochromes induced | | | |
|---|---|---|---|---|---|
| | | P-450 I A1 (P-450c) | P-450 I A2 (P-450d) | P-450 II A1 (P-450a) | P-450 II B1 & B2 (P-450b & e) |
| | | (ratio of induced/control) | | | |
| Phenobarbital (PB) | PB | 1 | 1 | 2 | 40 |
| 3-Methylcholanthrene (MC) | MC | 72 | 11 | 4 | 1 |
| Isosafrole | MC? | 19 | 22 | 2 | 13 |
| 2,2',4,4',5,5'-Hexachlorobiphenyl | PB | 1 | 1 | 2 | 73 |
| 3,3',4,4',5,5'-Hexachlorobiphenyl | MC | 43 | 40 | 8 | 1 |
| 2,3,3',4,4',5-Hexachlorobiphenyl | mixed | 43 | 10 | 6 | 47 |
| Aroclor 1254 | mixed | 50 | 20 | 4 | 45 |

Immature rats were pretreated with the different inducing agents and the specific cytochromes P-450 were quantified by radical immunodiffusion using monospecific antibodies (data from Conney, 1986).

Activation of 2-AAF to its ultimate carcinogen by N-hydroxylation, is induced by MC or by repeated dosage with 2-AAF. Treatment of rats with 2-AAF, unlike MC, did not induce P-448, nor did it induce NADPH-cytochrome c reductase, NADPH-cytochrome P-450 reductase, or amine oxidase, so that the nature of this type of enzyme induction is unknown (MALEJKA-GIGANTI et al., 1978).

Some inducing agents, such as the methylenedioxyphenyl compounds, safrole and isosafrole (THOMAS et al., 1983; IOANNIDES et al., 1985), hexachlorobenzene (LINKO et al., 1986) and many polyhalogenated biphenyls (Aroclors) (PARKINSON et al., 1983), and the drugs chlorpromazine and phenothiazine (THOMAS et al., 1983) act as mixed-type inducers increasing the synthesis of both cytochromes P-448 and P-450 (P-450 I and II) (see Table 2). A screening procedure for the characterization of different types of inducing agent, based on the O-dealkylation of a series of alkoxyresorufins, has been proposed; ethoxyresorufin O-deethylation (EROD) is highly selective for P-448 and pentoxyresorufin O-dealkylation is selective for PB-induced P-450s (BURKE et al., 1985).

## 2.2. Cytochrome P-450a (P-450 II A1)

Cytochrome P-450a is a constitutional enzyme in rat liver, developmentally and sex-specifically regulated. The enzyme develops early after birth and is then suppressed in the adult male rat. It is characterized by its testosterone $7\alpha$-hydroxylase activity and is induced by MC but not by PB (NAGATA et al., 1987) although other workers have also reported induction by PB (WAXMAN et al., 1985).

## 2.3. Cytochromes P-450$_{PB}$ (P-450 II B and C)

Phenobarbital (PB) induces two major forms of cytochrome P-450, namely P-450b (P-450 II B1) and P-450e (P-450 II B2). In addition a number of constitutive cytochromes P-450 only marginally induced by phenobarbital (PB$_{1a,b,c}$, PB$_{2a,b,c,d}$, or P-450 II C subfamily) have been isolated from liver of rats treated with PB. These have been characterized by their N-terminal amino acid sequences, immunological reactivities and their selective enzyme activities to a series of alkoxyresorufins (WOLF et al., 1986); the O-dealkylation of pentoxyresorufin is a rapid, sensitive and specific assay for determination of the induction of cytochromes P-450 by phenobarbital and similar inducing agents (LUBET et al., 1985; BURKE et al., 1985). Analysis of mRNA of rabbit liver identified three cytochromes P-450 of the P-450 II C subfamily, namely: PBc1 — not detectable in liver of untreated rabbits but induced in liver by

PB, PBc2 — present in control liver and kidney and induced 3-fold in both tissues by PB, and PBc3 — present in control liver only but not induced by PB (LEIGHTON and KEMPER, 1984); absence of PBc1 mRNA before treatment with PB indicates that an increase in gene transcription is involved in the PB induction of this enzyme.

Many other drugs, such as phenytoin and other anticonvulsants, warfarin (KLING et al., 1985) and proadifen (SKF 525-A) (BORNHEIM et al., 1983) induce the PB forms of P-450. Numerous pesticides and other xenobiotics also induce the PB forms of P-450 in various animal species including man. These include p,p'-DDT (1,1,1-trichloro-2,2-bis[p-chlorophenyl]ethane), its lipophilic metabolite DDE (1,1-bis[p-chlorophenyl]-2,2-dichloroethylene), chlordane (1,2,4,5,6,7,8,8-octachloro-3a,4,7,7a-tetrahydro-4,7-methanoindane), lindane (γ-hexachlorocyclohexane) and dieldrin, and the insecticides synergists, piperonyl butoxide and sesamex (FABACHER et al., 1980). DDE has a more persistent inducing effect than PB on rat liver P-450 and its mRNA (MORO-HASHI et al., 1984). However, from a study of the inductive effects of various pesticides on the various testosterone and benzo(a)pyrene hydroxylases it is evident that the organochlorine pesticides p,p'-DDE, dieldrin, heptachlor, chlordane, toxaphene and lindane induce not only the major PB-inducible cytochromes P-450b and P-450e, but also induce additional forms of the cytochrome (HAAKE et al., 1987).

PB may also induce cytochrome P-450 in primary cultures of rat hepatocytes, contrary to earlier reports which indicated that induction of cytochrome P-450 occurred only in vivo in the whole animal. The reason for this inconsistency is that PB induction of P-450 requires five days whereas MC induction of P-448 in cultured hepatocytes occurs within 48 hours (MICHALOPOULOS, 1976).

There are other major differences in the induction of the microsomal mixed-function oxidases by PB and the polycyclic hydrocarbons quite apart from the nature of the specific cytochromes P-450 induced. PB also induces cytochrome P-450 reductase and a variety of proteins and phospholipids, resulting in a marked proliferation of the endoplasmic reticulum. In contrast, the polycyclic hydrocarbons and other specific inducers of the P-448s yield no increase in the reductase and no significant proliferation of the endoplasmic reticulum. However, in dog, although PB induction was qualitatively and quantitatively similar to that in rat, induction by β-naphthoflavone resulted in smaller increases in P-448-dependent EROD activity than seen in rat and also increased NADPH-cytochrome c reductase activity, not observed in rat with P-448-type inducers (McKILLOP, 1985). Differences in the tissue distribution of enzyme induction also occur, and in rabbits PB increased only LM2 in liver and kidney, whereas TCDD increased both LM4 and LM6 in liver, kidney and lung (DEES et al., 1982).

A number of halogenated aromatic hydrocarbons are 'mixed' inducing agents increasing the activity of cytochromes P-448 (P-450 I), P-450 (P-450 II) or both, dependent on the positions of the halogen atoms and the degree of coplanarity of the congener molecule (see Table 2). The fungicide, hexachlorobenzene is a true 'mixed' type inducer, increasing both P-448 and P-450 (GUTKINA and MISKIN, 1986), whereas all lower chlorinated benzenes are PB-type inducers (LINKO et al., 1986). Of the polychlorinated biphenyls (Aroclors), the 2,3,4,4',5-pentachloro congener is a mixed inducing agent, whereas the 2,2',3,4,4',5-hexachloro and 2,2',3,4,4',5,6-heptachloro induce only the PB-inducible cytochromes P-450, (DENOMME et al., 1983), and 3,3',4,4',5-pentachloro induces only P-448 (OZAWA et al., 1979). Octachlorostyrene, a byproduct in the manufacture of many chlorinated hydrocarbons and a major environmental contaminant, induces only the PB-inducible forms of P-450 (HOLME and DYBING, 1982).

The sex-specific forms of cytochrome P-450 that have been isolated from male and female rats are responsible for the sex differences in microsomal oxidations in this species and which are induced by testosterone and estradiol respectively, and were unchanged or even decreased by treatment with phenobarbital, 3-methylcholanthrene, or the mixed inducing agent, the polychlorinated biphenyls, although testosterone 16$\alpha$-hydroxylase activity characteristic of the male P-450 was increased by the PB induction of cytochrome P-450b (P-450 II B1) (KAMATAKI, 1986).

## 2.4.  Cytochrome P-450$_{ALC}$ (P-450 II E)

Chronic administration of ethanol to rats is associated with the appearance of a novel form of cytochrome P-450 showing structural and catalytic properties different from those of the constitutive cytochromes P-450 and PB- and MC-inducible forms of the enzyme (JOLY et al., 1977). The rat enzyme, P-450j, is induced by ethanol and acetone, and has been isolated and characterized after induction with isoniazid (RYAN et al., 1985). This form of cytochrome P-450 (P-450$_{ALC}$ or P-450 3a) has been isolated also from rabbit liver after ethanol administration (KHANI et al., 1987). It is also induced by a variety of other structurally-unrelated chemicals including, in order of increasing potency, isoniazid, trichloroethylene, pyrazole, ethanol, imidazole and acetone (COON and KNOOP, 1987), is characterized by its high ability for the O$_2$-dependent oxidation of ethanol, other alcohols and aniline, and is also concerned in the oxidative activation of paracetamol, carbon tetrachloride and the carcinogen, dimethylnitrosamine. Cytochrome P-450$_{ALC}$ of rabbit liver is 100-fold more effective than either PB-induced P-450 (LM2) or MC-induced P-448 (LM4) in the reductive dehalogenation of CCl$_4$ (JOHANSSON and INGELMAN-SUNDBERG, 1985).

This alcohol-induced cytochrome exhibits tissue specificity in its induction and, in rabbit, the liver and kidney, but not lung and other extrahepatic tissues, is responsive to the specific inductive effects of alcohol (UENG et al., 1987).

Rat liver dimethylnitrosamine (DMN) N-demethylation is increased by fasting, due to the increase of a low $K_m$ form of the enzyme, which is also increased in diabetes and by treatment with acetone, ethanol, isopropanol and pyrazole (HONG et al., 1987). Fasting of rats for 24 and 48 hours caused 60 and 116% increases in the DMN N-demethylase which were accompanied by similar increases of cytochrome P-450j and of the corresponding mRNA, indicating that the enzyme induced by diabetes and fasting is probably identical with the cytochrome induced by ethanol (HONG et al., 1987).

## 2.5. Cytochromes P-450$_{PCN}$ (P-450 III)

Rats treated with the synthetic "catatoxic" steroid, pregnenolone-16$\alpha$-carbonitrile (PCN) are resistant to the toxicity of many drugs and chemicals, including carbon tetrachloride, and are less susceptible to the occurrence of DMN-induced cancer. This is attributed to the specific induction of a cytochrome P-450$_{PCN}$, a family of two or more proteins which have been isolated and purified from PCN-treated rats (ELSHOURBAGY and GUZELIAN, 1980; HOSTETLER et al., 1987). P-450$_{PCN}$ also exhibits aldrin epoxidase activity and in this respect is more active than the PB-induced P-450 (NEWMAN and GUZELIAN, 1983). P-450$_{PCN}$ is regulated by the endogenous inducer, corticosterone (HOSTETLER et al., 1987), and is induced in rats also by treatment with the synthetic steroids, dexamethasone and spironolactone, with the macrolide antibiotic, triacetyloleandomycin and the rifamycin antibiotic rifampicin (WRIGHTON et al., 1985), and with PB, chlordane and other PB-type inducers but not with MC (HEUMAN et al., 1982). Many other steroids, including glucocorticoids, androgens, estrogens and progestogens fail to induce P-450$_{PCN}$, and it has been demonstrated that induction is not associated with events mediated by the classical glucocorticoid receptor (HEUMAN et al., 1982). P-450$_{PCN}$ activity is not detectable in untreated rats and accounts for $< 2\%$ of the total constitutive cytochromes P-450, whereas after treatment with PCN the level of P-450$_{PCN}$ increases 20-fold (HEUMAN et al., 1982).

## 2.6. Cytochrome P-452 (P-450 IV)

Several structurally unrelated industrial plasticizers and hypolipidaemic drugs such as di(2-ethylhexyl)phthalate (DEHP), clofibrate (ethyl p-chlorophenoxy-isobutyrate), methyl clofenapate (methyl 2-[4-(p-chlorophenyl) phenoxy]-2-methylpropionate), and tibric acid (2-chloro-5-[dimethylpiperidinosulphonyl] benzoic acid), administered to rats and mice result in hepatomegaly, marked proliferation of hepatic peroxisomes, and a specific induction of liver microsomal cytochrome P-452 (HESS et al., 1965; SVOBODA et al., 1967; REDDY and KRISHNAKANTHA, 1975; and TAMBURINI et al., 1984).

Among the constitutive cytochromes P-450 are two or three enzymes which metabolize medium-chain-length fatty acids (C6—C12) by $\omega$ or $\omega - 1$ oxidation, producing primary and secondary alcohols which are further oxidized to dicarboxylic acids and $\omega - 1$ oxo-fatty acids respectively (ORTIZ DE MONTELLANO and REICH, 1984). The dicarboxylic acids are further catabolized by $\beta$-oxidation and although this is only a minor pathway of metabolism for unbranched fatty acids, except during starvation and ketosis, for branched chain or substituted fatty acids it is the major pathway. A laurate $\omega$-hydroxylase, cytochrome P-452, has been isolated from the liver of rats treated with clofibrate, and has been purified and characterized (TAMBURINI et al., 1984). P-452 is present in non-induced rat liver and kidney and its level is markedly increased (up to 30-fold) after administration of clofibrate, and probably other hypolipidaemic drugs and industrial ester plasticizers. P-452 has less than 35% similarity of cDNA nucleotide and amino acid sequences with cytochromes P-450c and P-450d (P-450 I), P-450e (P-450 II B2) and P-450$_{PCN}$ (P-450 III) indicating that it is a member of a new P-450 gene family and probably contains two or three genes (HARDWICK et al., 1987). This family of cytochromes P-450, present in a wide variety of tissues and cell types has also been shown to be involved in the metabolism of arachidonic acid (BAINS et al., 1985) and may thus be concerned in the biogenesis of the prostanoids. Simultaneous with the clofibrate-induced gene transcription of cytochrome P-452, transcription of the peroxisomal enzymes, fatty acyl-CoA oxidase and enoyl-CoA hydratase/3-hydroxyacyl CoA dehydrogenase, also occurs (REDDY et al., 1986).

The induction of cytochrome P-452 has been achieved in rat hepatocyte cultures and a good correlation has been obtained between induction in rats in vivo and in vitro (LAKE et al., 1986). DEHP and clofibrate induce similar forms of P-452, and for a range of hypolipidaemic drugs and phthalate monoesters a good correlation was obtained between the induction of the microsomal enzyme and peroxisomal enzyme activities. Structure activity relationships have shown that branched-chain esters were more potent inducing agents than straight-chain and octyl esters are more potent than hexyl esters (LAKE et al., 1986).

## 3.   Mechanisms of induction of the cytochromes P-450

The molecular mechanisms of induction of the cytochromes P-448 (P-450 I) have been studied in detail, especially by NEBERT and his colleagues (NEBERT and JENSEN, 1979) and have been shown to involve interaction of the inducing

---

agents with one or more cytosolic receptors to activate transcription, analogous to the mechanism of genomal regulation of protein/enzyme synthesis by the steroid hormones. No similar mechanisms have been established for the induction of the other families of cytochromes P-450, and no receptors have been identified. Indeed, because of the association of the PB-type inducers with simultaneous induction of the P-450 reductase and other enzymes, together with proliferation of the endoplasmic reticulum and cellular hypertrophy, it is likely that PB-induction involves a fundamentally different mechanism from that identified for P-448 induction.

## 3.1. Cytochromes P-448 (P-450 I)

In the induction of the cytochromes P-448, the inducing agent, or possibly a metabolite, binds to the cytosolic polycyclic aromatic hydrocarbon (Ah) receptor to be translocated into the nucleus where it effects genomal de-repression of the P-450 I genes and synthesis of the appropriate mRNAs and proteins by a mechanism, the details of which are still unknown. Isosafrole and other methylenedioxyphenyl inducers of the cytochromes P-448 do not appear to act via the Ah receptor (COOK and HODGSON, 1985), and evidence that the 'mixed' inducer, hexachlorobenzene, acts to induce P-448 by interaction with the Ah receptor is, at best, equivocal (LINKO et al., 1986). From computer graphic studies of the molecular dimensions and electronic structures of inducers of P-448 and specific substrates and inhibitors of these enzymes, the overall conformation of the binding site of the cytosolic receptor for enzyme induction would appear to be very similar to that of the active site of the enzyme (LEWIS et al., 1986). However, despite the similarity of the spatial dimensions of substrates and inducers of the cytochromes P-448, not all substrates are effective inducing agents (e.g. paracetamol), and not all potent inducing agents are readily metabolized by the enzyme (e.g. TCDD).

Genetic differences in the inducibility of the cytochromes P-448 have been extensively studied in mice (NEBERT, 1979). The B6 inbred mouse exhibits high inducibility of cytochrome P-448, while the D2 shows poor induction of this enzyme. MC at high doses induces both $P_1$-450 and $P_3$-450 in B6 mice but not in D2 mice, whereas high doses of the more potent inducer, TCDD, induces both isoenzymes in both strains of mice by increased gene transcription (GONZALEZ et al., 1984), indicating that the genetic differences in P-448 induction in mice are attributable to differences in affinities of the cytosolic Ah receptors for the inducing agents. Genetic differences in regulation of P-448 are seen in other species, and guinea pigs, a species relatively resistant to chemical carcinogens, are also resistant to induction by MC, in marked contrast to rats and mice (ABE and WATANABE, 1983).

Expression of the $P_1$-450 and $P_3$-450 genes in liver and extrahepatic tissues of mice treated with MC or TCDD indicate that although both genes are

**Table 3.** Molecular dimensions of cytochrome P-448/P-450 substrates/inducers
Data is from Lewis et al. (1986)

| Molecule | Length (Å) | Width (Å) | Depth (Å) | Area/depth² ($Å^{-1}$) | Cytochrome |
|---|---|---|---|---|---|
| Dibenzo(a,h)anthracene | 15.9 | 9.3 | 3.2 | 14.4 | P-448 |
| Benzo(a)pyrene | 13.6 | 9.0 | 3.2 | 11.9 | P-448 |
| N,N-Dimethylaminoazobenzene | 15.1 | 6.6 | 3.2 | 9.7 | P-448 |
| 2-Aminoanthracene | 12.9 | 7.3 | 3.2 | 9.2 | P-448 |
| 7-Ethoxyresorufin | 14.6 | 9.1 | 3.9 | 8.7 | P-448 |
| Ellipticine | 13.5 | 9.1 | 3.8 | 8.5 | P-448 |
| 4-Aminobiphenyl | 13.8 | 7.4 | 3.6 | 7.9 | P-448 |
| 3-Methylcholanthrene | 14.6 | 8.6 | 4.0 | 7.9 | P-448 |
| 1,2,7,8-Tetrachlorodibenzodioxin | 13.8 | 7.4 | 3.6 | 7.9 | P-448 |
| Trp-P-1 | 12.6 | 9.1 | 4.2 | 6.6 | P-448 |
| Zoxazolamine | 11.1 | 7.1 | 3.6 | 6.1 | P-448 |
| Benzene | 7.4 | 7.4 | 3.2 | 5.4 | P-448/P-450 |
| 2-Acetylaminofluorene | 14.4 | 7.3 | 4.6 | 5.0 | P-448 |
| Paracetamol | 11.6 | 7.3 | 4.2 | 4.8 | P-448 |
| β-Naphthoflavone | 13.5 | 8.3 | 5.3 | 4.0 | P-448 |
| Aflatoxin | 12.3 | 10.6 | 6.4 | 3.2 | P-448/P-450 |
| Clofibrate | 14.8 | 6.7 | 7.1 | 2.0 | P-452 |
| Hexobarbital | 9.0 | 7.5 | 6.2 | 1.7 | P-450 |
| Chlordane | 11.1 | 9.8 | 8.1 | 1.6 | P-450 |
| Rifampicin | 18.6 | 13.8 | 14.0 | 1.3 | P-450 |
| Phenobarbital | 10.1 | 7.3 | 8.1 | 1.1 | P-450 |
| Aldrin | 10.0 | 9.0 | 9.0 | 1.1 | P-450 |
| Diphenylhydantoin | 11.4 | 7.1 | 9.0 | 1.0 | P-450 |
| DDT | 13.6 | 8.1 | 11.5 | 0.8 | P-450 |

controlled by the same (Ah) receptor, striking tissue-specific differences in transcription and stabilization of mRNA affect the final mRNA concentrations. For example, for liver $P_1$-450, TCDD results in a 8-fold increase in transcription associated with a 27-fold increase in mRNA, and for kidney $P_3$-450 TCDD results in a 2-fold rise in transcription accompanied by a 12-fold increase in mRNA content (KIMURA et al., 1986). However, increases in mRNA do not always correlate with increases in cytochrome P-450 proteins; high doses of isosafrole to mice markedly increased hepatic $P_1$-450 and $P_3$-450 mRNAs but the corresponding catalytic activities were not raised (TUTEJA et al., 1986).

Structural requirements for the binding of inducing chemicals to the Ah cytosolic receptor and induction of the cytochromes P-448 have been studied

D. V. PARKE

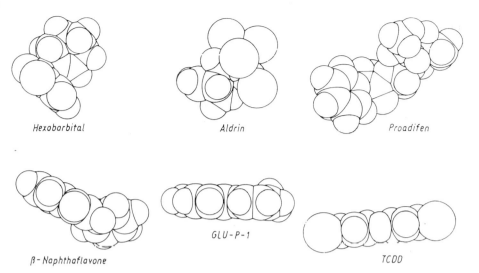

**Fig. 1.** Space-filled models of compounds exhibiting high specificity for cytochromes P-450 or cytochromes P-448.
Models were drawn using the PLUTO computer program and utilizing the following van der Waals radii to generate computer graphical plots of molecular geometries: carbon, 1.6 Å; nitrogen, 1.5 Å; oxygen, 1.4 Å; and hydrogen, 1.2 Å. Hexobarbital, aldrin and proadifen are specific substrates/inducers for cytochromes P-450, and $\beta$-naphthoflavone, Glu-P-1 and TCDD are specific substrates/inducers for cytochromes P-448 (PARKE et al., 1986).

with PCB congeners (PARKINSON et al., 1980). Potent inducers of P-448 were 3,3',4,4'-tetra, 3,4,4',5-tetra, 3,3',4,4',5-penta and 3,3',4,4',5,5'-hexa-chlorobiphenyls, and the structural requirements for induction of P-448 by PCBs were considered to be 'substitution at both para positions and at least two meta positions'. Mixed inducers of the MC-induced P-448s and the PB-induced P-450s require also the substitution of one ortho position, for example 2,3,3',4,4',5-hexachlorobiphenyl (PARKINSON et al., 1980). More detailed studies of the molecular orbitals of inducers have revealed that the type of cytochrome induced is governed by the spatial dimensions and electronic structures of the inducing agents (LEWIS et al., 1986; PARKE et al., 1986). Planar molecules fit the Ah receptor, and globular molecules are the preferred substrates and inducers of the PB-inducible forms of P-450. Table 3 shows the molecular dimensions of some substrates/inducers of the cytochromes P-448 (P-450 I) and of the PB-inducible cytochromes P-450 (P-450 II), and Figure 1 shows the molecular spatial conformations of typical substrates, inducers and inhibitors of the PB-inducible P-450s and P-448 cytochromes, respectively (PARKE et al., 1986). A correlation of the cytosolic receptor binding avidities and cytochrome P-448 induction (EROD activity) with molecular dimensions

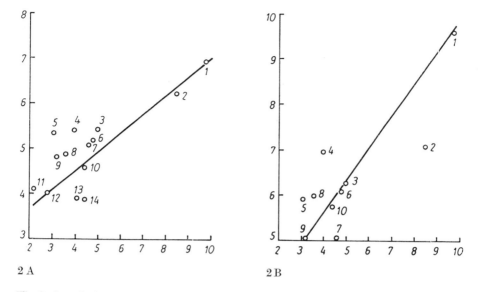

**Fig. 2.** Correlation plots of molecular dimensions of a series of polychlorinated biphenyls with (A) rat cytosolic TCDD receptor binding, and (B) cytochrome P-448 induction (EROD activity) in rat hepatoma cells.
Molecular dimensions (Area/Depth$^2$) were determined from crystal data or by the MINDO/3 method for determining optimal molecular geometries. Rat cytosolic TCDD receptor binding data ($-\log EC_{50}$), and P-448 induction (EROD activity) data ($-\log EC_{50}$) were obtained for: 1. 3,3′,4,4′,5-penta; 2. 3,3′,4,4′-tetra; 3. 2,3,4,4′,5-penta; 4. 2,3,3′,4,4′-penta; 5. 2,3,3′,4,4′,5′-hexa; 6. 2,3,3′,4,4′,5-hexa; 7. 2,3′,4,4′,5-penta; 8. 2′,3,4,4′,5-penta; 9. 2,3′,4,4′,5,5′-hexa; 10. 2,3,4,4′-tetra; 11. 2,2′,4,4′,5,5′-hexa; 12. 2,3′,4,4′,5′,6-hexa; 13. 2,2′,4,4′-tetra; and 14. 2,3,4,5-tetra (SAFE et al., 1985) (Figures from PARKE et al., 1986).

of a series of PCBs are shown in Figure 2 (PARKE et al., 1986). Correlation of cytosolic receptor binding avidities with induction of P-448 (EROD activity) of 29 polycyclic aromatic hydrocarbons further showed that (i) a common structural requirement for receptor binding or P-448 induction is the presence of a phenanthrene nucleus fused with at least one additional benzo ring, and (ii) there is no correlation between the binding avidities to the Ah receptor and the potential for induction of P-448 (PISKORSKA-PLISZCZNSKA et al., 1986).

Administration to rats of the mixed PCB inducer, Aroclor 1254, resulted in a biphasic (2−6 hour then 10−20 hour after dosing) sequential activation of cAMP-dependent protein kinase, induction of ornithine decarboxylase (ODC), activation of RNA polymerase I, and synthesis of cytochromes P-450 (COSTA et al., 1976). Further evidence with polycyclic aromatic hydrocarbon inducers indicated that in mice induction of ODC and cytochromes P-450 were independent processes and that ODC induction per se was not integral to cytochrome induction (RAUNIO and PELKONEN, 1983); furthermore, dibutyryl cAMP decreased the inductive effect of MC in rat liver (MESSNER, 1984).

Induction of benzo(a)pyrene hydroxylase (AHH) activity has been shown to occur in rat hepatocyte cultures, mediated by a riboflavin-mediated generation of superoxy anion ($O_2^-\cdot$) in the culture media (PAINE and McLEAN, 1974), or by exposure of mice to ionizing radiation (PRASAD et al., 1977). The reactive oxyradicals so formed act on histidine and other components of the medium to form stable cytochrome P-450/P-448 inducers (PAINE and FRANCIS, 1980). It is postulated that oxyradicals generated in vitro, or in vivo by decoupling of P-450 from its reductase with consequent leakage of oxyradicals, may play a central role in the induction of the cytochromes (PAINE, 1978). This is a tenable hypothesis, as substrates that are potent inducers (TCDD, PB) are poorly metabolized and hence result in activation of the P-450-oxygen high-spin complex which, because of difficulties in oxygenation of the substrate, generates superoxy anion, singlet oxygen and hence initiates cytochrome P-450 induction (PAINE, 1978).

## 3.2. Cytochromes P-450$_{PB}$ (P-450 II)

The earliest views on the mechanism of PB-induced induction of the cytochromes P-450 and proliferation of the endoplasmic reticulum inclined to an increase in the stabilization of mRNA or a decrease in its turnover, due to enhanced methylation, rather than to increased transcription (SMITH et al., 1976; LINDELL, 1977; REES, 1979). The three major mRNAs for PB-inducible cytochromes P-450 (P-450a, b and e, or P-450 II A1, B1 and B2) are under co-ordinate regulation, which is different from the regulation of the P-448 isoenzymes; the three cytochromes P-450 occur at very low levels in untreated rats but become major microsomal enzymes after PB treatment (PICKETT et al., 1983). Later, PB was shown unequivocally to increase the transcriptional activity of isolated rat hepatocytes up to 10-fold (HARDWICK et al., 1983a), and transcription of the P-450b gene in rat hepatocyte nuclei was increased 9-fold one hour after administration of PB to male rats (HARDWICK et al., 1983b). The mechanism of PB induction of cytochrome P-450 therefore probably involves enhancement of both translation and transcription (PHILLIPS et al., 1981). The extent of induction of P-450 produced by barbiturates is related to the plasma half-lives of the inducing agents; compounds with low rates of metabolism and long half-lives were the most potent inducing agents (IOANNIDES and PARKE, 1975).

Administration of the metabolites of PB, p- and m-hydroxyphenobarbital, to rats results only in proliferation of the endoplasmic reticulum, without induction of cytochrome P-450, indicating that these two inductive effects seen with PB may depend on different molecular mechanisms (CRESTEIL et al., 1980).

A study of polychlorinated biphenyl analogues of different spatial conformations indicates that the PB-inducible cytochromes P-450 are induced by the non-coplanar molecules, including 2,2′,3,4,4′,5-hexa and 2,2′,3,4,4′,5,6-penta-chlorobiphenyl. QSAR studies indicate that structural requirements for P-450 induction are substitution of the biphenyl molecule in both **para** positions, and at least two **meta** and two **ortho** positions (DENOMME et al., 1983).

# 4.  Developmental aspects of cytochrome P-450 induction

The cytochromes P-450 develop in the foetus; in foetal rodents the P-448s develop first, followed by development of the PB-cytochromes P-450 after birth (PARKE, 1984). PB, MC, and the PCB, Aroclor 1254, administered intraperitoneally to pregnant rats all increased benzo(a)pyrene hydroxylase (AHH) of foetal liver, whereas only MC and PCBs increased the AHH of placenta (ALVARES and KAPPAS, 1975). $\beta$-Naphthoflavone administered to pregnant rats resulted in a 250-fold induction in EROD activity and to an increase in P-450c (P-450 I A1), but not P-450d (P-450 I A2), in the hepatocytes and erythroblasts of foetal liver (SHIVERICK et al., 1986). Constitutive $P_1$-450 (P-450 I A1) mRNA, but not $P_3$-450 (P-450 I A2) mRNA, is increased in 7 day-old mouse embryos and $P_1$-450 gene activation can occur in the absence of induction by xenobiotics (KIMURA et al., 1987). A development lag in the inducibility of cytochrome P-450d, relative to that of P-450c, has been reported to occur in several species. From studies with foetal rat hepatocytes, it has been concluded that glucocorticoids act synergistically with polycyclic aromatic hydrocarbons to induce P-450c in foetal rat liver, and that this action is mediated through the classical type II glucocorticoid receptor (MATHIS et al., 1986).

Treatment of 0, 2 and 12-day old neonatal rats with PB induced hepatic cytochromes P-450b and P-450e, and MC induced P-450c, although there was immunological evidence that these neonatal cytochromes may differ somewhat from the adult enzymes (GULYAEVA et al., 1986).

Administration of PCBs to neonates via the maternal milk increased the AHH activity and cytochromes P-450 of the neonatal liver, PB similarly administered increased P-450 and ethylmorphine N-demethylase activity, but MC showed no induction of P-450 or of either enzyme activity, indicating that induction is dependent not only on the nature of the inducing agent but also on the route and method of administration and on the tissue examined (ALVARES and KAPPAS, 1975). Treatment of lactating rats with MC or $\beta$-naphthoflavone induced P-448 in rat liver and breast tissue, 2- and 3- to

10-fold respectively, and although the induced level in breast tissue was only 1% of that of liver this does indicate that carcinogens may be metabolized to more polar metabolites in breast tissue (RITTER et al., 1982). Administration of PB to neonates has been shown to 'imprint' or 'programme' the microsomal enzymes of the developed adult. After dosing 1- to 5-day-old rats with PB the microsomal enzymes showed expected increases at six days which largely disappeared at 22 and 50 days post partum. However, at 140 days the PB-treated rats of both sexes showed increased levels of P-450, the P-450 and cytochrome c reductases, mixed-function oxidase activities, glucuronyl transferase activity, and in vitro binding of benzo(a)-pyrene to DNA and in vivo binding of aflatoxin $B_1$ to hepatic DNA (BAGLEY and HAYES, 1985).

Induction of the microsomal mixed-function oxidases has also been demonstrated in human infants and anti-epileptic drug therapy with PB, phenytoin, primidone or carbamazepin during pregnancy led to marked enzyme induction in the neonates, as demonstrated by the [13]C-aminopyrine breath test (RATING et al., 1983).

Induction also occurs in old animals and although the liver mixed-function oxidase activities in senescent rats (100 weeks) are substantially lower than those in young rats they are readily induced by PB or $\beta$-naphthoflavone (KAO and HUDSON, 1980), MC or PCN (BIRNBAUM and BAIRD, 1978a), or PCBs (Aroclor 1254) (BIRNBAUM and BAIRD, 1978b), to levels similar to those seen in induced young rats.

## 5. Species differences in cytochrome P-450 induction

The families of cytochromes P-450, now so well characterized in mammalia, are not necessarily the same as those occurring in animal species of other classes, and consequently the regulation of biosynthesis of these enzymes may also vary. p,p'-DDT and p,p'-DDE administered to trout for four weeks did not result in any induction of the cytochromes P-450 or of the mixed-function oxidase enzymes, aldrin epoxidase or 7-ethoxycoumarin O-deethylase (ADDISON et al., 1977). In contrast, intraperitoneal or intramuscular injection of 5,6-benzoflavone, MC or other polycyclic aromatic hydrocarbons, to several species of marine fish increased their hepatic cytochrome P-450 contents and hydroxylation of benzo(a)pyrene, 7-ethoxycoumarin and 7-ethoxyresorufin (JAMES and BEND, 1980). This induction of mixed-function oxidases by inducers of cytochrome P-448 (P-450 I), but not by inducers of cytochromes P-450 (P-450 II) supports the subsequent conclusions that P-448-type enzymes predominate in fish and that the mammalian cytochromes P-450 developed subsequent to the evolution of terrestrial species (PARKE, 1985).

Administration of MC increased the cytochromes P-450, AHH and pnitrophenetole O-deethylase activities in liver microsomes of trout, frog, snake and rat, and increased aminopyrine N-demethylase in trout, but not frog, snake or rat. The P-450-type inhibitor, SKF 525-A inhibited the induced AHH activity in trout, snake and frog, but not rat, and the P-448-type inhibitor, α-naphthoflavone, inhibited the AHH activity in trout, frog and rat, but not snake, thus indicating that MC induces different types of cytochrome P-450 in different animal classes. Similarly, PB did not induce the cytochromes P-450, or the mixed-function oxidase activities in trout, frog or snake, even thought it accumulated in the livers of these species similar to that seen in rats (SCHWEN and MANNERING, 1982).

# 6.    Consequences of induction of the cytochromes P-450

The physiological roles of the cytochromes P-450 are so diverse and so critical to normal biochemical, cellular and tissue functions that it is perhaps surprising that induction of these enzymes does not result in more profound consequences than have been seen to occur. The reason for this is probably that most inducing agents are highly selective in the family of cytochromes P-450 they genomally derepress. Furthermore, with the exception of certain drugs, such as the anticonvulsants, the amounts of inducing agents normally ingested are small and, with the exception of tobacco smoke, TCDD and a few other halogenated chemicals, the amounts required to effect induction are greater than those to which man and animals are usually exposed. Among the more important consequences of induction of the cytochromes P-450 are the clinical effects associated with drug treatments, effects on intermediary metabolism, effects on chemical toxicity, and the possible effects on tumorigenesis.

## 6.1.  Clinical effects

The number of drugs of importance in the induction of the cytochromes P-450 are relatively few and are confined mostly to the anticonvulsants (phenobarbitone (PB), phenytoin (PT) and carbamazepine (CBZ)) and the antitubercular drug, rifampicin (BRECKENRIDGE, 1987). Few new drugs come onto the market which are potent enzyme-inducing agents, largely because of their potentially adverse effects in decreasing the therapeutic activities of other simultaneously administered drugs due to enhanced metabolic deactivation, their removal of the 'protective' action of oral contraceptive steroids, altered vitamin D metabolism with potential consequences of osteomalacia, and folate depletion resulting in macrocytic anaemia and teratogenesis (LABADARIOS et al., 1977).

Induction of cytochromes P-450 in man may be quantitatively monitored by determination of urinary excretion of D-glucaric acid or, preferably, of 6$\beta$-hydroxycortisol, by radioimmunoassay. Onset of induction has been shown to be more rapid with rifampicin than with PB, and considerable inter-subject variability in the time course of induction by different drugs has been demonstrated (BRECKENRIDGE, 1987). Different pathways of metabolism may be induced by different drugs, e.g. metabolism of antipyrine to 4-hydroxyantipyrine is induced by CBZ and PT, whereas metabolism to nor-antipyrine is increased by rifampicin. Induction is also dependent on age, for whereas rifampicin induces drug metabolism in young subjects it does not appear to have a similar effect in elderly patients (BRECKENRIDGE, 1987). Smoking also induces drug metabolism, e.g. of the $\beta$-adrenergic blocker drug, propranolol, but again the inductive effect of cigarette smoking has its greatest effect in young people.

The anticonvulsant drugs, PB, PT, CBZ, and primidone are among the most potent drug-inducing agents. Moreover, they are usually prescribed in combination and for long periods of time, and it is not surprising that such patients exhibit the highest degree of hepatic enzyme induction seen in man (PERUCCA, 1978). Patients treated with PB, PT, CBZ or primidone, either alone or in combination, showed high values of antipyrine clearance, and excreted larger amounts of D-glucaric acid or 6$\beta$-hydroxycortisol than controls. Significant correlations were found between the dosage of PB, PT, CBZ and primidone and both indices of enzyme induction, the relative potencies for antipyrine clearance at average therapeutic dose levels being PB (1.0), PT (0.92), CBZ (0.84) and primidone (0.82) (PERUCCA et al., 1984). Valproic acid, another anticonvulsant drug, is much less potent as an enzyme inducing agent in man.

The broad-spectrum antifungal drugs, miconazole and ketoconazole, inhibitors of P-450-mediated fungal sterol biosynthesis, are also inducers of the cytochromes P-450. Miconazole is more potent than ketoconazole as an inducing agent, and repeated intravenous administration to patients of therapeutic doses of miconazole, but not ketoconazole, increased the excretion of D-glucaric acid (LAVRIJSEN et al., 1986).

Clofibrate, an antilipidaemic drug, induces up to 30-fold the cytochrome P-452 (P-450 IV) that $\omega$-hydroxylates lauric acid and also results in peroxisomal proliferation, with increases in the peroxisomal lipid-metabolizing enzymes. This induction of P-452-mediated $\omega$-hydroxylation of medium length fatty acids, and their subsequent oxidation to dicarboxylic acids and $\beta$-oxidation, might account for the hypolipidaemic action of clofibrate. However, co-administration of the P-450 inhibitor, 1-aminobenzotriazole, which inhibits a major part of the cytochrome P-452 fatty acid hydroxylase induced by clofibrate, had no effect on the ability of clofibrate to lower serum triglyceride levels. This indicates that the antilipidaemic action of clofibrate is independent of the induction by clofibrate of cytochrome P-452 fatty acid hydroxylase activity (REICH and ORTIZ DE MONTELLANO, 1986).

Enzyme induction may also be the basis of therapeutic action, as in the case of aminoglutethimide which enhances the metabolic deactivation of estrogens. Aminoglutethimide is used in the treatment of advanced breast cancer as it inhibits aromatase and other adrenal cortical pathways of steroidogenesis. It is also a potent enzyme inducer, and appears to induce the P-450-dependent metabolism of endogenous estrogens, increasing the rate of estriol excretion, which may also be an important aspect of the effect of aminoglutethimide in decreasing the bioavailability of estrone sulfate, and of its therapeutic benefit in breast cancer (LONNING et al., 1986). With continuous treatment, aminoglutethimide induces its own metabolism, probably the N-hydroxylation, decreasing the half-life from 15 hours to 9 hours. Aminoglutethimide also induces a 3- to 5-fold increase in the clearance of warfarin, and similarly increases the clearance of theophylline, digitoxin, and medroxyprogesterone acetate, a steroid used in combination with aminoglutethimide in the treatment of breast cancer (KVINNSLAND et al., 1986).

A further possible example of enzyme induction as the basis of therapeutic action concerns the value of medroxyprogesterone acetate in the treatment of liver disease. The induction of liver enzymes and increased protein synthesis promoted by medroxyprogesterone accelerates recovery after liver damage by carbon tetrachloride and dimethylnitrosamine, and hence is recommended for treatment of hepatic injury (SAARNI et al., 1983).

## 6.2. Effects on intermediary metabolism

Induction of the cytochromes P-450 may also have effects on intermediary metabolism, including glucose metabolism, lipid metabolism, protein synthesis, and the metabolism of a number of vitamins including, particularly, folic acid and vitamin D.

In non-insulin-dependent diabetes mellitus, the conventional treatment of diet, exercise, sulphonylureas and metformin may sometimes give rise to hyperglycaemia and development of insulin resistance. PB type enzyme inducers increase insulin sensitivity in diabetics and can be used as an adjunct in treatment enabling insulin resistance to be overcome. The mechanism is believed to be associated with the increased hepatic microsomal mixed-function oxidase activity, and the consequent increased hepatic utilization of NADPH, and increased glucose uptake and utilization by the liver (SOTANIEMI et al., 1987).

Induction of P-450 by anticonvulsant drugs or ethanol also affects lipid metabolism, and decreases the mortality rate from coronary heart disease. The inducing agents, which increase the synthesis of hepatic proteins and phospholipids, also decrease serum low density lipoprotein (LDL) cholesterol, increase serum high density lipoprotein (HDL) cholesterol, and inhibit cholesterol accumulation in arterial walls and the formation of atherosclerotic plaque (LUOMA, 1987).

Inducers of the mixed-function oxidases, such as the anticonvulsant drugs, nikethamide, DDT and alcohol, result in liver enlargement, associated with hypertrophy, proliferation of the endoplasmic reticulum and increased protein synthesis. Chronic alcoholics have livers with normal histology but 40% greater in size than those of control subjects. Liver enlargement has been correlated with P-450 induction, is reversible when administration of the inducing agent is stopped, and is considered to be an adaptive, non-pathological process (PIRTTIAHO, 1987).

Long-term anticonvulsant therapy in epileptics is associated with osteomalacia and deficiency in vitamin D activity. Vitamin D is activated by metabolism to 25-hydroxyvitamin D by a liver microsomal enzyme which is further hydroxylated to 1,25-dihydroxyvitamin D in the kidney or deactivated by conjugation with glucuronic acid in liver and secreted in the bile. 1,25-dihydroxyvitamin D has high activity in promoting calcium absorption. Treated epileptics have low levels of 25-hydroxyvitamin D but normal levels of 1,25-dihydroxyvitamin D, and the osteomalacia may be associated with the increased biliary excretion of 25-hydroxyvitamin D, resulting from the drug-promoted enzyme induction (BRECKENRIDGE, 1987). Patients on prolonged treatment with anticonvulsant or phenothiazine drugs also show deficiency in folate, probably due to the increased rate of protein synthesis, and consequent high requirement for folate, associated with prolonged enzyme induction. This folate deficiency subsequently limits enzyme induction leading to impaired drug metabolism and consequent adverse drug reactions (LABADARIOS et al., 1977).


## 6.3.  Effects on chemical toxicity

Increased rates of metabolism of toxic chemicals, resulting from induction of the mixed-function oxidases, may lead to a decrease in toxicity, through acceleration of detoxication, or to an increase in toxicity because of increased rates of metabolic activation. The key to this enigma is the nature of the enzymes involved in detoxication and activation of the particular toxic chemical, and the nature of the cytochrome P-450 induced. Induction of rat liver enzymes with the PB-type inducers results in protection from the lethal effects of warfarin, meprobamate and strychnine, but increases the lethality of Schradan and carbon tetrachloride, and the hepatotoxicity of bromobenzene (CONNEY, 1986). Conversely, induction by the MC-type inducers affords protection against the lethality of zoxazolamine and the hepatotoxicity of bromobenzene, but increases paracetamol-induced liver necrosis and the lethality of dimethylnitrosamine (CONNEY, 1986; PHILLIPS et al., 1975). However, although treatment of rats with PB has little or no effect on detoxication of paracetamol (glucuronide conjugation) and also no effect on its metabolic activation (mercapturate excretion) it does lead to an acceleration of the hepatic necrosis, which is attributed to events not mediated through the cytochrome P-450 system (POULSEN et al., 1985). Similarly, the PB-enhanced hepatotoxicity of carbon tetrachloride is attributed to the simultaneous induction of the cytochrome P-450 reductase which does not occur with MC type inducing agents (SUAREZ et al., 1972). However, the protection given to rats against the acute toxicity of dimethylnitrosamine (DMN) by MC type induction, but not by PB-induction, cannot be explained simply in terms of differential induction of liver P-450-dependent DMN demethylase activity, and the hepatotoxicity of DMN probably involves degradation to formaldehyde by a multicomponent enzyme process (PHILLIPS et al., 1975).

The effects of PB- and MC-mediated enzyme induction on the pulmonary toxicity and acute lethality of 4-ipomeanol are similar, but for different reasons and can change the target organ of toxicity (STATHAM and BOYD, 1982). Both forms of enzyme induction decrease the concentration of ipomeanol in the lung and hence its pulmonary metabolism, covalent binding and pulmonary toxicity, by increasing the rate and extent of its metabolism in the liver, but whereas PB increases the detoxication of ipomeanol by conjugation with glucuronic acid and urinary excretion, MC increases the metabolic activation of the chemical in the liver, with consequent increase in covalent binding and hepatotoxicity (STATHAM and BOYD, 1982).

Pretreatment of rats with steroidal (PCN or spironolactone) or non-steroidal (PB, or phenytoin) inducing agents greatly changes the toxic responses to the various pyrrolizidine alkaloids. The toxicity of lasiocarpine (100 mg/kg) was greatly decreased (mortalities: control 100%; phenytoin, 50%; spironolactone, 40%; PB, 10%; PCN, 0%) but, in contrast, the toxicities of monocrotaline (90 mg/kg) (control, 5%; PB, phenytoin, PCN and spironolactone, 80−95%) and of heliotrine (300 mg/kg) (control, 25%; PB, 80%; PCN, 100%) were markedly increased. The hepatotoxicity of the pyrrolizidine alkaloids is considered to be due to their metabolic activation to pyrrole derivatives by the hepatic microsomal mixed-function oxidases, and PB and PCN markedly increase this metabolism to pyrroles with lasiocarpine (3- to 17-fold) and with heliotrine (6- to 30-fold). Although the increased toxicity correlates with increased metabolism to pyrroles with monocrotaline, heliotrine and other pyrrolizidine alkaloids, an inverse correlation was observed with lasiocarpine, indicating that an alternative mechanism may exist for lasiocarpine toxicity or that enzyme induction results in increased excretion of metabolites (TUCH-WEBER et al., 1974).

Apart from the well-known induction of specific families of cytochrome P-450 with PB (P-450 II, B1 & 2, C1–10) or MC (P-450 I), induction of other families of this enzyme is associated with changes in chemical toxicity. Selective induction of cytochrome P-450$_{ALC}$ (P-450 II E) by alcohol, acetone, isoniazid, etc., results in increased metabolic activation of dimethylnitrosamine, paracetamol and carbon tetrachloride to reactive intermediates and ultimate carcinogens, and probably accounts for the potentiation of toxicity/carcinogenicity of these chemicals by chronic ethanol consumption (KHANI et al., 1987). Similarly, chronic treatment of hamsters with alcohol increased the in vitro activation by liver microsomes of the aromatic amines 2-aminofluorene, 4-aminobiphenyl, benzidine and 2-acetamidofluorene to mutagens (IOANNIDES and STEELE, 1986). Induction of cytochrome P-450$_{PCN}$ (P-450IIIA) in rat liver by pregnenolone 16α-carbonitrile results in a decrease in the toxicity of carbon tetrachloride and dimethylnitrosamine (HARDWICK et al., 1983c). The peroxisomal proliferating, antilipidaemic drugs, clofibrate and ciprofibrate, which specifically induce cytochrome P-452 (P-450 IV), are associated with the formation of peroxide, lipid peroxidation, decreases in the enzymes protecting against autoxidative damage, hepatocellular necrosis and cancer, though which is cause and which is effect is unknown (GOEL et al., 1986).

Several instances of drug toxicity have been attributed to induction of the various cytochromes P-450. The anti-tubercular drug combination, isoniazid and rifampicin, may sometimes result in fulminant hepatitis, which is associated with the enzyme inductive effects of rifampicin (PESSAYRE et al., 1977), though both drugs are enzyme inducing agents, effecting the increase of different cytochromes P-450 and resulting in a complex interplay of mixed-function oxidases and pathways of drug detoxication and activation, especially as several other drugs were administered simultaneously. Similarly, enzyme induction of the cytochromes P-450 has been implicated in the hepatotoxicity of the muscle relaxant dantrolene which induces P-448 (P-450c), its own metabolic hydroxylation and its own toxicity (JAYYOSI et al., 1987), and in the hepatotoxicity of halothane which is metabolized to toxic reactive intermediates by a reductive P-450 pathway (McLAIN et al., 1979).

Potent enzyme inducing agents may also elicit effects on toxicity by their inhibition of P-450 enzymes, as inducing agents may function first as inhibitors before manifesting their inductive effects, and these alternate roles of inhibitor and inducer may even occur simultaneously with different P-450s or in different tissues. For example, TCDD, a potent inducer of P-448 which thereby enhances the mutagenicity/carcinogenicity of many P-448-activated carcinogens, also leads to toxic effects on the endocrine system, with decreased production of testosterone and corticosterone, due to TCDD inhibition of the P-450 enzyme, adrenal 21-hydroxylase (MEBUS and PIPER, 1986).

The potent anticonvulsant inducers, PB and phenytoin, result in megaloblastic anaemia in epileptic patients and to teratogenicity in their offspring, and in rats, as the consequence of folate deficiency, due to the high demands of PB type induction on protein synthesis and folate (LABADARIOS, 1975; LABADARIOS et al., 1977). Peripartum exposure of rats to maternally administered phenytoin also resulted in significant depression of the mixed-function oxidase activities in the 28 day neonates, and at four to five months these enzymes were still not normal and were much less responsive to the inductive effects of PB, indicating that exposure of neonates to inducing agents may disrupt the development of the hepatic monoxygenase system and its response to inducing agents in adult life, with consequent implications for enhanced chemical toxicity (SHAPIRO et al., 1986).

## 6.4. Induction and carcinogenesis

Many different kinds of carcinogens, such as polycyclic aromatic hydrocarbons, aromatic amines and amides, and azo compounds, are activated to the corresponding ultimate carcinogens by metabolic oxygenation by the cytochromes P-450; the most active of these being the cytochromes P-448 (P-450 I) and the cytochromes P-450$_{ALC}$ (P-450 II E) (IOANNIDES et al., 1984). Substrates and inducers of P-448 have been shown by computer graphics to be essentially rigid planar molecules characterized by small depth and large area/depth$^2$ ratios, in contrast to the substrates and inducers of the PB-induced cytochromes P-450 which are non-planar, globular molecules characterized by greater depth and by smaller area/depth$^2$ ratios (see Table 3) (LEWIS et al., 1986). The significance of the planar molecular conformation of many carcinogens is that oxygenation can occur in conformationally-hindered positions, thus yielding reactive intermediates which are resistant to detoxication by epoxide hydrolases and other Phase 2 conjugating enzymes, and so may interact with DNA and other vital cellular macromolecules (PARKE, 1987; IOANNIDES and PARKE, 1987).

The cytochromes P-448 are induced by a mechanism involving one or more cytosolic receptors (Ah receptors). Genetic differences in the inducibility of P-448, and hence susceptibility to tumorigenesis, have been extensively studied in mice (NEBERT and JENSEN, 1979). The B6 inbred mouse exhibits high inducibility of P-448 and high susceptibility to cancer, while the D2 mouse has poor inducibility of P-448 and low susceptibility to cancer. These genetic differences in mice are attributed to differences in the affinities of the Ah receptors for the inducing agents.

Inducers of P-448, such as TCDD, decrease the binding of epidermal growth factor (EGF) to its receptor, mediated by the increased formation of reactive intermediates of co-administered polycyclic aromatic hydrocarbons (KÄREN-LAMPI et al., 1983). EGF is one of a number of polypeptide growth factors which act by binding to transmembrane receptors thereby activating a protein tyrosine kinase, initiating a pleiotropic growth response which results in DNA transcription, protein synthesis and cell division. EGF and other growth factor receptors function to stimulate cell growth and play a key role in oncogenesis (STOSCHECK and KING, 1986). EGF markedly increased the growth in vitro of 81% of 186 human malignant tumour specimens derived from many different tissues (SINGLETARY et al., 1987). Furthermore, several oncogenes have been shown to regulate the same mRNA as growth factors or growth factor receptors (YAMAMOTO et al., 1986). Substrates/inducers of the cytochromes P-448 are therefore complete carcinogens in that when metabolically activated they can effect damage to DNA (genotoxic effects) and also effect changes in DNA regulation and cell division (epigenetic effects).

TCDD, the most potent inducer of P-448, is considered to be a co-carcinogen but not a carcinogen, whose mode of action is the induction of P-448 resulting in the greater activation of carcinogens such as 3-methylcholanthrene (KOURI et al., 1978). Similarly, in mammary cancer induced by 7,12-dimethylbenz(a)-anthracene (DMBA) in rats, the essential first step is the induction of P-448 and another cytochrome P-450, which then result in the activation of DMBA to the proximate carcinogen, with the initiation of malignancy (CHRISTOU et al., 1987).

The role of enzyme-inducing agents as 'promoters' of cancer has been a long-standing controversy. Inducers of the cytochromes P-448 unquestionably promote, and may also initiate, malignancy, but studies with inducers of the other forms of cytochromes P-450 have lead to equivocal results. The 'mixed' inducer, Aroclor 1254, a polychlorinated biphenyl mixture, with or without the accompanying polychlorinated dibenzofuran impurities, increased 4- to 5-fold the incidence of diethylnitrosamine (DEN)-induced hepatocellular carcinoma in rat (PRESTON et al., 1981). Similarly, PB at daily doses of 250 ppm or greater, enhanced the incidence of DEN-induced hepatocellular carcinoma in rats, thus showing a threshold for promotion of tumorigenesis; PB without DEN had no tumorigenic activity (PEREIRA et al., 1986). Thus, non-genotoxic carcinogens and tumour promoters, such as PB and dieldrin, exhibit a non-linear dose response, in contrast to the linear dose response of genotoxic carcinogens, and the promotional effectiveness of these non-genotoxic promoters appears to be related to their PB type enzyme induction potential. The liver tumour-promoting action in rats of a series of barbiturates was shown to be related to their enzyme-inducing potential and to correlate with their induction of the cytochromes P-450 and epoxide hydrolase, and with increased liver weight, in the following order of diminishing activity: phenobarbital > barbital > pentobarbital > amobarbital > hexobarbital > barbituric acid (NIMS et al., 1987). The significance to man of these findings in rat is uncertain as epidemiological studies of epileptic patients who had received large doses of PB and other anticonvulsant enzyme-inducing drugs for long periods of time have shown no increase of tumours. Unfortunately, many of these tumour promotion studies have been based on unreliable end-points, such as $\gamma$-glutamyltransferase ($\gamma$-GT)-positive lesions and other pre-malignant foci, instead of frank malignancy, and a recent study in rats showed that whereas PB, DDT and BHT, but not nafenopin (NAF), administered chronically after DEN initiation, enhanced the development of premalignant ($\gamma$-GT-positive) lesions at 3 to 14 weeks, only PB, DDT and NAF were found to enhance the development of frank liver carcinoma (PRÉAT et al., 1986). However, although PB promotes carcinogenesis when administered after DEN treatment, it decreases the carcinogenic effect when administered simultane-ously with DEN (BARBASON et al., 1986).

The induction of cytochromes P-450, other than P-448, may be involved in the initiation of malignancy. Cytochrome P-450$_{ALC}$ (P-450 II E) induced by alcohol, acetone, isoniazid or trichloroethylene, or by fasting (HONG et al., 1987) is responsible for the activation of the hepatocarcinogen, dimethyl-nitrosamine (KHANI et al., 1987). Cytochrome P-452 (P-450 IV), is induced by the hypolipidaemic drugs, clofibrate, ciprofibrate, the phthalate esters and other chemicals which give rise to peroxisomal proliferation, and is associated with, though is possibly independent of, the generation of peroxide and hepatocellular carcinoma (GOEL et al., 1986). These peroxisome proliferators constitute a novel class of non-mutagenic hepatocarcinogens all of which induce a similar pleiotropic response consisting of hepatomegaly, and induction of peroxisomal and microsomal enzymes (RAO and REDDY, 1987). However, as these observations have been confined mostly to high dosage of peroxisomal proliferators to rodents and probably would be dependent on the rate of tissue oxygen uptake, which is 10-fold higher in rodents than man, the low therapeutic doses for humans are unlikely to be associated with risk of malignancy (BERGE et al., 1984).

Atherosclerosis, in addition to malignant tumours, has been associated with mutagenic events. Inbred strains of mice with different susceptibilities to malignancy (Ah responsive and Ah non-responsive), when treated with MC, developed atherosclerotic lesions which were significantly greater in the Ah-responsive mice, indicating that the carcinogen MC can initiate atherosclerotic lesions and that the extent is dependent on metabolic activation by the cyto-chromes P-448 (PAIGEN et al., 1986).

# 7.  References

ABE, T. and M. WATANABE (1983), Mol. Pharmac. **23**, 258—264.
ADDISON, R. F., M. E. ZINCK, and D. E. WILLIS (1977), Comp. Biochem. Physiol. **57**, 39—43.
ALVARES, A. P. and A. KAPPAS (1975), FEBS Lett. **50**, 172—174.
ASTRÖM, A. and J. W. DE PIERRE (1985), Carcinogenesis **6**, 113—120.

BAGLEY, D. M. and J. R. HAYES (1985), Biochem. Pharmacol. **34**, 1007—1014.

BAINS, S. K., S. M. GARDINER, K. MANNWEILER, D. GELLETT, and G. G. GIBSON (1985), Biochem. Pharmacol. **34**, 3221—3229.

BARBASON, H., C. H. MORMONT, S. MASSART, and B. BOUZAHZAH (1986), Eur. J. Cancer Clin. Oncol. **22**, 1073—1078.

BERGE, R. K., L. H. HOSØY, A. AARSLAND, O. M. BAKKE, and M. FARSTAD (1984), Toxicol. Appl. Pharmacol. **73**, 35—41.

BILIMORIA, M. H. and D. J. ECOBICHON (1980), Toxicology **15**, 83—89.

BIRNBAUM, L. S. and M. B. BAIRD (1978a), Exp. Geront. **13**, 299—303.

BIRNBAUM, L. S. and M. B. BAIRD (1978b), Exp. Geront. **13**, 469—477.

BOOBIS, A. R., D. W. NEBERT, and J. S. FELTON (1977), Mol. Pharmacol. **13**, 259—268.

BORNHEIM, L. M., P. G. PETERS, and M. R. FRANKLIN (1983), Chem.-Biol. Interactions **47**, 45—55.

BRADLOW, J. A., L. H. GARTHOFF, N. E. HURLEY, and D. FIRESTONE (1980), Fd. Cosmet. Toxicol. **18**, 627—635.

BRECKENRIDGE, A. (1987), in: Enzyme Induction in Man, (E. A. SOTANIEMI and R. O. PELKONEN, eds.), Taylor and Francis, London. 201—209.

BURKE, M. D., S. THOMPSON, C. R. ELCOMBE, J. HALPERT, T. HAAPARANTA, and R. T. MAYER (1985), Biochem. Pharmacol. **34**, 3337—3345.

CHRISTOU, M., C. J. MOORE, M. N. GOULD, and C. R. JEFCOATE (1987), Carcinogenesis **8**, 73—80.

CONNEY, A. H., E. C. MILLER, and J. A. MILLER (1956), Cancer Res. **16**, 450—459.

CONNEY, A. H. (1986), Life Sci. **39**, 2493—2518.

COOK, J. C. and E. HODGSON (1985), Chem.-Biol. Interactions **54**, 299—315.

COON, M. J. and D. R. KNOOP (1987), Arch. Toxicol. **60**, 16—61.

COSTA, M., E. R. COSTA, C.-A. MANEN, I. G. SIPES, and D. H. RUSSELL (1976), Mol. Pharmacol. **12**, 871—878.

CREAVEN, P. J. and D. V. PARKE (1966), Biochem. Pharmacol. **15**, 7—16.

CRESTEIL, T., J.-L. MAHU, P. M. DANSETTE, and J.-P. LEROUX (1980), Biochem. Pharmacol. **29**, 1127—1133.

CRESTEIL, T., E. Le PROVOST, J. P. LEROUX, and P. LESCA (1982), Biochem. Biophys. Res. Commun. **107**, 1037—1045.

DEES, J. H., B. S. S. MASTERS, U. MULLER-EBERHARD, and E. F. JOHNSON (1982), Cancer Res. **42**, 1423—1432.

DEGAWA, M., M. KOJIMA, T. MASUKO, T. HISHIMUNA, and Y. HASHIMOTO (1985), Biochem. Biophys. Res. Commun. **133**, 1072—1077.

DEGAWA, M., M. KOJIMA, Y. SATO, and Y. HASHIMOTO (1986), Biochem. Pharmacol. **35**, 3565—3570.

DENOMME, M. A., S. BANDIERA, I. LAMBERT, L. COPP, L. SAFE, and S. SAFE (1983), Biochem. Pharmacol. **32**, 2955—2963.

ELSHOURBAGY, N. A. and P. S. GUZELIAN (1980), J. Biol. Chem. **255**, 1279—1285.

FABACHER, D. L., A. P. KULKARNI, and E. HODGSON (1980), Gen. Pharmacol. **11**, 429—435.

GELBOIN, H. V. (1983), N. Engl. J. Med. **309**, 105.

GOEL, S. K., N. D. LALWANI, and J. K. REDDY (1986), Cancer Res. **46**, 1324—1330.

GOLDSTEIN, J. A. and P. LINKO (1984), Mol. Pharmacol. **25**, 185—191.

GONZALEZ, F. J., R. H. TUKEY, and D. W. NEBERT (1984), Mol. Pharmacol. **26**, 117 to 121.

GONZALEZ, F. J., S. KIMURA, and D. W. NEBERT (1985), J. Biol. Chem. **260**, 5040—5049.

GULYAEVA, L. F., V. M. MISHIN, and V. V. LYACHESLAV (1986), Int. J. Biochem. **18**, 829—834.

GUTKINA, N. I. and V. M. MISKIN (1986), Chem.-Biol. Interactions **58**, 57—68.

HAAKE, J., M. KELLEY, B. KEYS, and S. SAFE (1987), Gen. Pharmacol. **18**, 165—169.
HARDWICK, J. P., F. SCHWALM, and A. RICHARDSON (1983a), Biochem. J. **210**, 599—606.
HARDWICK, J. P., F. J. GONZALEZ, and C. B. KASPER (1983b), J. Biol. Chem. **258**, 8081—8085.
HARDWICK, J. P., F. J. GONZALEZ, and C. B. KASPER (1983c), J. Biol. Chem. **258**, 10182—10186.
HARDWICK, J. P., B.-J. SONG, E. HUBERMAN, and F. J. GONZALEZ (1987), J. Biol. Chem. **262**, 801—810.
HESS, R., W. STAUBLI, and W. REISS (1965), Nature (London) **208**, 856—858.
HEUMAN, D. M., E. J. GALLAGHER, J. L. BARWICK, N. A. ELSHOURBAGY, and P. S. GUZELIAN (1982), Mol. Pharmacol. **21**, 753—760.
HOLME, J. A. and E. DYBING (1982), Biochem. Pharmacol. **31**, 2523—2529.
HONG, J., J. PAN, F. J. GONZALEZ, H. V. GELBOIN, and C. S. YOUNG (1987), Biochem. Biophys. Res. Commun. **142**, 1077—1083.
HOSTETLER, K. A., S. A. WRIGHTON, P. KREMERS, and P. S. GUZELIAN (1987), Biochem. J. **245**, 27—33.
HSIA, M. T. S. and B. L. KREAMER (1979), Res. Commun. Chem. Pathol. Pharmacol. **25**, 319—331.
IOANNIDES, C. and D. V. PARKE (1975), J. Pharm. Pharmacol. **27**, 739—746.
IOANNIDES, C., P. Y. LUM, and D. V. PARKE (1984), Xenobiotica **14**, 119—138.
IOANNIDES, C., M. DELAFORGE, and D. V. PARKE (1985), Chem.-Biol. Interactions **53**, 303—311.
IOANNIDES, C. and C. M. STEELE (1986), Chem.-Biol. Interactions **59**, 129—139.
IOANNIDES, C. and D. V. PARKE (1987), Biochem. Pharmacol. **36**, 4197—4207.
IWASAKI, K., P. Y. LUM, C. IOANNIDES, and D. V. PARKE (1986), Biochem. Pharmacol. **35**, 3879—3884.
JAMES, M. O. and J. R. BEND (1980), Toxicol. Appl. Pharmacol. **54**, 117—133.
JAYYOSI, Z., M. TOTIS, H. SOUHAILI, C. GOULON-GINET, M. H. LIVERTOUX, A. M. BATT, and G. SIEST (1987), Biochem. Pharmacol. **36**, 2481—2487.
JOHANSSON, I. and M. INGELMAN-SUNDBERG (1985), FEBS Lett. **183**, 265—269.
JOLY, J.-G., J.-P. VILLENEUVE, and P. MAVIER (1977), Alcoholism: Clin. Exper. Res. **1**, 17—20.
KAMATAKI, T., K. MAEDA, M. SHIMADA, and R. KATO (1986), J. Biochem. **99**, 841—845.
KAO, J. and P. HUDSON (1980), Biochem. Pharmacol. **29**, 1191—1194.
KÄRENLAMPI, S. O., H. J. EISEN, O. HANKINSON, and D. W. NEBERT (1983), J. Biol. Chem. **258**, 10378—10383.
KHANI, S. C., P. G. ZAPHIROPOULOS, V. S. FUJITA, T. D. PORTER, D. R. KOOP, and M. J. COON (1987), Proc. Natl. Acad. Sci. U.S.A. **84**, 638—642.
KIMURA, T., M. KODAMA, C. NAGATA, T. KAMATAKI, and R. KATO (1985), Biochem. Pharmacol. **34**, 3375—3377.
KIMURA, S., F. J. GONZALEZ, and D. W. NEBERT (1986), Mol. Cell Biol. **6**, 1471—1477.
KIMURA, S., J.-C. DONOVAN, and D. W. NEBERT (1987), J. Exp. Pathol. **3**, 61—74.
KLING, L., L. WOLFGANG, and K. J. NETTER (1985), Biochem. Pharmacol. **34**, 85—91.
KOURI, R. E., T. H. RUDE, R. JOGLEKAR, P. M. DANSETTE, D. M. JERINA, S. A. ATLAS, I. S. OWENS, and D. W. NEBERT (1978), Cancer Res. **38**, 2777—2783.
KRIJGSHELD, K. R. and T. E. GRAM (1984), Biochem. Pharmacol. **33**, 1951—1956.
KVINNSLAND, S., P. E. LØNNING, and P. M. UELAND (1986), Breast Cancer Research and Treatment **7** (Suppl.), 73—76.
LABADARIOS, D. (1975), Studies on the effects of drugs on nutritional status. Ph. D. Thesis, University of Surrey.
LABADARIOS, D., J. W. T. DICKERSON, D. V. PARKE, E. G. LUCAS, and G. H. OBUWA (1977), Br. J. Clin. Pharmacol. **4**, 167—173.

---

LAKE, B. G., T. J. B. GRAY, and S. D. GANGOLLI (1986), Environ. Hlth. Persp. **67**, 283—290.

LAVRIJSEN, K., J. VAN HOUDT, D. THIJS, W. MEULDERMANS, and J. HEYKANTS (1986), Biochem. Pharmacol. **35**, 1867—1878.

LEIGHTON, J. K. and B. KEMPER (1984), J. Biol. Chem. **259**, 11165—11168.

LEWIS, D. F. V., C. IOANNIDES, and D. V. PARKE (1986), Biochem. Pharmacol. **35**, 2179—2185.

LIEM, H. H., U. MULLER-EBERHARD, and E. F. JOHNSON (1980), Mol. Pharmacol. **18**, 565—570.

LINDELL, T. J., R. ELLINGER, J. T. WARREN, D. SUNDHEIMER, and A. F. O'MALLEY (1977), Mol. Pharmacol. **13**, 426—434.

LINDESKOG, P., T. HAAPARANTA, M. NORGÅRD, H. GLAUMANN, T. HANSSON, and J.-A. GUSTAFSSON (1986), Arch. Biochem. Biophys. **244**, 492—501.

LINKO, P., H. N. YEOWELL, T. A. GASIEWICZ, and J. A. GOLDSTEIN (1986), J. Biochem. Toxicol. **1**, 95—107.

LONNING, P. E., S. KVINNSLAND, T. THORSEN, and D. EKSE (1986), Breast Cancer Research and Treatment **7** (Suppl.), 77—82.

LOTLIKAR, P. D., R. N. PANDEY, M. S. CLEARFIELD, and S. M. PAIK (1984), Toxicol. Lett. **21**, 111—118.

LUBET, R. A., R. T. MAYER, J. W. CAMERON, R. W. NIMS, M. D. BURKE, T. WOLFF, and F. P. GUENGERICH (1985), Arch. Biochem. Biophys. **238**, 43—48.

LUOMA, P. V. (1987), in: Enzyme Induction in Man, (E. A. SOTANIEMI and R. O. PELKONEN, eds.), Taylor and Francis, London. 231—242.

MALEJKA-GIGANTI, D., R. C. McIVER, A. L. GLASEBROOK, and H. R. GUTMANN (1978), Biochem. Pharmacol. **27**, 61—69.

MATHIS, J. M., R. A. PROUGH, R. N. HINES, E. BRESNICK, and E. R. SIMPSON (1986), Arch. Biochem. Biophys. **246**, 439—448.

McDANELL, R., A. E. M. McLEAN, A. B. HANLEY, R. K. HEANEY, and G. R. FENWICK (1987), Fd. Chem. Toxicol. **25**, 363—368.

McKILLOP, D. (1985), Biochem. Pharmacol. **34**, 3137—3142.

McLAIN, G. E., I. G. SIPES, and B. R. BROWN Jr. (1979), Anesthesiology **51**, 321—326.

MEBUS, C. A. and W. N. PIPER (1986), Biochem. Pharmacol. **35**, 4359—4362.

MESSNER, B., J. STILL, and J. BERNDT (1984), Pesticide Biochem. Physiol. **21**, 283—289.

MICHALOPOULOS, G., C. A. SATTLER, G. L. SATTLER, and H. C. PITOT (1976), Science **193**, 907—909.

MIZOKAMI, K., T. NOHMI, M. FUKUHARA, A. TAKANAKA, and Y. OMORI (1986), Biochem. Biophys. Res. Commun. **139**, 466—472.

MOROHASHI, K.-I., H. YOSHIOKA, K. SOGAWA, Y. FUJII-KURIYAMA, and T. OMURA (1984), J. Biochem. **95**, 949—957.

NAGATA, K., T. MATSUNAGA, J. GILLETTE, and H. V. GELBOIN (1987), J. Biol. Chem. **262**, 2787—2793.

NEBERT, D. W. and N. M. JENSEN (1979), Crit. Rev. Biochem. **6**, 401—437.

NEBERT, D. W. and F. J. GONZALEZ (1987), Ann. Rev. Biochem. **56**, 945—93.

NEBERT, D. W., M. ADESNIK, M. J. COON, R. W. ESTABROOK, F. J. GONZALEZ, F. P. GUENGERICH, I. C. GUNSALUS, E. F. JOHNSON, B. KEMPER, W. LEVIN, I. R. PHILLIPS, R. SATO, and M. R. WATERMAN (1987), DNA **6**, 1—11.

NEWMAN, S. L. and P. S. GUZELIAN (1983), Biochem. Pharmacol. **32**, 1529—1531.

NIMS, R. W., D. E. DEVOR, J. R. HENNEMAN, and R. A. LUBET (1987), Carcinogenesis **8**, 67—71.

NIWA, A., K. KUMAKI, and D. W. NEBERT (1975), Mol. Pharmacol. **11**, 399—408.

ORTIZ DE MONTELLANO, P. R. and N. O. REICH (1984), J. Biol. Chem. **259**, 4136 to 4141.

OZAWA, N., S. YOSHIHARA, K. KAWANO, Y. OKADA, and H. YOSHIMURA (1979), Biochem. Biophys. Res. Commun. **91**, 1140—1147.

PAIGEN, B., P. A. HOLMES, A. MORROW, and D. MITCHELL (1986), Cancer Res. **46**, 3321—3324.

PAINE, A. J., and A. E. M. MCLEAN (1974), Biochem. Biophys. Res. Commun. **58**, 482—486.

PAINE, A. J. (1978), Biochem. Pharmacol. **27**, 1805—1813.

PAINE, A. J., and J. E. FRANCIS (1980), Chem.-Biol. Interactions **30**, 343—353.

PARKE, D. V. (1984), In: Toxicology and the Newborn, Eds. S. KACEW and M. J. REASOR. Elsevier, Amsterdam. Pp. 1—31.

PARKE, D. V. (1985), Marine Environ. Res. **17**, 97—100.

PARKE, D. V., C. IOANNIDES, and D. F. V. LEWIS (1986), in: Toxicology in Europe in the Year 2000, FEST Suppl., (C. M. Hodel, ed.), Elsevier, Cambridge. Pp. 14—18.

PARKE, D. V. (1987), Arch. Toxicol. **60**, 5—15.

PARKINSON, A., L. ROBERTSON, L. SAFE, and S. SAFE (1980), Chem.-Biol. Interactions **30**, 271—285.

PARKINSON, A., A. H. SAFE, L. W. ROBERTSON, P. E. THOMAS, D. E. RYAN, L. M. REIK, and W. LEVIN (1983), J. Biol. Chem. **258**, 5967—5976.

PEREIRA, M. A., S. L. HERREN-FREUND, and R. E. LONG (1986), Cancer Lett. 305 to 311.

PERUCCA, E. (1978), Pharmac. Ther. **2**, 285—314.

PERUCCA, E., E. HEDGES, K. A. MAKKI, M. RUPRAH, J. F. WILSON, and A. RICHENS (1984), Br. J. Clin. Pharmacol. **18**, 401—410.

PESSAYRE, D., M. BENTATA, C. DEGOTT, O. NOUEL, J.-P. MIGUET, B. RUEFF, and J.-P. BENHAMOU (1977), Gastroenterology **72**, 284—289.

PHILLIPS, J. C., C. E. HEADING, B. G. LAKE, S. D. GANGOLLI, and A. G. LLOYD (1975), Fd. Cosmet. Toxicol. **13**, 611—617.

PHILLIPS, I. R., E. A. SHEPHARD, F. MITANI, and B. R. RABIN (1981), Biochem. J. **196**, 839—851.

PHILLIPSON, C. E., P. M. M. GODDEN, P. Y. LUM, C. IOANNIDES, and D. V. PARKE (1984), Biochem. J. **221**, 81—88.

PICKETT, C. B., R. L. JETER, R. WANG, and A. Y. H. LU (1983), Arch. Biochem. Biophys. **225**, 854—860.

PIRTTIAHO, H. I. (1987), in: Enzyme Induction in Man, (E. A. SOTANIEMI and R. O. PELKONEN, eds.), Taylor and Francis, London. Pp. 125—138.

PISKORSKA-PLISZCZYNSKA, J., B. KEYS, S. SAFE, and M. S. NEWMAN (1986), Toxicology Lett. **34**, 67—74.

POLAND, A. and E. GLOVER (1974), Mol. Pharmacol. **10**, 349—359.

POULSEN, H. E., A. LERCHE, and N. T. PEDERSEN (1985), Pharmacology **30**, 100 to 108.

PRASAD, N., R. PRASAD, J. THORNBY, S. C. BUSHONG, L. B. NORTH, and J. E. HARRELL (1977), Cancer Res. **37**, 3771—3773.

PRÉAT, V., J. DE GERLACHE, M. LANS, H. TAPER, and M. ROBERFROID (1986), Carcinogenesis **7**, 1025—1028.

PRESTON, B. D., J. P. VAN MILLER, R. W. MOORE, and J. R. ALLEN (1981), J. Natl. Cancer Inst. **66**, 509—515.

RATING, D., E. JÄGER-ROMAN, H. NAU, W. KUHNZ, and H. HELGE (1983), Pediat. Pharmacol. **3**, 209—218.

RAO, M. S. and J. K. REDDY (1987), Carcinogenesis 8, 631—636.

RAUNIO, H. and O. PELKONEN (1983), Cancer Res. **43**, 782—786.

REDDY, J. K. and T. P. KRISHNAKANTHA (1975), Science (Wash. DC) **190**, 787 to 789.

REDDY, J. K., K. GOEL, R. NEMALI, J. CARRINO, G. LAFFLER, M. REDDY, J. SPERBECK, T. OSUMI, T. HASHIMOTO, N. D. LALWNI, and M. S. RAO (1986), Proc. Natl. Acad. Sci. U.S.A. **83**, 1747—1751.

REES, D. E. (1979), Gen. Pharmacol. **10**, 341—350.

REICH, N. O. and P. R. ORTIZ DE MONTELLANO (1986), Biochem. Pharmacol. **35**, 1227—1233.

REMMER, H. and B. ALSLEBEN (1958), Klin. Wochenschrift **36**, 332—333.

RITTER, C. L. and D. MALEJKA-GIGANTI (1982), Biochem. Pharmacol. **31**, 239—247.

ROBERTSON, L. W., A. PARKINSON, S. BANDIERA, and S. SAFE (1981), Chem.-Biol. Interactions **35**, 13—24.

ROBINSON, R. C., K.-C. CHENG, S. S. PARK, H. V. GELBOIN, and F. K. FRIEDMAN (1986), Biochem. Pharmacol. **35**, 3827—3830.

RYAN, D. E., L. RAMANATHAN, S. IIDA, P. E. THOMAS, M. HANIU, J. E. SHIVELY, C. S. LIEBER, and W. LEVIN (1985), J. Biol. Chem. **260**, 6385—6393.

SAARNI, H. U., J. STENGÅRD, N. T. KÄRKI, and E. A. SOTANIEMI (1983), Biochem. Pharmacol. **32**, 1075—1081.

SAFE, S., S. BANDIERA, T. SAWYER, B. ZMUDZKA, G. MASON, M. ROMKES, M. A. DE-NOMME, J. SPARLING, A. B. OKEY, and T. FUJITA (1985), Environ. Health Perspect. **61**, 21—33.

SCHWEN, R. J. and G. J. MANNERING (1982), Comp. Biochem. Physiol. **71B**, 445—453.

SERABJIT-SINGH, C. J., J. R. BEND, and R. M. PHILPOT (1985), Mol. Pharmacol. **28**, 72—79.

SHAPIRO, B. H., G. M. LECH, and R. M. BARDALES (1986), J. Pharmac. Exp. Ther. **238**, 68—75.

SHIVERICK, K. T., A. G. KVELLO-STENSTROM, W. H. DONNELLY, A. S. SALHAB, J. A. GOLDSTEIN, and M. O. JAMES (1986), Biochem. Biophys. Res. Commun. **141**, 299—305.

SINGLETARY, S. E., F. L. BAKER, G. SPITZER, S. L. TUCKER, B. TOMASOVIC, W. A. BROCK, J. A. AJANI, and A. M. KELLY (1987), Cancer Res. **47**, 403—406.

SMITH, S. J., D. K. LIU, T. B. LEONARD, B. W. DUCEMAN, and E. S. VESELL (1976), Mol. Pharmacol. **12**, 820—831.

SOTANIEMI, E. A., J. T. LAHTELA, and J. STENGÅRD (1987), in: Enzyme Induction in Man, (E. A. SOTANIEMI and R. O. PELKONEN, eds.), Taylor and Francis, London. Pp. 219—230.

STATHAM, C. N. and M. R. BOYD (1982), Biochem. Pharmacol. **31**, 3973—3977.

STOSCHECK, C. M. and L. E. KING (1986), Cancer Res. **46**, 1030—1037.

SUAREZ, K. A., G. P. CARLSON, G. C. GULLER, and N. FAUSTO (1972), Toxicol. Appl. Pharmacol. **23**, 171—177.

SVOBODA, D., H. GRADY, and D. AZARNOFF (1967), J. Cell. Biol. **35**, 127—152.

TAMBURINI, P., H. A. MASSON, S. K. BAINS, R. J. MAKOWSKI, B. MORRIS, and G. G. GIBSON (1984), Eur. J. Biochem. **139**, 235—246.

THOMAS, P. E., L. M. REIK, D. E. RYAN, and W. LEVIN (1983), J. Biol. Chem. **258**, 4590—4598.

TUCHWEBER, B., K. KOVACS, M. V. JAGO, and T. BEAULIEU (1974), Res. Commun. Chem. Pathol. Pharmacol. **7**, 459—480.

TUTEJA, N., F. J. GONZALEZ, and D. W. NEBERT (1986), Biochem. Pharmacol. **35**, 718—720.

UENG, T.-H., F. K. FRIEDMAN, H. MILLER, S. S. PARK, H. V. GELBOIN, and A. P. ALVARES (1987), Biochem. Pharmacol. **36**, 2689—2691.

WAXMAN, D. J., G. A. DANNAN, and F. P. GUENGERICH (1985), Biochemistry **24**, 4409—4417.

WELCH, R. M., Y. E. HARRISON, A. H. CONNEY, P. J. POPPERS, and M. FINISTER (1968), Science (Wash. D.C.) **160**, 541—542.

WOLF, C. R., S. SEILMAN, F. OESCH, R. T. MAYER, and M. D. BURKE (1986), Biochem. J. **240**, 27—33.

WONG, T. K., B. A. DOMIN, P. E. BENT, T. E. BLANTON, M. W. ANDERSON, and R. M. PHILPOT (1986), Cancer Res. **46**, 999—1004.

WRIGHTON, S. A., P. MAUREL, E. G. SCHUETZ, P. B. WATKINS, B. YOUNG, and P. S. GUZELIAN (1985), Biochemistry **24**, 2171—2178.

YAMAMOTO, T., S. IKAWA, T. AKIYAMA, K. SEMBA, N. NOMURA, N. MIYAJIMA, T. SAITO, and K. TOYOSHIMA (1986), Nature, Lond. **319**, 230—234.

# Chapter 2
# The P450 Gene Superfamily

D. W. Nebert and F. J. Gonzalez

The evolution, structure and regulation of genes in the P450 superfamily have been extensively reviewed through early 1987 (NEBERT and GONZALEZ, 1987). The purpose of this Chapter is to update this information through spring of 1988. For many of the numerous trivial names for the same P450 protein, the reader is referred to the 1987 review and references therein. The references to many findings described in this Chapter can be found in the bibliography of NEBERT and GONZALEZ (1987) and NEBERT et al. (1989a).

# 1. P450 protein sequence comparisons and evolution

The divergence of distinct P450 gene families from a common ancestor has been estimated with the aid of computer algorithms for sequence alignment and phylogenetic tree construction (NEBERT et al., 1987; NEBERT and GONZALEZ, 1987; NELSON and STROBEL, 1987; NEBERT et al., 1989b). From such estimations a nomenclature system has been proposed, with Roman numerals and capital letters denoting families and subfamilies, respectively (NEBERT

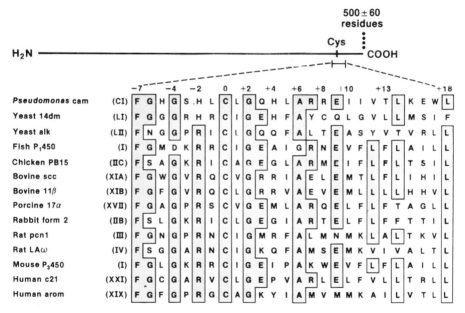

**Fig. 1.** Diagram of the linear P450 protein and approximate location of the highly conserved cysteinyl-containing peptide involved in the heme-binding site among ten eukaryotic species and one prokaryote. Positive and negative numbers refer to amino acid positions downstream and upstream, respectively, from the cysteine that binds the heme iron.

et al., 1987). The nomenclature for human and mouse chromosomal loci (P450 genes) has recently been proposed (NEBERT et al., 1989a).

If one examines the P450 amino acid sequences by global alignment programs, it has been established that, although there are many dissimilarities among the proteins, there is a 26-residue region near the carboxy-terminus of all P450 proteins that exhibits a high degree of similarity (Fig. 1). These data strongly suggest that all P450 genes — ranging from bacteria to humans — have diverged from the same ancestral gene, which probably existed more than 2 billion years ago (Fig. 2). Presently in our data base, we have more than 100 P450 amino acid sequences, derived from the unique genes of 17 eukaryotes and two prokaryotes. Each gene produces a single protein (the enzyme), and the superfamily comprises at least 20 families with the II

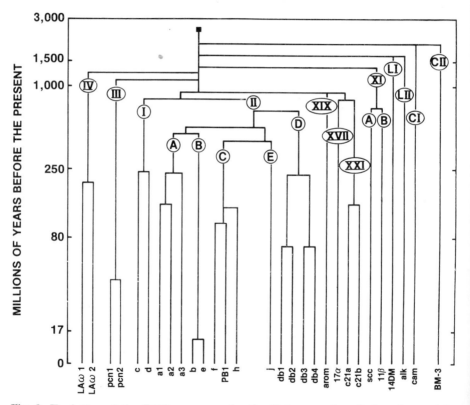

**Fig. 2.** Evolution of the P450 gene superfamily. Data are primarily based on rat for families I through IV, human for families XI, XVII, XIX, and XXI, two yeast species for families LI and LII, and two bacterial species for families CI and CII. The data were compiled from all P450 full-length cDNA or protein sequences available to us as of the end of May, 1988.

D. W. NEBERT; F. J. GONZALES

family having eight subfamilies and the XI family having two subfamilies (Fig. 2).

With use of the FASTP algorithm (WILBUR and LIPMAN, 1983), the amino acid sequence of a P450 protein from one gene family exhibits $\leq 37\%$ resemblance to that from any other family, the amino acid sequences of two or more P450 proteins within the same subfamily are $\geq 60\%$ similar, and the amino acid sequence of a P450 from a particular subfamily gene subfamily is $38\%$ to $59\%$ identical to that from any other subfamily in the same family.

All mammals contain at least ten P450 gene families. One can appreciate from Figure 2 that there has been a "burst" of new P450 genes during the past 800 million years. Family II is known to have a minimum of 26 gene duplications (NELSON and STROBEL, 1987). The emergence of a new gene would represent a gene duplication event, followed by divergence due to mutations and, ultimately, fixation of a gene having in some instances selective advantages during evolutionary pressures. The emergence of many new genes, particularly in family II, most likely reflects "animal-plant warfare" that has been carried on during the past 800 million years. As animals diverged from plants and began to ingest plants, the plants enhanced their own survival by synthesizing new noxious metabolites to make them less attractive. In return, animals responded by developing new enzymes to process the new noxious plant metabolites. Twentieth-century drugs are often derived from plant metabolites and always resemble in chemical structure such metabolites. It is therefore not surprising that animals possess sufficiently diverse P450 enzymes to handle essentially any newly synthesized drug or chemical: the P450 superfamily in animals has evolved under this kind of evolutionary pressure for at least 800 million years.

One other appreciation of this evolutionary scheme (Fig. 2) is worthy of mention. Since the human-rodent split approximately 80 million years ago, it is entirely possible that different mammalian species will have varying numbers of functional P450 genes in their corresponding subfamily. Four such examples have already been well documented. (a) The rat IID subfamily has five active genes, while the human IID subfamily has only three (GONZALEZ and NEBERT, in press). (b) Rabbit forms 1 and 1-88 in the IIC subfamily have diverged after the rabbit-rodent separation, meaning that one of these two genes does not exist in the rat or mouse (or human). (c) Rat b and e diverged approximately 12 million years ago, whereas the rat-mouse split occurred about 17 million years ago; this means that the orthologue of rat b (or e) does not exist in the mouse (or rabbit or human). (d) The rabbit IIE subfamily has two members that arose via gene duplication after the rabbit-rodent divergence (KHANI et al., 1988), meaning that one of these two genes does not exist in the rat or mouse (or human). These observations and evolutionary explanation for differences in drug metabolism between, for example, human and rat might have important ramifications for drug companies to consider: the degree of toxicity, teratogenicity or carcinogenicity of a new drug determined in a rat may not be similar to that in the human — due to the presence or absence of a particular P450 gene in one of the two species.

## 2. The *CYP1* gene family

Family I in rat, rabbit, mouse, and human has two genes that are inducible by polycyclic hydrocarbons, benzoflavones and tetrachlorodibenzo-p-dioxin (TCDD; "dioxin"). In response to the ingestion of flavones, many of which are found in plants, we believe that, more than 500 million years ago, animals developed this P450 induction system, complete with a receptor for plant flavones (NEBERT and GONZALEZ, 1987). All of the CYP1 genes examined to date have seven exons, and, of the ten mammalian P450 families, only the CYP1 genes have a nontranslated first exon. Rat, rabbit, mouse and human are known to have two active CYP1 genes: CYP1A1 and CYP1A2. These correspond to rat P450c and P450d, respectively, rabbit forms 6 and 4, respectively, and mouse or human $P_1450$ and $P_3450$, respectively. The mouse $P_2450$ protein was found to be an allelic variant of the C57BL/6N mouse $P_3$ gene and differs in only one amino acid residue, plus two nucleotide changes in the 3′ nontranslated region (KIMURA and NEBERT, 1986).

The human $P_1$ full-length cDNA and complete gene and flanking regions have been isolated and sequenced in several laboratories. An increased risk of cigarette smoking-induced bronchogenic carcinoma appears to be associated with the high $P_1$ inducibility phenotype (KOURI et al., 1982). Although restriction fragment length polymorphisms (RFLPs) have been detected with the human $P_1$ cDNA (JAISWAL and NEBERT, 1986; SPURR et al., 1987), no correlation between enhanced $P_1$ inducibility and any of these RFLP patterns has been found to date (JAISWAL et al., 1985; JAISWAL and NEBERT, 1986). The phenotype of $P_1$ enzyme inducibility can be correlated with the $P_1$ mRNA content of lymphocytes cultured with mitogen and polycyclic hydrocarbon inducer (JAISWAL et al., 1985). Interestingly, Epstein-Barr virus-transformed B cell lines from individual patients (WAITHE et al., 1986) do not retain their $P_1$ inducibility phenotype.

The human $P_3$ full-length cDNA (JAISWAL et al., 1986; 1987) and complete gene and flanking regions (K. IKEYA et al., 1989; QUATTROCHI et al., 1986) have also been sequenced. Since the $P_3$ protein actively metabolizes 2-acetyl-aminofluorence and aminobiphenyls (KADLUBAR and HAMMONS, 1987), which are potent carcinogens in laboratory animals, the human orthologue should also be of great interest in clinical cancer genetics studies.

The trout $P_1$ cDNA has also been sequenced (HEILMANN et al., 1988). The deduced protein is statistically significantly more similar to mammalian P450 IA1 than IA2 proteins, and there is no Southern blot or Northern blot evidence for a second CYP1 gene in trout. These data suggest that the CYP1A1 gene is the older and that the CYP1A2 gene arose in land animals via a duplication event sometime after the separation of trout and mammalian predecessors.

Somatic cell genetics has been used to identify genes necessary for the mouse $P_1450$ [aryl hydrocarbon hydroxylase (AHH)] induction response. AHH-inducible cells grown in the presence of benzpyrene generate lethal amounts of toxic metabolites. By means of resistance to benzpyrene toxicity, therefore, clones deficient in AHH induction have been selected from the mouse hepatoma line Hepa-1 (HANKINSON, 1979) and shown to be mutational in origin and stable in phenotype (HANKINSON, 1981). The recessive mutant clones have been assigned to at least four complementation groups called A, B, C, and D. The $P_1450$ cDNAs from two Group A mutants have been sequenced and found to be defective (KIMURA et al., 1987b), confirming that the $P_1$ structural gene is defective in Group A variants ($P_1^-$). Groups B, C, and D represent regulatory mutant cell lines. Mutations in Group B result in less than 10% of the aromatic hydrocarbon (Ah) receptor levels (LEGRAVEREND et al., 1982) normally found in the parent Hepa-1 wild-type ($wt$) and are designated receptorless ($r^-$). Group B variants probably include mutations in the Ah receptor gene(s). Mutations in Group C affect the normal nuclear translocation of the inducer-receptor complex (LEGRAVEREND et al., 1982) and are designated nuclear translocation-defective ($nt^-$). Another possibility is that this mutation affects chromatin binding of the Ah receptor and that the line be designated $cb^-$ for chromatin-binding-defective. The Group D mutant appears to lack the Ah receptor, which can be restored by 5-azacytidine or sodium butyrate treatment but not mutagen treatment. Miller and coworkers have independently isolated and characterized $r^-$ and $nt^-$ benzpyrene-resistant AHH$^-$ mutant lines (MILLER and WHITLOCK, 1981; MILLER et al., 1983).

Expression vectors containing mouse and human $P_1$ upstream sequences and the chloramphenicol acetyltransferase (CAT) gene in $wt$, $nt^-$, and $P_1^-$ stable transformants in our laboratories have provided evidence for (a) a promoter containing a TATA box, (b) a negative control element between 400 and 800 bases upstream from the mRNA cap site, and (c) a dioxin-inducible enhancer between 900 and 1,100 bases upstream from the cap site. Both the negative element and the positive element are dependent upon a functional Ah receptor for metabolism-deficient derepression and induction by foreign chemicals, respectively (NEBERT and JONES, 1989).

$P_1$ sequences between 400 and 600 bases upstream from the cap site are involved in a negative autoregulatory loop controlling constitutive gene expression (GONZALEZ and NEBERT, 1985). Absence of $P_1$ metabolism leads to the lack of a putative repressor binding to the negative control region; consequently, constitutive $P_1$ mRNA concentrations are exceedingly high in Group A mutants (HANKINSON et al., 1985). Interestingly, this metabolism-deficient-dependent derepression of constitutive activity appears to require the Ah receptor, since no augmentation of control activity is seen in the $nt^-$ mutant containing 368, 823, or 1,646 bp of upstream sequences. The interaction of the receptor-dependent, metabolism-dependent, and promoter regions of the $P_1$ upstream sequences is thus regarded as extremely complicated.

At least three *cis*-acting drug regulatory elements (DREs) — with a consensus sequence of 5'-(G/C)N(T/G)(A/G)GCTGGG-3' — were identified in the rat CYP1A1 flanking region at approximately 1.0, 1.5 and 3.3 kb upstream from the mRNA cap site (SOGAWA et al., 1986). Two xenobiotic responsive elements (XREs) — with a consensus sequence of 5'-C(G/C)T(G/C)(C/T)-T(G/C)TCACGC(T/A)A-3' — were identified between 1,000 and 1,100 bases upstream from the rat CYP1A1 cap site (FUJISAWA-SEHARA et al., 1987). Both the DREs and the XREs are present at similar sites in the mouse and human CYP1A1 upstream regions (NEBERT and JONES 1989), and it is likely that inverted repeats of both these elements are involved in the positive regulation (induction by foreign chemicals), as well as the negative regulation (derepression in the absence of a functional CYP1A1 gene product) of this gene. Further elucidation of these complex regulatory mechanisms will be relevant to gaining a better understanding of polycyclic hydrocarbon carcinogenesis.

# 3.    The *CYP2* gene family

This group of genes was originally called the "phenobarbital-inducible family." It is now clear that most phenobarbital-inducible genes reside within the IIB and IIC subfamilies. All of the CYP2 genes examined to date contain nine exons.

## 3.1.    The *CYP2A* subfamily

Rat P450a is highly specific for testosterone $7\alpha$-hydroxylase activity. Since three rat P450a cDNA sequences have now been characterized (NEBERT et al., 1989a), we shall refer to them as IIA1, IIA2 and IIA3 or, more simply, a1, a2 and a3, and the genes as CYP2A1, CYP2A2 and CYP2A3. The a1 cDNA from nontreated rat liver has been isolated and sequenced and found to be markedly induced by 3-methylcholanthrene but not phenobarbital (NAGATA et al., 1987).

The rat a1 protein is 52% similar to the rat b and e proteins. When rat e cDNA was used for screening a human liver cDNA library (PHILLIPS et al., 1985), the partial cDNA sequence of the gene called P450(1) was obtained and presumed to be a member of the IIB subfamily. Further analysis showed that this gene belongs to the IIA subfamily (NEBERT et al., 1987). The amino acid sequence of P450(1) (331 residues only) is 60% and 51% similar to rat a and e, respectively. All human-rodent orthologues to date exhibit between 71% and 80% similarity in amino acid sequence (NEBERT et al., 1987); hence, P450(1) is in the subfamily IIA but is not the orthologue of rat a. NAGATA and GONZALES (unpublished) have recently isolated the same P450a cDNA that PHILLIPS et al. (1985) described; this protein is 65%, 72% and 78% similar to the rat a1, a2 and a3 proteins, respectively, indicating that the human gene is probably orthologous to the rat CYP2A3 gene.

Rat liver a2 cDNA expresses testosterone 15α- and 6β-hydroxylase activities and also produces one other major, as yet unidentified, testosterone metabolite (MATSUNAGA et al., 1988). These activities are in contrast to the highly specific testosterone 7α-hydroxylase reaction of a1 (NAGATA et al., 1987). Rat a3 appears to be specifically expressed in lung and is also inducible by 3-methylcholanthrene (GONZALES, 1988). The cDNA for a male-specific testosterone 15α-hydroxylase in mouse liver and kidney was recently reported (SQUIRES and NEGISHI, 1988); the deduced protein was found to be 69%, 65% and 94% similar to the rat a1, a2 and a3 proteins, respectively. It is therefore possible that the mouse 15α gene is the orthologue of the rat CYP2A3 gene. The rat CYP2A1 and CYP2A2 genes contain nine exons (GONZALEZ, 1988).


### 3.2. The *CYP2B* subfamily

It became clear that two genes in the rat IIB subfamily exist because the two gene products (REIK et al., 1985), as well as both genes (SUWA et al., 1985), have been well characterized. Sharing 14 amino acid differences, the b and e proteins exhibit 97% identity. A third rat IIB gene, CYP2B3, has recently been identified (LABBE et al., 1988). At least one phenobarbital-inducible gene has been partially characterized in the mouse (STUPANS et al., 1984).

Numerous reports of liver P450 cDNA sequences from phenobarbital-treated rats and rabbits have described deduced proteins having between two and 11 amino acid differences (NEBERT and GONZALEZ, 1987 and references therein; GASSER et al., 1988; IMAI et al., 1988). Such microheterogeneity is believed to represent allelic variants of the two or three genes in the IIB subfamily. There is probably a very extensive polymorphism among rabbits that are not highly inbred. Whether rabbit form 2 is the orthologue of rat b or e has not been established.

A chicken gene (clone PB15; gene A) has been sequenced and shown to be induced by phenobarbital, allylisopropylacetamide, and 3,5-diethoxy-carbonyl-1,4-dihydrocollidine (HOBBS et al., 1986). Two genes (A and B) are transcriptionally active, producing mRNAs of 3.2 and 2.5 kb, respectively (B. K. MAY, personal communication). Of interest, the percent amino acid resemblance between chicken gene A and rat CYP2 genes is significantly higher for the IIC subfamily (57%) than the IIB subfamily (49%). These chicken genes are now designated as the two members of the CYP2H subfamily.

## 3.3. The *CYP2C* subfamily

Subfamily IIC has the largest number of members in all vertebrate species studied to date. Although most of the genes in this subfamily are constitutively expressed, several appear to be phenobarbital-inducible (LEIGHTON et al., 1984). The characterization of four closely related clones from rabbit liver — PBc1, PBc2, PBc3, and PBc4, (LEIGHTON et al., 1984) — have now been completed (LEIGHTON and KEMPER, 1984; GOVIND et al., 1986). One of these — PBc2, also called P450K — has been shown to be responsible for renal $\omega-1$ hydroxylation of lauric acid in rabbit kidney (FINLAYSON et al., 1986). The PBc3 and PBc4 genes are identical with rabbit form 3b and form 1-88, respectively. Rabbit form 1, particularly high for progesterone 21-hydroxylase (JOHNSON and GRIFFIN, 1985) and benzpyrene hydroxylase activity (RAUCY and JOHNSON, 1985), exhibits 94% similarity with the 1-88 protein, indicating that the two genes diverged long after the rabbit-rodent and rabbit-human speciations; therefore, the rat, mouse and human would not be expected to have orthologues to both of these genes (NEBERT et al., 1987). Rat f, g and i appear to represent one subset, and rat PB1 and h another subset, within the IIC subfamily (Fig. 2). Either rat PB1 or h probably corresponds to the genes 1 and 1-88 that arose in the rabbit via a duplication event. Human form 1 (TUKEY et al., 1985) and P450mp (UMBENHAUER et al., 1987) are members of the human IIC subfamily. KIMURA and coworkers (1987a) reported the sequence of two human liver P450 genes having relatively abundant mRNA concentrations and designated IIC1 and IIC2. Their IIC1 and IIC2 requences are 98% and 99% similar to form 1 and mp, respectively, strongly suggesting that IIC1 and IIC2 are allelic variants of the form 1 and mp genes, respectively. It is presently impossible to determine which laboratory animal CYP2C genes represent the orthologues of human IIC1 and IIC2. Hence, the IIC subfamily is complicated because of its large number of genes.

## 3.4. The *CYP2D* subfamily

A human debrisoquine 4-hydroxylase polymorphism was described (IDLE and SMITH, 1979) in which the "extensive metabolizer" (EM phenotype) is able to break down the antihypertensive drug 10 to 200 times better than the "poor metabolizer" (PM phenotype). The metabolism of more than 20 other drugs appears to be related to this same polymorphism (GONZALEZ et al., 1988b). Two rat IID sequences, db1 and db2, exhibit 73% amino acid resemblance (GONZALEZ et al., 1987). The db1 protein has high debrisoquine 4-hydroxylase and bufuralol 1'-hydroxylase activities, whereas the db2 enzyme does not. The catalytic specificity of db2 is unknown.

The antibody to rat db1 was used to quantitate immunochemically the human db1 enzyme and isolate the human db1 cDNA (GONZALEZ et al., 1988b). When cDNAs from EM and PM individuals were sequenced, three variants representing aberrant splicing defects were identified: variant *a* exhibited the retention of intron 5; variant *b*, retention of intron 6; and variant *b'*, loss of the first three exons plus the 3' half of exon 6. These splicing errors lead to immunologically undetectable, or very low, amounts of the human db1 protein (GONZALEZ et al., 1988b). Among a Nigerian population with cancer of the liver and gastrointestinal tract (IDLE et al., 1981) and in a study of 245 cigarette smokers with bronchogenic carcinoma (AYESH et al., 1984), it appears that a disproportionately greater number of the EM phenotype exhibit these malignancies.

With use of the human CYP2D6 cDNA as a probe, RFLP patterns have been found with six restriction endonucleases (GONZALEZ et al., 1988c). To this end, a recent study was undertaken, in which RFLP patterns from lymphocyte DNA were compared among 53 unrelated people and five families having at least one PM individual; about 75% of all PM individuals could be grouped into two haplotypes representing two mutant alleles of the CYP2D1 gene (SKODA et al., 1988). It should be possible to use this approach to determine individual risk for idiosyncratic reactions to certain prescribed drugs, as well as risk of certain types of malignancy.

Rat db3, db4 and db5 sequences have recently been determined, indicating that the rat has five functional CYP2D genes, at least three of which are expressed in liver and kidney (GONZALEZ and NEBERT, 1990). A male-specific testosterone 16α-hydroxylase from mouse liver (WONG et al., 1987) shows 82%, 72%, 78% and 70% similarity in amino acid sequence to rat db1, db2, db3 and db4, respectively, and less than 45% similarity to any other CYP2 gene. Since 18% divergence is more than twice that seen for any other orthologous P450 gene between rats and mice, we conclude that the mouse 16α and rat db1 are not orthologues. Moreover, it is unexpected that a male-specific testosterone hydroxylase gene, known to occur in the rat IIA and IIC subfamilies but not IID subfamily, would be present in the mouse IID subfamily.

---

## 3.5. The *CYP2E* subfamily

At least one P450 protein, designated 3a in rabbit and j in rat and human, appears to be induced by such chemicals as ethanol, imidazole, acetone, trichloroethylene and pyrazole. The increased protein induction by ethanol and these other chemicals occurs by a posttranscriptional mechanism (SONG et al., 1986). The rat gene and flanking regions have been sequenced (UMENO et al., 1988). Interestingly, the rabbit possesses two very closely related CYP2E genes, both of which are expressed in liver (KHANI et al., 1988). These genes have diverged after the rabbit-rodent split, meaning that both genes will not be present in rat (or mouse or human). The CYP2E1 gene product, the P450j enzyme, metabolizes aniline, alcohols and certain nitrosamines (YANG et al., 1985). Due to the possible role of IIE1 in the metabolism of certain dietary and environmental nitrosamines, it may be possible to correlate certain RFLP patterns with enhanced risk of cancer caused by these agents. An RFLP for the CYP2E1 gene on chromosome 10 has been reported (McBRIDE et al., 1987).

## 3.6. The *CYP2H* subfamily

A human lung CYP2F and a rat olfactory-specific CYP2G c-DNA have been characterized (NEBERT et al., 1989a).
A chicken gene (clone PB15; gene A) has been sequenced and shown to be induced by phenobarbital, allylisopropylacetamide, and 3,5-diethoxycarbonyl-1,4-dihydrocollidine (HOBBS et al., 1986). Two genes (A and B) are transcriptionally active, producing mRNAs of 3.2 and 2.5 kb, respectively, and represents two members of the IIH subfamily.

## 4. The *CYP3* gene family

A cDNA clone was isolated from pregnenolone 16α-carbonitrile (PCN)-treated rats (HARDWICK et al., 1983); the sequence was called pcn1 (GONZALEZ et al., 1985) and revealed a protein with 88% similarity to that encoded by rat pcn2 (GONZALEZ et al., 1986). Among the best activities carried out by products of the steroid-inducible P450 III gene family is testosterone 6β-hydroxylase (GONZALEZ et al., 1986). The first 17 NH$_2$-terminal residues of rat pcn1 and pcn2 are identical. Interestingly, phenobarbital induces rat pcn1 and pcn2, whereas PCN induces the pcn1 gene but not the pcn2 gene (GONZALEZ et al., 1986).

Rat pcn1 and pcn2 have diverged since the mammalian radiation, so it is highly unlikely that both genes will exist in the human. Three distinctly different human P450 III cDNA clones have been isolated and sequenced in three laboratories. These have been designated HLp (MOLOWA et al., 1986), P450nf, for nifedipine oxidation (BEAUNE et al., 1986), and pcn1 (GONZALEZ et al., 1988a). The cDNA clone pcn1 has been shown to express nifedipine oxidase activity in COS-1 cells (GONZALEZ et al., 1988a). Because the amino acid sequences of HLp and nf are 97% similar, the possibility exists that they could be allelic variants of the same gene (NEBERT et al., 1987). However, there are numerous differences between the 3′ nontranslated regions of pcn1, compared with both HLp and nf, suggesting that these indeed represent three different gene products. It remains to be determined whether HLp or nf catalyzes nifedipine oxidation and how many CYP3 genes are actually expressed in human liver.

Macrolide antibiotics, such as triacetyloleandromycin and griseofulvin, induce at least one gene of the P450 III family in the rat (WRIGHTON et al., 1985) and rabbit (DALET et al., 1988). It is unclear which human CYP3 gene will be the one induced by macrolide antibiotics, or if more than one human CYP3 gene will be induced by these agents. Clinical studies of the human orthologues of these CYP3 genes, such as correlation of RFLP patterns with gene expression, may be important for predicting macrolide antibiotic response or toxicity caused by such drugs. The CYP3A1 locus is linked to the COL1A2 gene on chromosome 7 (BROOKS et al., 1988).

A house fly P450 cDNA was recently isolated with the use of a polyclonal antibody developed against phenobarbital-treated house fly P450 (R. FEYER-EISEN et al., 1989). Interestingly, the amino acid sequence is most similar to the mammalian III family, although the degree of divergence can be used as a criterion for assignment to a new P450 gene family. This insect P450 gene is the first member of the CYP6 family.

## 5. The *CYP4* gene family

The serum lipid lowering agents and peroxisome proliferators (e.g. clofibrate) form a novel class of hepatocellular tumor promoters in rodents (REDDY et al., 1980) and induce a protein (P450 LAω) having high specificity toward lauric acid ω-hydroxylation (ORTON and PARKER, 1982). The LAω cDNA encodes a protein having an amino acid sequence $\leq 33\%$ similar to that encoded by P450 genes in any of the other seven mammalian families (HARDWICK et al., 1987). A second gene, designated p-2, was shown to be inducible in rabbit lung during pregnancy and to express prostaglandin ω-hydroxylase activity (MATSUBARA et al., 1987). Rat liver LAω and rabbit lung p-2 exhibit 74% similar amino acid sequences and could be the orthologues of one another. There probably exist several other members of this relatively ancient mammalian P450 IV family (Fig. 2). A second rat liver CYP4 gene has been characterized; both genes examined to date contain thirteen exons (GONZALEZ, 1988).

## 6. P450 gene families involved in steroidogenesis

Congenital adrenal hyperplasia may involve a defect in any of the four P450 enzymes involved in adrenal steroidogenesis (NEW et al., 1983). The 21-hydroxylase defect can be manifest as either a simple virilizing form or a life-threatening salt-wasting form. The incidence of 21-hydroxylase deficiency is far greater than the incidence of the other three enzyme deficiencies combined. These rarer forms of congenital adrenal hyperplasia — 20,22-desmolase (cholesterol side-chain cleavage; scc) deficiency, 17α-hydroxylase deficiency, and 11β-hydroxylase deficiency — can also be serious and sometimes lethal. Mechanisms of gene conversions, unequal crossovers, polymorphisms, mutations and, occasionally, large deletions have been shown to be responsible for 21-hydroxylase deficiency (SCHNEIDER et al., 1986; HARADA et al., 1987; JOSPE et al., 1987; MATTESON et al., 1987; RODRIGUES et al., 1987; SHIROISHI et al., 1987; AMOR et al., 1988; HIGASHI et al., 1988; MILLER, 1988) and the 20,22-desmolase defect (HAUFFA et al., 1985). It should be noted that both P450scc and 11β-hydroxylase are mitochondrial proteins, whereas the 21-hydroxylase and 17α-hydroxylase are microsomal enzymes, and all four are encoded by nuclear genes.

### 6.1. The *CYP11* family

At the present time there are two members of the XI family, and both genes code for mitochondrial proteins. The scc and 11β-hydroxylase genes are members of the XIA and XIB subfamilies, respectively. Bovine scc and 11β-hydroxylase proteins have both been shown to undergo posttranslational processing whereby an NH₂-terminal peptide of at least two dozen amino acids is removed to form the mature P450 protein (MATOCHA and WATERMAN, 1985). The mitochondrial P450 proteins are more similar than the microsomal P450 proteins are to the bacterial P450 proteins. Both observations — posttranslational processing and resemblance to prokaryotic proteins — are common properties of mitochondrial proteins.

Bovine scc precursor is 521 residues and the mature enzyme is 481 residues long (MOROHASHI et al., 1984). The human scc precursor is 522 residues and the mature enzyme is 482 residues long (MOROHASHI et al., 1987a). The bovine 11β precursor is 504 residues and the mature enzyme is 479 amino acids in length (CHUA et al., 1987; MOROHASHI et al., 1987b). The CYP11A1 gene contains nine exons (MOROHASHI et al., 1987a), while the CYP11B1 gene structure has not yet been reported.

## 6.2. The *CYP17* family

Human and bovine family XVII consists of a single gene encoding an enzyme with both $17\alpha$-hydroxylase and 17,20-lyase activities important during steroidogenesis. The CYP17 gene is divided into eight exons; of the seven introns, five interrupt the gene at precisely the same location as CYP21 gene introns (PICADO-LEONARD and MILLER, 1988). Because of this similarity in gene structure — except that the first and fourth introns of CYP21 have been deleted in CYP17 — it is possible that these two families are distantly related (see Fig. 2).

## 6.3. The *CYP19* family

The "aromatase system" is involved in the conversion of androgens to estrogens. A defect in P450arom regulation, leading to marked gynecomastia with normal male genitalia, has been shown to represent aromatase activity 10 to 50 times higher than normal (HEMSELL et al., 1977; BERKOVITZ et al., 1985). Because aromatase activity represents aromatization of the steroid A ring, which includes oxidation and loss of carbon-19, the P450 gene family has been named XIX (NEBERT et al., 1989a). The cDNA probe hybridizes to several sizes of mRNA in a number of cells and tissues including ovarian granulosa cells, testicular Sertoli cells, adipose cells, placental syncytiotrophoblasts, hypothalamic cells and blastocysts (EVANS et al., 1986; 1987). The partial amino acid sequence (about 60 residues missing) deduced from the cDNA has been reported (SIMPSON et al., 1987; CHEN et al., 1988). Interestingly, both laboratories obtained the identical cDNA that encodes only the COOH-terminal 419 amino acids, suggesting that mRNA secondary structure may pose a problem for the reverse transcriptase.

The CYP19 gene structure is not yet known. Although Southern blot analysis is consistent with at least two genes, it is possible that one is a pseudogene. Northern blot analysis indicates three sizes of mRNA from human placenta (3.0, 2.7 and 2.4 kb) and two sizes from human adipose stromal cells (3.5 and 3.0 kb), suggesting either differential processing or products from more than one gene (EVANS et al., 1986; 1987).

## 6.4. The *CYP21* family

The human 21-hydroxylase gene is localized in the HLA region of chromosome 6 very near the human complement C4A gene. Two 21-hydroxylase genes were found, each located near the 3′ end of one of the two C4 genes (WHITE et al., 1985). A tandem arrangement in the human (5'-C4A-C21A-C4B-C21B-3') is similar to that in the mouse, which has Slp (sex-limited protein) in place of the C4A gene. With deletion mutants from patients with various types of 21-hydroxylase deficiency, it was concluded that the human C21B gene is functional but the C21A gene is not (WHITE et al., 1985), whereas the murine C21A gene is active and the C21B gene is not (PARKER et al., 1985). The proposed human loci nomenclature and the chromosomal arrangement is 5'-C4A-CYP21P-C4B-CYP21-3', and the proposed mouse loci nomenclature and the chromosomal arrangement is 5'-Slp-Cyp21-C4-Cyp21p-3' (NEBERT et al., 1989a). The CYP21 genes contain ten exons.

One or several small deletions, mostly in exon 3 of the human CYP21P gene (HIGASHI et al., 1986; WHITE et al., 1986), and a deletion of all of exon 2 of the mouse Cyp21p gene (CHAPLIN et al., 1986), account for the nonfunctional gene of the pair of 21-hydroxylase genes in either species. Of interest, the upstream regulatory sequences of the mouse Cyp21p "inactive" gene are functional (CHAPLIN et al., 1986). That there is one active and one inactive C21 gene in close proximity of one another in both human and mouse, and that the regulatory region of the inactive mouse gene is functional (CHAPLIN et al., 1986), suggest a possible important role in regulation or other advantage in evolutionary survival.

## 7. The *CYP51* gene family

The structural gene and flanking regions for lanosterol 14$\alpha$-demethylase (14DM) from *Saccharomyces cerevisiae* revealed an open reading frame of 530 amino acids and is regarded as the first member of family LI (KALB et al., 1987). As with most yeast genes, the CYP51 gene has no introns. A possible TATA promoter element exists more than 260 bases upstream from the initiation codon. The 14DM protein is more closely related to mammalian P450 proteins than to bacterial P450 protein. The orthologous gene from *Candida tropicalis* ATCC750 was recently sequenced; the deduced LI protein of 528 amino acids exhibits 66% resemblance to the *S. cerevisiae* 14DM (NEBERT et al., 1989a).

## 8. The *CYP52* gene family

When yeast is grown in the presence of n-alkanes, the P450 enzyme that is induced (P450alk) in turn aids in the catabolism of the n-alkane, such as tetradecane, as an energy source (SANGLARD et al., 1986). As the first enzyme in the pathway for alkane assimilation, alk is essential for the growth of *C. tropicalis* on this carbon source. The structural gene and flanking regions for alk from *C. tropicalis* have been sequenced (NEBERT et al., 1989a). The deduced protein is 524 residues long. With use of the Wilbur-Lipman alignment program (WILBUR and LIPMAN, 1983), the deduced protein exhibits $< 30\%$ amino acid similarity to that of all other P450 gene families, including LI. This sequence has thereby been assigned as the first member of family LII, and the proposed name for the gene is CYP52.

## 9.   The *CYP101* gene family

P450cam from *Pseudomonas putida* is induced by, and in turn metabolizes, camphor. The deduced protein from the recent cDNA sequence (UNGER et al., 1986) has 415 amino acids and is regarded as the first member of family CI. Prokaryotic P450 metabolism requires three components: the flavoprotein NADH-P450 oxidoreductase, the nonheme iron-sulfur moiety putidaredoxin, and the P450 protein. Interestingly, the mitochondrial P450 catalytic activity requires three components: the flavoprotein oxidoreductase, the iron-sulfur moiety adrenodoxin, and the P450 protein (NEBERT and GONZALEZ, 1987, and references therein). Because P450cam is soluble rather than membrane-bound, purification of sufficient quantities for studying the crystal structure has led to a detailed understanding of the substrate-free and substrate-bound forms of this bacterial enzyme (POULOS et al., 1986). On the basis of hydropathy profiles of 34 aligned P450 proteins and other structural information, a 3-dimensional model has been proposed for microsomal P450 proteins that is similar to the 3-dimensional structure of P450cam (NELSON and STROBEL, 1988).

## 10.   The *CYP102* gene family

In the presence of NADPH and atmospheric oxygen, a single polypeptide of approximately 120 kilodaltons in *Bacillus megaterium* can catalyze the hydroxylation of long-chain fatty acids without the requirement of any other protein (WEN and FULCO, 1987). The polypeptide appears to comprise a NADPH-P450 oxidoreductase subunit and a P450 subunit (NARHI and FULCO, 1987). The P450 moiety, BM-3, is markedly induced by barbiturates, perhaps by interaction with a repressor, i.e. derepression (WEN and FULCO, 1987). The sequence of the deduced BM-3 protein (NEBERT et al., 1989a) indicates that this gene is the first member of family CII.

## 11.   Criteria for a single gene product

Many anti-P450 antibodies recognize P450 proteins that are more than $60\%$ similar, and many cDNA probes can detect P450 mRNA sequences that are at least $55\%$ similar. This empirical observation becomes especially important for any P450 families in which two or more genes are expressed (NEBERT and GONZALEZ, 1987). This potential problem should be taken into consideration when new P450 cDNAs are isolated or when antibodies — even antibodies that react with single bands on Western blots — are used for studying correlations with certain catalytic activities. For example, P450pcn1 and pcn2 are both present in phenobarbital- and PCN-treated rat liver microsomes, both proteins comigrate on sodium dodecyl sulfate polyacrylamide gels, and antibodies against pcn1 are known to inhibit pcn2 catalytic activity. The pcn2 gene was only identified via cDNA cloning and sequencing (GONZALEZ et al., 1986).

Previous studies on rat microsomal P450-mediated debrisoquine 4-hydroxylase never indicated the possible presence of four other closely related P450 protein in the rat P450 IID subfamily. Both chromatographic and cDNA sequence data have clearly established that the rat db1 protein exists in the presence of four other homologous proteins in the same subfamily (GONZALEZ and NEBERT, 1990). These proteins can only be resolved via gel electrophoresis under unusual running conditions. Moreover, the antibody against db1 protein cross-reacts with the db2 protein, and the purified proteins comigrate on conventional gels (GONZALEZ et al., 1987). In fact, five closely related proteins in the rat IID subfamily have now been characterized.

In summary, any report in which a P450 protein is quantitated by Western blot analysis, without additional supporting data, must be viewed with caution. A single protein on a gel or immunoblot, or a single $NH_2$-terminal sequence, should never be used as the sole criterion for a monospecific antibody or a purified protein. The best criteria include isolation, sequencing, and expression of the cDNA. Any report in which a P450 mRNA is quantitated by Northern blot or dot blot using a near full-length cDNA probe, without additional supporting data, should also be viewed with caution. A cDNA encoding one P450 can react with at least two mRNA sequences of identical size on Northern blots, which is known to happen with pcn1 and pcn2 (GONZALEZ et al., 1986) and with db1 and db2 (GONZALEZ et al., 1987). Probes utilizing specific 3' nontranslated regions or specific oligonucleotides should therefore be used to distinguish between highly homologous mRNA sequences.


## 12. Regulation of P450 gene expression

Many P450 genes have been shown to be under complex, and highly unique, control during development and under the influence of sex-specific signals. The reader is referred to NEBERT and GONZALEZ (1978) and GONZALEZ (1988) for a discussion of this topic in depth.

By far the most common mechanism for increases in the P450 enzyme involves transcriptional activation of the gene. Genes of all eight mammalian families are known to respond to exogenous inducers or endogenous signals in this way.

In addition, several P450 genes have been shown to be regulated post-transcriptionally. For example, treatment of rabbits with the antibiotic tri-acetyloleandromycin results in an increase in 3c (P450 III) mRNA without any rise in transcription (DALET et al., 1986). Treatment of rats with dexa-methasone leads to marked increases in b (IIB) mRNA in the absence of any increase in the transcriptional rate (SIMMONS et al., 1987). Treatment of rats with ethanol, acetone or 4-methylpyrazole causes a rapid increase in the j protein (IIE) in the absence of any increase in j mRNA, suggesting a post-translational induction process (SONG et al., 1986). On the other hand, rats rendered diabetic exhibit a 10-fold increase in j mRNA in the absence of an increase in transcription (SONG et al., 1987). Treatment of mice with TCDD leads to a complex combination of transcriptional and posttranscriptional effects on constitutive and inducible expression of the $P_1$ and $P_3$ genes (family I), as well as striking tissue-specific differences in these effects (KIMURA et al., 1986).

The XIA, XIB, XVII, XIX and XXI genes are transcriptionally activated by certain endogenous signals (i.e. ACTH, prostaglandins, FSH, hCG, gastrin, prolactin, LH) through the action of cyclic AMP (NEBERT and GONZALEZ, 1987, and references therein). It appears that cAMP does not stimulate P450 gene transcription directly but rather by way of an intermediate protein. Candidates for this protein might include the sterol carrier protein-2 (TRZECIAK et al., 1987) and the steroidogenesis activator polypeptide (PEDERSEN and BROWNIE, 1987).

Insulin-like growth factors (IGFs) are polypeptides important for cell proliferation and growth and appear to influence steroidogenesis P450 gene expression. IGF-II mRNA accumulation in human cultured placental cells, granulosa cells and fetal adrenal cells treated with dibutyryl cAMP is regulated in parallel with scc mRNA (VOUTILAINEN and MILLER, 1987). The IGF-II gene is regulated differently in fetal testis than it is in fetal adrenal, placenta or adult granulosa cells (VOUTILAINEN and MILLER, 1988). IGF-II might therefore behave in an autocrine or paracrine fashion to stimulate adrenal and ovarian growth in response to ACTH and gonadotropins, respectively.

More is known about regulation of the CYP1A1 gene than that of any other mammalian P450 gene. The reader is referred to NEBERT and GONZALEZ (1987) and NEBERT and JONES (1989) for detailed reviews on this subject, as well as Section 2 of this Chapter for a brief summary. Induction involving foreign chemical inducers acts in a positive manner by way of an inducible enhancer element about 1 kilobase upstream from the mRNA cap site. A negative DNA element, dependent upon a functional CYP1A1 gene product (the enzyme), is located about 500 bases upstream from the cap site. Both positive regulation (induction) and release of negative regulation (derepression

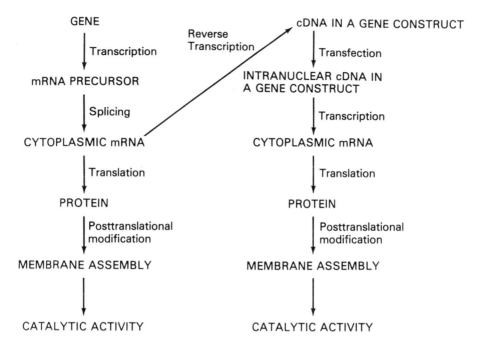

**Fig. 3.** Sequence of events by which a P450 cDNA in an expression vector should lead to enzymic activity.

in the absence of $P_1$ metabolism) lead to increases in the rate of transcriptional activation of the CYP1A1 gene and require a functional Ah receptor. Derepression, which can occur in the absence of a foreign chemical inducer, presumably requires binding of an endogenous ligand to the Ah receptor.

## 13.  P450 expression vectors

In many instances of P450 cDNA clones so far isolated, the true function of the particular gene product in the intact animal remains uncertain. Expression of the cDNA or genomic clone in yeast, mammalian cultured cells, or transgenic mice (i.e. expression of one enzyme activity and not another) might resolve this uncertainty. The experiment involves the insertion of a P450 cDNA (Fig. 3) or complete gene (Fig. 4) into an appropriate expression vector, having upstream control sequences and downstream termination (and splicing) signals, and transfection (or electroporation or microinjection) of that construct into a suitable host eukaryotic cell. The net result includes transcription of the cDNA or gene, translation of the mature mRNA, and

D. W. Nebert; F. J. Gonzales

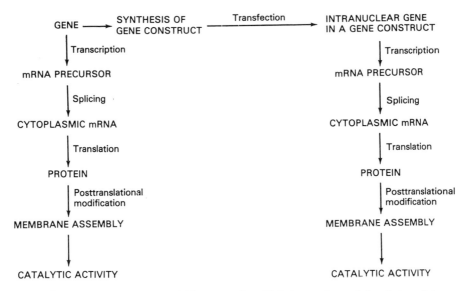

**Fig. 4.** Sequence of events by which a complete P-450 gene (containing introns) in an expression vector should lead to enzymic activity.

incorporation of the ultimate P450 protein into the membrane with the correct catalytic activity.

In the first successful experiments reported, the translated region of rat c cDNA was placed in the yeast expression vector pAAH5 and transfected into transformed yeast; benzpyrene hydroxylase activity was expressed when purified rat NADPH-P450 oxidoreductase was added in vitro (OEDA et al., 1985). This group then constructed a chimeric cDNA encoding the first 518 residues of rat P450c linked to all but the first 56 residues of rat NADPH-P450 oxidoreductase (MURAKAMI et al., 1987). The 130-kDa protein displayed P450c enzymic activity in the absence of exogenously added NADPH-P450 oxidoreductase. This elegant genetic engineering study demonstrates the interaction among the FAD and FMN binding sites of the oxidoreductase and the substrate- and heme-binding sites of the P450 protein.

Further studies of P450 expression vectors include the expression of: steroid $17\alpha$-hydroxylase and 17,20-lyase activities in transformed monkey kidney COS-1 cells with the pcD vector containing bovine XVII cDNA (ZUBER et al., 1986); the transcribed mouse CYP21 gene driven by the CYP21P promoter in Y1 adrenocortical tumor cells (CHAPLIN et al., 1986); benzpyrene hydroxylase activity in yeast with the pAAH5 vector containing chimeric $P_1$ cDNA sequences from mouse Hepa-1 wild-type and $P_1^-$ mutant lines (KIMURA et al., 1987b); and testosterone $7\alpha$-hydroxylase and lauric acid $\omega$-hydroxylase activities in yeast with the pAAH5 vector containing rat IIA1 cDNA (NAGATA et al., 1987) and IVA1 cDNA (HARDWICK et al., 1987), respectively. Bufurolol 1'-hydroxylase activity has been produced in COS-1 cells with the SV40- and adenovirus-based p91023(B) vector containing human db1 cDNA (GONZALEZ et al., 1988b), and nifedipine oxidase activity in COS-1 cells with the p91023(B) vector containing human pcn1 cDNA (GONZALEZ et al., 1988a). A vaccinia virus vector has been used to produce high levels of mouse $P_1$ and $P_3$ catalytic activities (BATTULA et al., 1987). Expression of the P450cam (CI) protein and enzymic activity in *Escherichia coli* (KOGA et al., 1985; UNGER et al., 1986) and *Pseudomonas putida* (KOGA et al., 1985) has also been reported. Following a triple transfection of COS-1 cells with plasmids containing bovine scc, adrenodoxin and 17α-hydroxylase cDNA sequences (ZUBER et al., 1988), it was shown that all three gene products can function in these cells: cholesterol can be converted to pregnenolone in mitochondria, followed by 17α-hydroxy-pregnenolone and dehydroepiandrosterone formation in the endoplasmic reticulum. Interestingly, although cAMP-dependent regulation of steroido-genesis is not present in the monkey kidney fibroblast COS-1 cells, the cellular mechanisms necessary to support both mitochondrial and microsomal steroid hydroxylations appear not to be tissue-specific (ZUBER et al., 1988).

## 14. Future prospects

A better understanding of the gene structure, evolution and regulation of the P450 gene superfamily has resulted from the literal explosion of molecular biologic studies in this research field during the past decade. A greater under-standing about P-450 gene evolution has led to a proposal for a simplified, universal nomenclature system (NEBERT et al., 1987; 1989a; 1989b). It is anti-cipated that P450 mutants, in both the human population and laboratory ani-mals, will enhance our knowledge of the enzyme active-site. For example, a sing-le base mutation in the mouse CYP1A1 gene, leading to Arg-245 → Pro, severely affects benzpyrene hydroxylation (KIMURA et al., 1987b). A single base change in the CYP21 gene, leading to Ile-172 → Asn, causes decreases in 21-hydroxy-lase activity (AMOR et al., 1988). A 4-base repeat at amino acid 480 out of a total of 508 in the CYP17 gene, leading to a shift in reading frame and a truncated protein three amino acids shorter, is a cause of 17α-hydroxylase deficiency (KAGIMOTO et al., 1988). Expression vectors, coupled with site-directed mutagenesis, will provide another means for dissecting the substrate-binding, heme-binding and flavoprotein-binding sites of the P-450 molecule. Lastly, the identification and characterization of cis-acting DNA regulatory elements, and the cloning of genes that encode trans-acting regulatory proteins that interact with these elements, will lead to a better understanding of constitutive and inducible P450 gene expression, as well as developmental-, sex- and tissue-specific expression of these important enzymes.

# Acknowledgements

We thank our colleagues — especially David R. NELSON and Michael R. WATERMAN — for valuable discussions and the critical reading of this manuscript.

# 15. References

AMOR, M., K. L. PARKER, H. GLOBERMAN, M. I. NEW, and P. C. WHITE (1988), Proc. Natl. Acad. Sci. U.S.A. 85, 1600—1604.

AYESH, R., J. R. IDLE, J. C. RITCHIE, M. J. CROTHERS, and M. R. HETZEL (1984), Nature 313, 169—170.

BATTULA, N., J. SAGARA, and H. V. GELBOIN (1987), Proc. Natl. Acad. Sci. U.S.A. 84, 4073—4077.

BEAUNE, P. H., D. R. UMBENHAUER, R. W. BORK, R. S. LLOYD, and F. P. GUENGERICH (1986), Proc. Natl. Acad. Sci. U.S.A. 83, 8064—8068.

BERKOVITZ, G. D., A. GUERAMI, T. R. BROWN, P. C. MacDONALD, and C. J. MIGEON (1985), J. Clin. Invest. 75, 1763—1769.

BROOKS, B. A., O. W. McBRIDE, C. T. DOLPHIN, M. FARRALL, P. J. SCAMBLER, F. J. GONZALEZ, and J. R. IDLE (1988), Am. J. Hum. Genet. 43, 280—284.

CHAPLIN, D. D., L. G. GALBRAITH, J. G. SEIDMAN, P. C. WHITE, and K. L. PARKER (1986), Proc. Natl. Acad. Sci. U.S.A. 83, 9601—9605.

CHEN, S., M. J. BESMAN, R. S. SPARKES, S. ZOLLMAN, I. KLISAK, T. MOHANDAS, P. F. HALL, and J. E. SHIVELY (1988), DNA 7, 27—38.

CHUA, S. C., P. SZABO, A. VITEK, K.-H. GRZESCHIK, M. JOHN, and P. C. WHITE (1987), Proc. Natl. Acad. Sci. U.S.A. 84, 7193—7197.

DALET, C., J. M. BLANCHARD, P. GUZELIAN, J. BARWICK, H. HARTLE, and P. MAUREL (1986), Nucl. Acids Res. 14, 5999—6015.

DALET, C., P. CLAIR, M. DAUJAT, P. FORT, J. M. BLANCHARD, and P. MAUREL (1988), DNA 7, 39—46.

EVANS, C. T., D. B. LEDESMA, T. Z. SCHULZ, E. R. SIMPSON, and C. R. MENDELSON (1986), Proc. Natl. Acad. Sci. U.S.A. 83, 6387—6391.

EVANS, C. T., C. Jo CORBIN, C. T. SAUNDERS, J. C. MERRILL, E. R. SIMPSON, and C. R. MENDELSON (1987), J. Biol. Chem. 262, 6914—6920.

FEYEREISEN, R., J. F. KOENER, D. E. FARNSWORTH, and D. W. NEBERT (1989) Proc. Natl. Acad. Sci. U.S.A. 86, 1465—1469.

FINLAYSON, M. J., B. KEMPER, N. BROWNE, and E. F. JOHNSON (1986), Biochem. Biophys. Res. Commun. 141, 728—733.

FUJISAWA-SEHARA, A., K. SOGAWA, M. YAMANE, and Y. FUJII-KURIYAMA (1987), Nucl. Acids Res. 15, 4179—4191.

GASSER, R., M. NEGISHI, and R. M. PHILPOT (1988), Mol. Pharmacol. 32, 22—30.

GONZALEZ, F. J. and D. W. NEBERT (1985), Nucl. Acids Res. 13, 7269—7288.

GONZALEZ, F. J., D. W. NEBERT, J. P. HARDWICK, and C. B. KASPER (1985), J. Biol. Chem. 260, 7435—7441.

GONZALEZ, F. J., B.-J. SONG, and J. P. HARDWICK (1986), Mol. Cell. Biol. 6, 2969 to 2976.

GONZALEZ, F. J., T. MATSUNAGA, K. NAGATA, U. A. MEYER, D. W. NEBERT, J. PASTEWKA, K. KOZAK, J. R. GILLETTE, H. V. GELBOIN, and J. P. HARDWICK (1987), DNA 6, 149—161.

GONZALEZ, F. J. (1988) Pharmacol. Rev. 40, 243—288.

GONZALEZ, F. J., B. J. SCHMID, M. UMENO, O. W. McBRIDE, J. P. HARDWICK, U. A. MEYER, H. V. GELBOIN, and J. R. IDLE (1988a), DNA 7, 79—86.

GONZALEZ, F. J., R. C. SKODA, S. KIMURA, M. UMENO, U. M. ZANGER, D. W. NEBERT, H. V. GELBOIN, J. P. HARDWICK, and U. A. MEYER (1988b), Nature **331**, 442—446.
GONZALEZ, F. J., F. VILBOIS, J. P. HARDWICK, O. W. MCBRIDE, D. W. NEBERT, H. V. GELBOIN, and U. A. MEYER (1988c), Genomics **2**, 174—179.
GONZALEZ, F.. J and D. W. NEBERT (1990) Trends Genet., in press.
GOVIND, S., P. A. BELL, and B. KEMPER (1986), DNA **5**, 371—382.
HANKINSON, O. (1979), Proc. Natl. Acad. Sci. U.S.A. **76**, 373—376.
HANKINSON, O. (1981), Somat. Cell. Genet. **7**, 373—388.
HANKINSON, O., R. D. ANDERSEN, B. BIRREN, F. SANDER, M. NEGISHI, and D. W. NEBERT (1985), J. Biol. Chem. **260**, 1790—1795.
HARADA, F., A. KIMURA, T. IWANAGA, K. SHIMOZAWA, J. YATA, and T. SASAZUKI (1987), Proc. Natl. Acad. Sci. U.S.A. **84**, 8091—8094.
HARDWICK, J. P., F. J. GONZALEZ, and C. B. KASPER (1983), J. Biol. Chem. **258**, 10182—10186.
HARDWICK, J. P., B.-J. SONG, E. HUBERMAN, and F. J. GONZALEZ (1987), J. Biol. Chem. **262**, 801—810.
HAUFFA, B. P., W. L. MILLER, M. M. GRUMBACH, F. A. CONTE, and S. L. KAPLAN (1985), Clin. Endocrinol. **23**, 481—493.
HEMSELL, D. L., C. D. EDMAN, J. F. MARKS, P. K. SIITERI, and P. C. MACDONALD (1977), J. Clin. Invest. **60**, 455—464.
HEILMANN, L. J., Y.-Y. SHEEN, S. W. BIGELOW, and D. W. NEBERT (1988), DNA **7**, 379—387.
HIGASHI, Y., H. YOSHIOKA, M. YAMANE, O. GOTOH, and Y. FUJII-KURIYAMA (1986), Proc. Natl. Acad. Sci. U.S.A. **83**, 2841—2845.
HIGASHI, Y., A. TANAE, H. INOUE, and Y. FUJII-KURIYAMA (1988), Am. J. Hum. Genet. **42**, 017—025.
HOBBS, A. A., L. A. MATTSCHOSS, B. K. MAY, K. E. WILLIAMS, and W. H. ELLIOTT (1986), J. Biol. Chem. **261**, 9444—9449.
IDLE, J. R. and R. L. SMITH (1979), Drug Metab. Rev. **9**, 301—317.
IDLE, J. R., A. MAHGOUB, T. P. SLOAN, R. L. SMITH, C. O. MBANEFO, and E. A. BABA-BUNMI (1981), Cancer Lett. **11**, 331—338.
IKEYA, K., A. K. JAISWAL, R. A. OWENS, J. E. JONES, D. W. NEBERT, and S. KIMURA (1989) Mol. Endocrinol. **3**, 1399—1408.
IMAI, Y., M. KOMORI, and R. SATO (1988), Biochemistry **27**, 80—88.
JAISWAL, A. K., F. J. GONZALEZ, and D. W. NEBERT (1985), Nucl. Acids Res. **13**, 4503—4520.
JAISWAL, A. K. and D. W. NEBERT (1986), Nucl. Acids Res. **14**, 4376.
JAISWAL, A. K., D. W. NEBERT, and F. J. GONZALEZ (1986), Nucl. Acids Res. **14**, 6773—6774.
JAISWAL, A. K., D. W. NEBERT, O. W. MCBRIDE, and F. J. GONZALEZ (1987), J. Exp. Pathol. **3**, 1—17.
JOHNSON, E. F. and K. J. GRIFFIN (1985), Arch. Biochem. Biophys. **237**, 55—64.
JOSPE, N., P. A. DONOHOUE, C. VAN DOP, R. H. MCLEAN, W. B. BIAS, and C. J. MIGEON (1987), Biochem. Biophys. Res. Commun. **142**, 798—804.
KADLUBAR, F. F. and G. J. HAMMONS (1987), in: Mammalian Cytochromes P-450, (F. P. GUENGERICH, ed.), CRC Press Inc., Boca Raton, Florida, Vol. II, 81—130.
KAGIMOTO, M., J. S. D. WINTER, K. KAGIMOTO, E. R. SIMPSON, and M. R. WATERMAN (1988), Mol. Endocrinol. **2**, 564—570.
KALB, V. F., C. W. WOODS, T. G. TURI, C. R. DEY, T. R. SUTTER, and J. C. LOPER (1987), DNA **6**, 529—537.
KHANI, S. C., T. D. PORTER, V. FUJITA, and M. J. COON (1988), J. Biol. Chem. **263**, 7170—7175.

KIMURA, S. and D. W. NEBERT (1986), Nucl. Acids Res. **14**, 6765—6766.
KIMURA, S., F. J. GONZALEZ, and D. W. NEBERT (1986), Mol. Cell. Biol. **6**, 1471—1477.
KIMURA, S., J. PASTEWKA, H. V. GELBOIN, and F. J. GONZALEZ (1987a), Nucl. Acids Res. **15**, 10053—10054.
KIMURA, S., H. H. SMITH, O. HANKINSON, and D. W. NEBERT (1987b), EMBO J. **6**, 1929—1933.
KOGA, H., B. RAUCHFUSS, and I. C. GUNSALUS (1985), Biochem. Biophys. Res. Commun. **130**, 412—417.
KOMORI, M., Y. IMAI, S. TSUNASAWA, and R. SATO (1988), Biochemistry **27**, 73—80.
KOURI, R. E., C. E. MCKINNEY, D. J. SLOMIANY, D. R. SNODGRASS, N. P. WRAY, and T. L. MCLEMORE (1982), Cancer Res. **42**, 5030—5037.
LABBE, D., A. JEAN, and A. ANDERSON (1988), DNA **7**, 253—260.
LEGRAVEREND, C., R. R. HANNAH, H. J. EISEN, I. S. OWENS, D. W. NEBERT, and O. HANKINSON (1982), J. Biol. Chem. **257**, 6402—6407.
LEIGHTON, J. K. and B. KEMPER (1984), J. Biol. Chem. **259**, 11165—11168.
LEIGHTON, J. K., B. A. DEBRUNNER-VOSSBRINCK, and B. KEMPER (1984), Biochemistry **23**, 204—210.
MATOCHA, M. F. and M. R. WATERMAN (1985), J. Biol. Chem. **260**, 12259—12265.
MATSUBARA, S., S. YAMAMOTO, K. SOGAWA, N. YOKOTANI, Y. FUJII-KURIYAMA, M. HANIU, J. E. SHIVELY, O. GOTOH, E. KUSUNOSE, and M. KUSUNOSE (1987), J. Biol. Chem. **262**, 13366—13371.
MATSUNAGA, T., K. NAGATA, E.J. HOLSZTYNSKA, D.P. LAPENSON, A.J. SMITH, R. KATO, H. V. GELBOIN, D. J. WAXMAN, and F. J. GONZALEZ (1988), J. Biol. Chem. **263**, 17995—18002.
MATTESON, K. J., J. A. PHILLIPS III, W. L. MILLER, B.-C. CHUNG, P. J. ORLANDO, H. FRISCH, A. FERRANDEZ, and I. M. BURR (1987), Proc. Natl. Acad. Sci. U.S.A. **84**, 5858—5862.
MCBRIDE, O. W., M. UMENO, H. V. GELBOIN, and F. J. GONZALEZ (1987), Nucl. Acids Res. **15**, 10071.
MILLER, A. G. and J. P. WHITLOCK, Jr. (1981), J. Biol. Chem. **256**, 2433—2437.
MILLER, A.G., D. ISRAEL, and J.P. WHITLOCK, Jr. (1983), J. Biol. Chem. **258**, 3523—3527.
MILLER, W. L. (1988), Am. J. Hum. Genet. **42**, 4—7.
MOLOWA, D. T., E. G. SCHUETZ, S. A. WRIGHTON, P. B. WATKINS, P. KREMERS, G. MENDEZ-PICON, G. A. PARKER, and P. S. GUZELIAN (1986), Proc. Natl. Acad. Sci. U.S.A. **83**, 5311—5315.
MOROHASHI, K., Y. FUJII-KURIYAMA, Y. OKADA, K. SOGAWA, T. HIROSE, S. INAYAMA, and T. OMURA (1984), Proc. Natl. Acad. Sci. U.S.A. **81**, 4647—4651.
MOROHASHI, K., K. SOGAWA, T. OMURA, and Y. FUJII-KURIYAMA (1987a), J. Biochem. **101**, 879—887.
MOROHASHI, K., H. YOSHIOKA, O. GOTOH, Y. OKADA, K. YAMAMOTO, T. MIYATA, K. SOGAWA, Y. FUJII-KURIYAMA, and T. OMURA (1987b), J. Biochem. (Tokyo) **102**, 559—568.
MURAKAMI, H., Y. YABUSAKI, T. SAKAKI, M. SHIBATA, and H. OHKAWA (1987), DNA **6**, 189—197.
NAGATA, K., T. MATSUNAGA, J. R. GILLETTE, H. V. GELBOIN, and F. J. GONZALEZ (1987), J. Biol. Chem. **262**, 2787—2793.
NARHI, L. O. and A. J. FULCO (1987), J. Biol. Chem. **262**, 6683—6690.
NEBERT, D. W. and F. J. GONZALEZ (1987), Annu. Rev. Biochem. **56**, 945—993.
NEBERT, D. W., M. ADESNIK, M. J. COON, R. W. ESTABROOK, F. J. GONZALEZ, F. P. GUENGERICH, I. C. GUNSALUS, E. F. JOHNSON, B. KEMPER, W. LEVIN, I. R. PHILLIPS, R. SATO, and M. R. WATERMAN (1987), DNA **6**, 1—11.
NEBERT, D. W. and J. E. JONES (1989), Int. J. Biochem. **21**, 243—252.

NEBERT, D. W., D R NELSON, M. ADESNIK, M. J. COON, R. W. ESTABROOK, F. J. GON-
ZALEZ, F. P. GUENGERICH, I. C. GUNSALUS, E. F. JOHNSON, B. KEMPER, W. LEVIN,
I. R. PHILLIPS, R. SATO, and M. R. WATERMAN (1989a) DNA 8, 1—13.

NEBERT, D. W., D. R. NELSON, and R. FEYEREISEN (1989b) Xenobiotica 19, 1149 to
1160.

NELSON, D. R. and H. W. STROBEL (1987), Mol. Biol. Evol. 4, 572—593.

NELSON, D. R. and H. W. STROBEL (1988), J. Biol. Chem. 263, 6038—6050.

NEW, M. I., B. DUPONT, K. GRUMBACH, and L. S. LEVINE (1983), in: The Metabolic
Basis of Inherited Disease, (J. B. STANBURY, J. B. WYNGAARDEN, D. S. FREDRICKSON,
J. L. GOLDSTEIN, and M. S. BROWN, eds.), McGraw-Hill, New York, 973—1000.

OEDA, K., T. SAKAKI, and H. OHKAWA (1985), DNA 4, 203—210.

ORTON, T. C. and G. L. PARKER (1982), Drug. Metab. Disp. 10, 110—115.

PARKER, K. L., D. D. CHAPLIN, M. WONG, J. G. SEIDMAN, J. A. SMITH, and B. P.
SCHIMMER (1985), Proc. Natl. Acad. Sci. U.S.A. 82, 7860—7864.

PEDERSEN, R. C. and A. C. BROWNIE (1987), Science 236, 188—190.

PHILLIPS, I. R., E. A. SHEPHARD, A. ASHWORTH, and B. R. RABIN (1985), Proc. Natl.
Acad. Sci. U.S.A. 82, 983—987.

PICADO-LEONARD, J. and W. L. MILLER (1987), DNA 6, 439—448.

POULOS, T. L., B. C. FINZEL, and A. J. HOWARD (1986), Biochemistry 25, 5314—5322.

QUATTROCHI, L. C., U. R. PENDURTHI, S. T. OKINO, C. POTENZA, and R. H. TUKEY
(1986), Proc. Natl. Acad. Sci. U.S.A. 83, 6731—6735.

RAUCY, J. L. and E. F. JOHNSON (1985), Mol. Pharmacol. 27, 296—301.

REDDY, J. K., D. L. AZARNOFF, and C. E. HIGNITE (1980), Nature 283, 397—398.

REIK, L. M., W. LEVIN, D. E. RYAN, S. L. MAINES, and P. E. THOMAS (1985), Arch.
Biochem. Biophys. 242, 365—382.

RODRIGUES, N. R., I. DUNHAM, C. Y. YU, M. C. CARROLL, R. R. PORTER, and R. D.
CAMPBELL (1987), EMBO J. 6, 1653—1661.

SANGLARD, D., O. KÄPPELI, and A. FIECHTER (1986), Arch. Biochem. Biophys. 251,
276—286.

SCHNEIDER, P. M., M. C. CARROLL, C. A. ALPER, C. RITTNER, A. S. WHITEHEAD, E. J.
YUNIS, and H. R. COLTEN (1986), J. Clin. Invest. 78, 650—657.

SHIROISHI, T., T. SAGAI, S. NATSUUME-SAKAI, and K. MORIWAKI (1987), Proc. Natl.
Acad. Sci. U.S.A. 84, 2819—2823.

SIMMONS, D. L., P. MCQUIDDY, and C. B. KASPER (1987), J. Biol. Chem. 262, 326—332.

SIMPSON, E. R., EVANS, C. T., CORBIN, C. J., POWELL, F. E., LEDESMA, D. B., and C. R.
MENDELSON (1987), Mol. Cell. Endocrinol. 52, 267—272.

SKODA, R. C., F. J. GONZALEZ, A. DEMIERRE, and U. A. MEYER (1988), Proc. Natl.
Acad. Sci. U.S.A. 85, 5240—5243.

SOGAWA, K., A. FUJISAWA-SEHARA, M. YAMANE, and Y. FUJII-KURIYAMA (1986), Proc.
Natl. Acad. Sci. U.S.A. 83, 8044—8048.

SONG, B.-J., H. V. GELBOIN, S. S. PARK, C. S. YANG, and F. J. GONZALEZ (1986), J.
Biol. Chem. 261, 16689—16697.

SONG, B.-J., T. MATSUNAGA, J. P. HARDWICK, S. S. PARK, R. L. VEECH, C. S. YANG,
H. V. GELBOIN, and F. J. GONZALEZ (1987), Mol. Endocrinol. 1, 542—547.

SPURR, N. K., A. C. GOUGH, K. STEVENSON, and C. R. WOLF (1987), Nucl. Acids Res.
15, 5901.

SQUIRES, E. J. and NEGISHI, M. (1988), J. Biol. Chem. 263, 4166—4171.

STUPANS, I., T. IKEDA, D. J. KESSLER, and D. W. NEBERT (1984), DNA 3, 129—138.

SUWA, Y., Y. MIZUKAMI, K. SOGAWA, and Y. FUJII-KURIYAMA (1985), J. Biol. Chem.
260, 7980—7984.

TRZECIAK, W. H., E. R. SIMPSON, T. J. SCALLEN, G. V. VAHOUNY, and M. R. WATER-
MAN (1987), J. Biol. Chem. 262, 3713—3717.

TUKEY, R. H., S. OKINO, H. BARNES, K. J. GRIFFIN, and E. F. JOHNSON (1985), J. Biol. Chem. **260**, 13347—13354.

UMENO, M., B. J. SONG, C. KOZAK, H. V. GELBOIN, and F. J. GONZALEZ (1988), J. Biol. Chem. **263**, 4956—4962.

UMBENHAUER, D. R., M. V. MARTIN, R. S. LLOYD, and F. P. GUENGERICH (1987), Biochemistry **26**, 1094—1099.

UNGER, B. P., I. C. GUNSALUS, and S. G. SLIGAR (1986), J. Biol. Chem. **261**, 1158—1163.

VOUTILAINEN, R. and W. L. MILLER (1987), Proc. Natl. Acad. Sci. U.S.A. **84**, 1590 to 1594.

VOUTILAINEN, R. and W. L. MILLER (1988), DNA **7**, 9—15.

WEN, L.-P., and A. J. FULCO (1987), J. Biol. Chem. **262**, 6676—6682.

WHITE, P. C., D. GROSSBERGER, B. J. ONUFER, D. D. CHAPLIN, M. I. NEW, B. DUPONT, and J. L. STROMINGER (1985), Proc. Natl. Acad. Sci. U.S.A. **82**, 1089—1093.

WHITE, P. C., M. I. NEW, and B. DUPONT (1986), Proc. Natl. Acad. Sci. U.S.A. **83**, 5111—5115.

WILBUR, W. J. and D. J. LIPMAN (1983), Proc. Natl. Acad. Sci. U.S.A. **80**, 726—730.

WONG, G., K. KAWAJIRI, and M. NEGISHI (1987), Biochemistry **26**, 8683—8690.

WRIGHTON, S. A., P. MAUREL, E. G. SCHUETZ, P. B. WATKINS, B. YOUNG, and P. S. GUZELIAN (1985), Biochemistry **24**, 2171—2178.

YANG, C. S., Y. Y. TU, D. R. KOOP, and M. J. COON (1985), Cancer Res. **45**, 1140—1145.

ZUBER, M. X., E. R. SIMPSON, and M. R. WATERMAN (1986), Science **234**, 1258—1261.

ZUBER, M. X., J. I. MASON, E. R. SIMPSON, and M. R. WATERMAN (1988), Proc. Natl. Acad. Sci. U.S.A. **85**, 699—703.

# Chapter 3

# The Induction of Cytochrome P-450c by Polycyclic Hydrocarbons Proceeds Through the Interaction of a 4S Cytosolic Binding Protein

E. Bresnick and W. H. Houser[1]

[1] The author's research activities have been supported by grants from the N.I.H.,
ESO3980 and CA36106. The Eppley Institute is a National Cancer Institute-designated
Laboratory Cancer Research Center, supported in part by CA36727.

# 1.  Introduction and hypothesis

The cytochrome P-450-dependent mixed function oxidases represent a family of isoenzymic hemoproteins that play a major role in the biotransformation of a number of endogenous fatty acids and steroids as well as of xenobiotics (CONNEY et al., 1967; GILLETTE et al., 1972). The elimination of these hydrophobic substances is necessary for the continued well-being of the organism. The cytochromes P-450 facilitate the oxidative metabolism of xenobiotics by enhancing their hydrophilicity. During the course of these chemical reactions, however, many xenobiotics are converted to highly-reactive electrophilic compounds that interact with the cellular macromolecular material, e.g., DNA, and provoke a pathological response such as cancer. Consequently, the cytochromes P-450 have generated considerable research interest.

The response of these hemoproteins to a large group of chemically-unrelated substances represents a unique feature of the cytochrome P-450 group. Thus, animals exposed to pesticides, barbiturates, environmental contaminants including polycyclic aromatic hydrocarbons, certain organic solvents, defined steroids, polychlorinated biphenyls, and some synthetic flavonoids respond by inducing select members of the cytochrome P-450 group (CONNEY et al., 1967; WATERMAN and ESTABROOK, 1983; IVERSEN et al., 1986). In particular, the exposure of rodents to polycyclic hydrocarbons such as benzo(a)pyrene (BP) or 3-methylcholanthrene (3MC) leads to an induction of principally two of these hemoproteins, cytochromes P-450c[1] and P-450d, in a number of tissues. The manner by which this elevation in cytochrome P-450c or P-450d occurs has not been definitively established although the mechanism of induction is under study in several laboratories (KUMARA et al., 1984; JONES et al., 1985; HOUSER et al., 1985).

Based upon the then emerging mechanism of induction of estrogen-sensitive proteins by this hormone, the senior author (BRESNICK, 1966) postulated that polycyclic hydrocarbons might also increase the biosynthesis of certain cytochrome P-450-dependent enzyme systems through the intervention of a "receptor", i.e., a protein which would interact with the polycyclic hydrocarbon in a high affinity and saturable manner and induce cytochrome P-450 synthesis. The postulated mechanism is presented in Figure 1.

It was POLAND's laboratory (POLAND et al., 1979) that first demonstrated the specific, high-affinity and saturable binding of a dioxin derivative to hepatic cytosolic proteins. This dioxin, 2,3,7,8-tetrachlorodibenzo-p-dioxin (TCDD), is also a potent inducer of a cytochrome P-450-dependent mono-

---

[1] We have adopted the nomenclature proposed by LEVIN and his colleagues (RYAN et al., 1979) in which the major rat cytochrome P-450 induced by 3MC is referred to as cytochrome P-450c; the minor form, as cytochrome P-450d. Cytochromes P-450c and P-450d correspond to the mouse cytochromes $P_1$-450 and $P_3$-450 and to the rabbit forms 6 and 4, respectively.

---

oxygenase, aryl hydrocarbon hydroxylase (AHH), in rodents (GUENTHNER and NEBERT, 1977; OKEY et al., 1979). Indeed, the toxicity of TCDD and its congeners in murine strains correlated with the binding of these components to the cytosolic binding protein (POLAND and GLOVER, 1980).

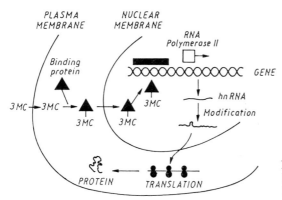

**Fig. 1.** Induction of cytochrome P-450c by polycyclic hydrocarbons. 3MC, 3-methylcholanthrene.

Additional evidence has since been reported on the saturable, high-affinity interaction of polycyclic hydrocarbons, such as 3MC, to proteins present in rat liver cytosol (GUENTHNER and NEBERT, 1977; OKEY et al., 1979; TIERNEY et al., 1980; HEINTZ et al., 1981; OKEY and VELLA, 1982; ZYTKOVICZ, 1982; COLLINS and MARLETTA, 1984).

In addition to high-affinity and saturable binding to a "receptor", the resultant ligand-protein complex must translocate into the nucleus as depicted in Figure 1. Such a temperature-dependent translocation into the nucleus has been demonstrated in vitro (OKEY et al., 1979; TIERNEY et al., 1980; HEINTZ et al., 1981; TUKEY et al., 1982; GREENLEE and POLAND, 1979; OKEY et al., 1980).

In a more recent study from our laboratory, the binding protein has been tested for its ability to interact with specific regions of the cytochrome P-450c gene (HOUSER et al., 1986). Such interaction does indeed occur, fulfilling yet another requirement of the hypothesis. Whether this interaction does stimulate the transcription of the cytochrome P-450c gene as demanded by the hypothesis has yet to be affirmed.

In this review, the purification of the polycyclic hydrocarbon-binding protein, its properties and its interaction with a defined region of the cytochrome P-450c will be discussed.

E. BRESNICK; W. H. HOUSER

# 2.   Isolation and properties of the polycyclic hydrocarbon-binding protein from rat hepatic cytosol

The purification procedure for obtaining the binding protein, i.e., the "receptor", from rat liver has been published (HOUSER et al., 1985). The purification scheme is briefly outlined in Table 1.

**Table 1.** Isolation of the polycyclic hydrocarbon-binding protein from rat liver

| Step | Specific activity (cpm $\times$ $10^{-3}$ mg$^{-1}$) | Total activity (%) | Overall purification |
|---|---|---|---|
| $10^5 \times$ g Supernate | 2.0 | 100 | 1 |
| Sepharose CM chromatography | 5.6 | 168 | 2.7 |
| Sephacryl S200 chromatography | 31.4 | 203 | 15.4 |
| Phenyl sepharose chromatography | 1,086 | 497 | 533 |
| DEAE Sepharose chromatography | 5,095 | 341 | 2,499 |
| Affi-gel-chromatography | 12,500 | 102 | 6,130 |

Data obtained from HOUSER et al. (1985).

The end result of the isolation procedure is the preparation of a protein which is or is near homogeneous. During the isolation from rat hepatic cytosol, a number of interfering substances are removed thus yielding an apparent increase in total activity, e.g., phenyl sepharose chromatography effects a 2.5 fold increase in total activity. The overall purification is in excess of 6,000-fold. SDS gel electrophoresis indicates two major bands as present. We believe these to represent different protomeric units of a dimeric polycyclic hydrocarbon-binding protein.

## 2.1.   Properties of the polycyclic hydrocarbon-binding protein

The substrate specificity of the binding protein has been assayed by using the magnitude of the inhibition by various polycyclic hydrocarbons and other agents of the interaction of labeled 3MC to the protein. These data which have previously been published (TIERNEY et al., 1983) are presented in Table 2.

Structures with at least three fused benzene rings were required in order to inhibit the binding of 3MC to the cytosolic protein thus suggesting a structural requirement for ligand-"receptor" interaction. Equally efficacious as in-

**Table 2.** Ability of various substances to inhibit the binding of 3MC to partially-purified "receptor"

| Competitor | Inhibition of specific 3MC binding (%) |
|---|---|
| 1. Polycyclic hydrocarbons | |
| Naphthalene, 2,3-benzanthracene, or pentacene | 0—11 |
| Phenanthrene or anthracene | 18—21 |
| Coronene | 41 |
| Pyrene<br>1,2-Benzanthracene, triphenylene, or chrysene | 73—79 |
| Benzo(a)pyrene, benzo(e)pyrene, or 3-methylcholanthrene | 91—100 |
| 2. Steroidal hormones | |
| 17$\beta$-Estradiol. dihydrotestosterone, progesterone, pregnenolone-16$\alpha$-carbonitrile, or dexamethasone | 0—9 |
| 3. Other | |
| Phenobarbital or isosafrole | 0 |
| Araclor 1254 or 7,8-benzoflavone | 9—13 |
| 5,6-Benzoflavone | 62 |

Abstracted from TIERNEY et al. (1983). Partially-purified rat hepatic binding protein was incubated with 5 pmol of $^3$H-3MC plus a 100-fold excess of the competitor for 18 hr at 4°. Sucrose density gradient analysis was used for the assessment of the amount of specific binding to the protein.

hibitors, and presumably as substrates for this binding protein, were benzo(a)-pyrene and benzo(e)pyrene. Triphenylene, 1,2-benzanthracene and pyrene were slightly less active in this regard.

The polycyclic hydrocarbon-binding protein exhibited little if any affinity for estrogenic, androgenic, progestational, or glucocorticoid hormones. Furthermore, the inducers phenobarbital and isosafrole were unable to inhibit the binding of 3MC to the protein. In this class, only the inducer, 5,6-benzoflavone ($\beta$-naphthoflavone), exerted any appreciable activity.

The affinity of the rat hepatic cytosolic fraction for polycyclic hydrocarbons has been examined by Scatchard analysis (TIERNEY et al., 1980; HOLDER et al., 1981). An example of such an analysis with benzo(a)pyrene as the substrate is offered in Figure 2. From these data, a $K_D$ of 2.5 nM and a binding capacity with the crude cytosolic fraction of 770 fmol/mg can be calculated. These results are in close agreement with those obtained for the binding of 3MC to the cytosolic protein (TIERNEY et al., 1980).

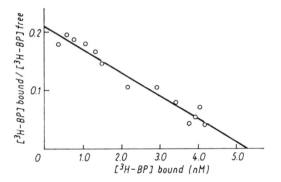

**Fig. 2.** Scatchard analysis of ³H-benzo(a)pyrene binding to the rat hepatic cytosolic fraction.
The data have been obtained from the publication of HOLDER et al. (1981). Rat liver cytosol (1 mg/ml) was incubated with various concentrations of radioactively-labeled benzo(a)pyrene in the absence or presence of a 160-fold excess of unlabeled polycyclic hydrocarbon. The specific binding was determined as the difference between these values and has been plotted according to SCATCHARD (1949).

In the mouse, high-affinity binding of TCDD and some polycyclic hydrocarbons has been reported to be associated with an 8S cytosolic protein (GUENTHNER et al., 1977; OKEY et al., 1979; NEBERT et al., 1984; OKEY et al., 1984). These studies have suggested the presence of the 8S TCDD-binding protein as a prerequisite for the induction of AHH. We were interested in determining the sedimentation constant of the polycyclic hydrocarbon-binding protein that is present in rat hepatic cytosol. Our previous studies (TIERNEY et al., 1980; HEINTZ et al., 1981) as well as those of others (ZYTKOVICZ, 1982; COLLINS and MARLETTA, 1984) have shown that this binding protein sediments at approximately 4S. An example of sucrose density gradient analysis of this binding is indicated in Figure 3.

The binding protein for polycyclic hydrocarbons exhibits a sedimentation constant of 4S (Fig. 3). As stated above, it has been reported that the TCDD binding occurs to an 8S protein species. We had been concerned with the possibility that the 4S binding protein may, in fact, also interact in a specific, saturable and high-affinity manner with TCDD and that the 4S protein may represent an artifactual subunit of the 8S protein. These possibilities have been tested recently (HOUSER et al., 1986). In this study, we have reaffirmed

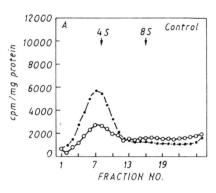
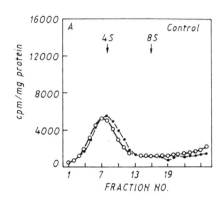

**Fig. 3**                    **Fig. 4**

**Fig. 3.** Sedimentation analysis of benzo(a)pyrene binding to rat hepatic cytosol.
These data are taken from HOUSER et al. (1986). One to 2 mg of rat hepatic cytosolic fraction was incubated for 60 min at 4° with $^3$H-benzo(a)pyrene in the absence (●—●) or presence (○—○) of a 200 — fold excess of unlabeled polycyclic hydrocarbon. An aliquot of the mixture was then subjected to analysis of a 5—20% sucrose gradient that was centrifuged at 55,000 rpm for 2 hr in a vertical rotor. The gradient was fractionated and the samples were counted. The positions of authentic 4S and 8S markers are indicated on the figure as arrows.

**Fig. 4.** Lack of effect of TCDBF on the binding of $^3$H-benzo(a)-pyrene to 4S rat hepatic cytosolic protein.
The results are abstracted from HOUSER et al. (1986). The $10^5 \times g$ supernate was prepared from rat liver and was incubated with the labelled polycyclic hydrocarbon in the presence (○—○) or absence (●—●) of a 200 fold excess of TCDBF. An aliquot of the reaction mixtures was then subjected to density analysis on a 5—20% sucrose gradient. The 4S and 8S sedimentation markers are indicated on the figure as arrows.

that a) no appreciable 8S protein exists in rat hepatic cytosol that interacts in a high-affinity manner with polycyclic hydrocarbons, i.e., benzo(a)pyrene, and b) the addition of a 200-fold excess of the dioxin congener, tetrachloro-dibenzofuran (TCDBF), does **not** inhibit the interaction of the polycyclic hydrocarbon to the 4S protein. These results are shown in Figure 4.

We have concluded from these studies that at least two binding proteins must exist, a) an 8S species that has high affinity for dioxins such as TCDD and

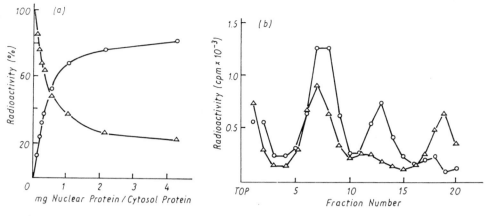

**Fig. 5 A.**          **Fig. 5 B.**

**Fig. 5.** 'Translocation' of the ligand-charged cytosol to the nucleus. The data were obtained from HOLDER et al. (1981). The transfer of $^3$H-benzo(a)pyrene from charged cytosolic 4S binding protein to isolated hepatic nuclei is indicated as a function of increasing amount of nuclei (Fig. 5 A). $\triangle—\triangle$, residual radioactivity in supernate; o—o, nuclear radioactivity. After incubation of the charged cytosol with isolated rate hepatic nuclei at 4° for 1 hr, an aliquot of the reaction mixture was subjected to density gradient analysis on 5—20% sucrose overlaid on a 2.3 M sucrose cushion (Fig. 5 B). The direction of sedimentation is from left to right with the 4S and nuclear regions indicated by the arrows. $^3$H-Benzo(a)pyrene-charged cytosol, o—o; nuclear incubation, $\triangle—\triangle$.

TCDBF and which is found in mouse hepatic cytosol, and b) a 4S species that has high affinity for polycyclic hydrocarbons such as benzo(a)pyrene and 3-methylcholanthrene and which is found in rat hepatic cytosol.

## 3.    Translocation into the nucleus

A second requirement imposed by the hypothesis (see page 64) is the translocation of the 'charged'-binding protein into the nucleus and the subsequent interaction of this protein (or the ligand) with chromatin. The ability of the polycyclic hydrocarbon charged 4S binding protein to 'translocate' into the nucleus has been tested previously in this laboratory (TIERNEY et al., 1980; HOLDER et al., 1981). An example of these results is presented in Figure 5. $^3$H-Benzo(a)pyrene-complexed 4S protein was incubated with increasing amounts of isolated rat hepatic nuclei and the level of radioactivity that is transferred to the nuclei as well as the remaining radioactivity in the cytosolic fractions was determined. As indicated in Figure 5 A, an increasing amount of $^3$H-benzo(a)pyrene became associated with the nuclei. In a com-

---

panion experiment (Fig. 5 B), the incubation of ³H-benzo(a)pyrene-charged cytosol with nuclei and sucrose density sedimentation analysis of the reaction resulted in a transfer of the radioactivity from the cytosolic to the nuclear compartment.

## 4. Molecular biology of the cytochrome P-450c gene

The last aspect in the definition of the role played by the 4S polycyclic hydrocarbon-binding protein in the induction of cytochrome P-450c is a demonstration of the increased transcription of the cytochrome P-450c gene as a

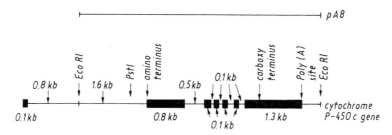

**Fig. 6.** Structure of the rat cytochrome P-450c gene.
The 5' and 3'-ends of the gene are indicated on the left and right, respectively, of the figure. The exons are depicted as solid bars while the introns are represented as the thine lines between them. The positions of the DNA sequence which are responsible for the N- and C-termini of cytochrome P-450c are indicated in the figure. The sizes of each exon and intron are presented in the figure. The relationship of the genomic clone, pA8, to the cytochrome P-450c gene is also shown.

result of the interaction of the 'receptor' with the latter. Required for this demonstration is the DNA sequence of the cytochrome P-450c gene as well as of the 5'-flanking regions.

The sequence for a near-full length cDNA clone which codes for cytochrome P-450c was first published by YABUSAKI et al. (1984). The structure and sequence for the cytochrome P-450c gene was subsequently reported by SOGAWA et al. (1984) and from our laboratory (HINES et al., 1985). The structure of the cytochrome P-450c gene is indicated in Figure 6.

The cytochrome P-450c gene spans over 6 kbp of information with two highly conserved regions, HR1 and HR2, as proposed by SOGAWA et al. (1984), representing possible heme binding sites located within exons 2 and 7, respectively. DNA sequence which is almost identical to SV40 core enhancer region is found in intron 1 and in the 5'-flanking region. Furthermore, some unusual stretches consisting of alternating base pairs, $(TC)_{33}$, $(TG)_{23}$, $(AG)_{19}$, and

E. BRESNICK; W. H. HOUSER

$(TG)_{14}$, are also found in regions upstream from the gene and in intron 1. Finally, the consensus sequence, TCTCCT, which has been suggested as contributing to the binding of the glucocorticoid receptor to DNA (KARIN et al., 1984) occurs three times within approximately 300 bp in intron 1 of the gene. This structural feature is particularly important in view of the observation by MATHIS et al. (1986) that glucocorticoids act synergistically with polycyclic hydrocarbons in increasing the level of cytochrome P-450c expressed in fetal rat liver.

## 5. Interaction of the 4S binding protein with the cytochrome P-450c gene

As indicated previously, verification of the hypothesis requires the interaction of the 4S polycyclic hydrocarbon-binding protein with the cytochrome P-450c gene. Once we had determined the fine structure of the latter, we believed that some of the unusual DNA sequences in intron 1 and in the 5'-flanking

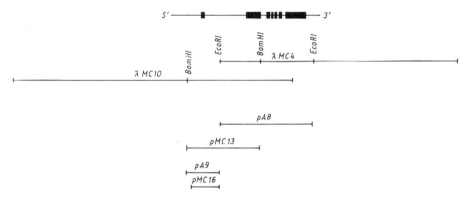

**Fig. 7.** The rat cytochrome P-450c gene and various subclones. The rat cytochrome P-450c gene includes 6.7 kbp of information (HINES et al., 1985). pA8, pMC13, and pA9 are subclones of this gene.

regions might serve as potential binding sites for the 4S protein. This aspect has been experimentally tested in the studies of HOUSER et al. (1985).

The ability of the 4S binding protein to interact with specific regions of the cytochrome P-450c gene was investigated using a modified streptomycin nitrocellulose-binding assay (SPELSBERG, 1983). The potential target molecules for the binding protein included: pA8, a 5.4 kbp of rat genomic information which contained all of the coding information for the rat cytochrome P-450c gene as well as the 3′ half of intron 1; pMC13, which contained a 4.3 kbp insert of exon 1, intron 1, part of exon 2 as well as 887 bp of 5′ information; and pA9 which contained 1.9 kbp of DNA including exon 1, the 5′ half of intron 1 as well as 887 bp of 5′ upstream information. The relationships among these plasmids are depicted in Figure 7.

**Table 3.** Interaction of the 4S binding protein with various labeled DNA preparations

| DNA Source | % Radioactivity associated in the complex | | |
|---|---|---|---|
| | Receptor (R) | DNA | R + DNA |
| Total rat liver | $0.04 \pm 0.03$ | $3.3 \pm 0.3$ | $25.9 \pm 8.6$ |
| pUC8 | $0.3 \pm 0.4$ | $2.5 \pm 0.5$ | $12.0 \pm 1.9$ |
| pA8 | $0.1 \pm 0.05$ | $2.8 \pm 1.2$ | $8.2 \pm 1.1$ |
| pA9 | $0.3 \pm 0.01$ | $6.1 \pm 2.5$ | $23.5 \pm 6.2$ |
| pMC13 | $0.3 \pm 0.3$ | $3.7 \pm 2.0$ | $22.3 \pm 6.2$ |

These results were abstracted from the publication of HOUSER et al. (1985). [35]S-nucleo-tide-labeled DNA was incubated with purified benzo(a)pyrene charged 4S protein and the amount of its complexation was determined by a streptomycin-nitrocellulose-binding assay. pUC8, plasmid vector; pA8, pA9, pMC13, see Figure 7 for their definition.

The results of our experiments are presented in Table 3. When total rat liver that had been end-labeled with a [35]S-deoxyribonucleotide was incubated with the partially-purified 4S binding protein (charged with benzo(a)pyrene), greater than 25% of the radioactivity became associated with a complex that bound to nitrocellulose (Table 3.). This complex was dependent upon added binding protein. The plasmid vector, pUC8, was used as a negative control; approximately 12% of the labelled DNA became associated in a complex that was not filterable through nitrocellulose. A portion of the cyto-chrome P-450c gene, pA8, also gave comparable (in fact, less) binding to the partially-purified binding protein. However, when plasmids containing more 5′-upstream regions were incubated with the 4S protein, considerable binding resulted, i.e., equivalent to that observed with total rat liver DNA. This is apparent with both pA9 and pMC13, both of which share the same 5′-end. pA9 and pMC13 contain the bases from −887 to +999 and −887 to +3,447, respectively, of the cytochrome P-450c gene. Consequently, in regard to interaction with the 4S binding protein, an important region of the cytochrome P-450c gene must be somewhere between −887 and +999 bp. The importance of this region is further reinforced by footprinting experiments, indicated below. It is also germane to mention that WHITLOCK's laboratory has recently published the existence of a binding region for TCDD on the mouse cytochrome $P_1$-450 gene somewhere between $-1,580$ and $-1,310$ bp (JONES et al., 1985).

## 5.1. Exonuclease 'footprinting' of the interaction of the 4S protein with the cytochrome P-450c gene

We have employed the exonuclease 'footprinting' technique to further define the interaction of the 4S binding protein with the cytochrome P-450c gene. Using the method of ELBRECHT et al. (1985), we have 3'-end labeled pMC16 (see Fig. 7.) as well as 5'-end labelled this plasmid. After treating pMC16 with either BamH1 or EcoR1, a $-600$ to $+1,003$ DNA fragment is produced which is now labeled on only its 3' or 5' end, respectively. The latter is incubated with the 4S binding protein under the conditions as described by ELBRECHT et al. (1985). The 3'-end labeled and 5'-end labeled DNA fragments were then digested with lambda exonuclease or $T_4$ DNA polymerase and exonuclease III, respectively. After subjecting the resultant material to DNA sequencing gel electrophoresis and radioautography, the region of the DNA that has been protected by the binding protein can be easily ascertained by comparison with a comparable DNA fragment to which the protein has not been added. A typical result from this experiment is shown in Figure 8.

These results have been interpreted as indicating protection afforded by the 4S binding protein to two regions of the cytochrome P-450c gene (summarized in Fig. 9), at approximately 200 and 400 bp upstream from exon 1. These regions are within the area of the gene suggested by the streptomycin-nitrocellulose-binding assay as being important.

## 6. Transfection experiments with genetic reconstructs

Further preliminary evidence as to the importance of the upstream regions in the regulation of cytochrome P-450c gene expression is provided by a series of genetic reconstruction experiments in which the chloramphenicol acetyl transferase (CAT) gene is employed as the reporter.

The strategy of our fusion experiments was to use a plasmid that had been constructed by ROSENTHAL et al. (1983), pA10cat, which includes the ampicillin resistance gene for selection, a TATA region, the CAT gene, GC-rich 21 bp repeats from SV40, an origin of replication, and the 72 bp SV40 sequence containing enhancer elements. We have modified pA10cat by converting a HindIII site just 5' to the CAT cartridge to a Bg1II site and elimination of a Bg1II fragment which contains the TATA region and enhancer. The resultant

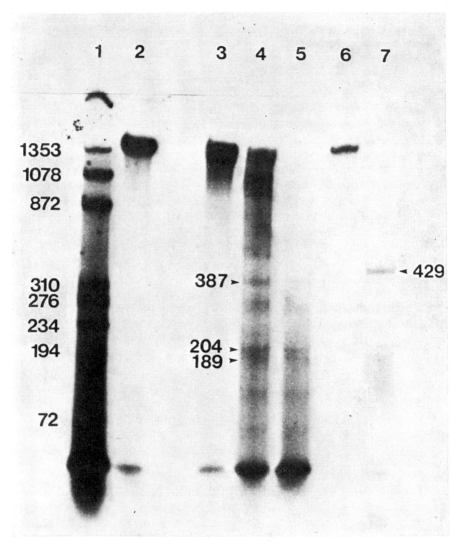

**Fig. 8.** Exonuclease 'footprinting' with pMC16 and the 4S binding protein.
pMC16 was end labeled on its 3′- and 5′termini using T4 DNA polymerase, and alpha-$^{32}$p-deoxyribonucleotide triphosphate or calf intestinal phosphatase, T4 polynucleotide kinase and gamma-$^{32}$P ATP, respectively. After incubating this DNA with either BamH1, a 3′- labeled −600 to +1,003 DNA fragment was obtained. To these fragments, 4S binding protein was added as indicated in the figure, and after a period of incubation, the DNA was treated with DNA polymerase and exonuclease III. The resultant material was subjected to gel electrophoresis under conditions of DNA sequencing, and the gel was radioautographed. Lanes 1, standards (sizes are given in bp); 2, pMC16; 3, pMC16 + 2.5 µl boiled "receptor"; 4, pMC16 + 2.5 µl "receptor"; 6, pMC16 + 5 µl boiled "receptor"; and 7, pMC16 + 5 µl receptor. The protected regions are indicated in bp.

E. Bresnick; W. H. Houser

construct, pD10cat is opened with BglII, and an appropriate fragment from pMC 13 which bears the TATA box, some 5'-upstream regions, exon I, and parts of intron I, is inserted (see Fig. 10). The final constructs include a) pMCcat in which the putative regulatory regions are in the appropriate order as in pD10cat and b) pMC'cat in which the insert from pMC13 has been placed in reverse sequence. These two plasmids along with the positive control, pRSVcat (which contains a strong promoter driving CAT expression), were transfected

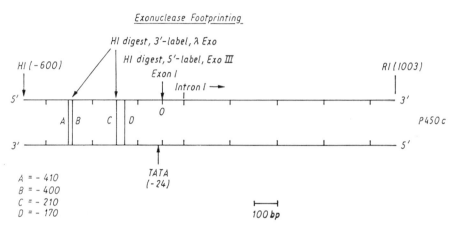

**Fig. 9.** Reconstruction of the interaction of pMC16 with the 4S binding protein as determined by exonuclease footprinting. HI, BamHI-treated 3'- and 5'-end labeled pMC16; RI, EcoRI-treated material; lambda exo, labda exonuclease; Exo III, exonuclease III. The positions of the starts for exon 1, intron 1, and the TATA box are indicated in the figure. The areas protected by the 4S binding protein are indicated as A, B, C, and D.

into RL-PR-C rat hepatocytes and into H4IIE rat hepatoma cells. Both the hepatocytes and the hepatoma cells are inducible for cytochrome P-450c and have 4S binding protein activity. As a negative control, the cells were transfected with pD10cat. 3-Methylcholanthrene (3MC) was added to the cells at $10^{-5}$ M, and after 24 hr the transient expression of CAT activity was determined by the method of GORMAN et al. (1982) as well as the steady-state level of the mRNA for cytochrome P-450c (by dot blot analysis). The latter is shown in Figure 11. The level of mRNA for cytochrome P-450c in control liver or hepatoma cells was negligible. Substantial induction took place in both cell types after 3MC treatment (Fig. 11). As is apparent from Figure 12., CAT expression is also significantly affected by 3MC treatment. The proper order of the regulatory sequences is required since pMC'cat + 3MC does not lead to CAT expression in either of the cell types. Secondly, 3MC is required

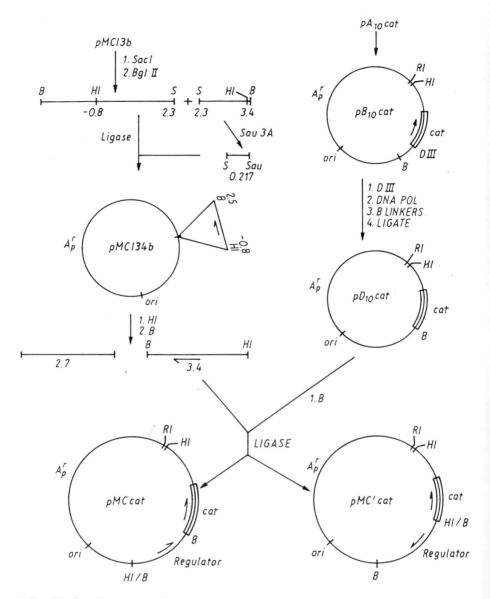

**Fig. 10.** Genetic construction of pMCcat and pMC'cat. B, Bgl II; RI, EcoRI; HI BamHI; S, SacI; Apr, ampicillin resistance gene; POL, DNA polymerase I.

E. Bresnick; W. H. Houser

for the appropriate expression of the information in pMCcat (e.g., A 1 and 2, B 1 and 2 in Fig. 12.). Consequently, we can conclude from these studies that the regulatory sequences governing the 3MC-dependent expression of a target gene are present in the region from the middle of intron 1 to $-887$ bp. The importance of this region as demonstrated in these gene fusion experiments is further reinforcement for the studies presented in this report.

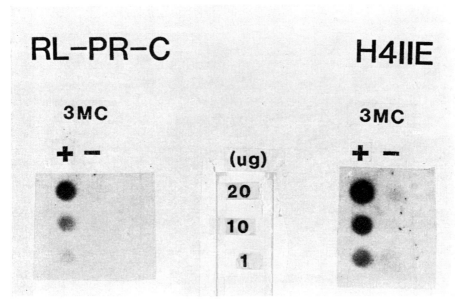

Fig. 11. Cytochrome P-450c mRNA in hepatocytes and hepatoma cells in response to 3-methylcholanthrene. Total RNA was extracted from RL-PR-C rat hepatocytes and from H4IIE rat hepatoma cells 24 hr after treatment with $10^{-5}$ M 3MC $(+)$ or the vehicle alone $(-)$. The RNA was dotted on nitrocellulose and the amount of cytochrome P-450c mRNA was determined after hybridization to nick-translated pA8. The amount of RNA used is indicated as μg.

Additional confirmation for the importance of this region derives from some deletion experiments we have performed. In these experiments, pMCcat has been treated with the restriction endonucleases, BstEII or MstII. These enzymes removed the region, $-238$ to $-660$, and $-95$ to $-665$, respectively. The resultant plasmids were used in transfection experiments and CAT expression was determined as described above. In neither case, was a 3MC-responsive CAT expression demonstrable (see Fig. 12 A, B).

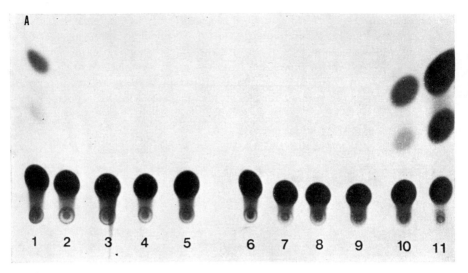

Fig. 12 A

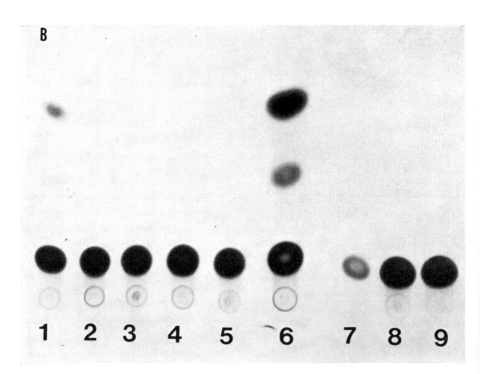

Fig. 12 B.

E. Bresnick; W. H. Houser

# 7. Summary

In this report, we have presented a series of studies that impact on the premise that polycyclic hydrocarbons such as benzo(a)pyrene are able to induce cytochrome P-450c only after interaction in a saturable and high-affinity manner with a rat hepatic cytosolic protein, a "receptor". We have purified this protein and demonstrated some of its properties which are in concert with the proposed mechanism.

We and others have subsequently isolated the gene for cytochrome P-450c and determined its fine structure and DNA sequence. A number of interesting and potentially-important structural features are noted in this gene, particularly in the 5'-flanking region and in intron 1. We have since demonstrated that the 4S polycyclic hydrocarbon-binding protein interacts in a saturable manner with a portion of this gene. This binding results in a unique footprinting pattern that suggests interaction at −200 and −400 bp of the gene.

We have used genetic constructions to demonstrate the importance of this region of the cytochrome P-450c gene. In these investigations, the chloramphenicol acetyl transferase gene was used as a reporter and the suspected regulatory sequences were inserted upstream from the latter. After transfection of these constructs, into both a 'normal' rat hepatocyte and rat hepatoma cell line, we then challenged the system with 3-methylcholanthrene. As a negative control, we employed a construct in which the rat sequences had been inserted in inverted fashion. This did not respond positively in regard to CAT expression upon challenge with the polycyclic hydrocarbon. On the other hand, when these sequences were inserted in appropriate fashion, a 3-methylcholanthrene-inducible expression of CAT was observed in both cell lines.

---

<

**Fig. 12.** CAT expression in RL-PR-C hepatocytes (A) and in hepatoma cells (B) in response to 3-methylcholanthrene. RL-PR-C (A) and H4IIE (B) were transfected by the protoplast fusion technique (YOAKUM, 1984) with either pRSCcat, pMCcat, pMC'cat, or pD10cat. 3MC (10 μM) was added to the cells and CAT expression was determined 24 hr later in extracts from these cells. The radioautographs show 3 spots with the acetylated chloramphenicol indicated as the area furthest away from the origin.

A. 1, pMCcat + 3MC; 2, pMCcat − 3MC; 3, pMC'cat + 3MC; 4, pMC'cat − 3MC;
5, BstEII deletion from pMCcat + 3MC; 6, BstEII deletion from pMCcat − 3MC;
7, MstII deletion from pMCcat + 3MC; 8, MstII deletion from pMCcat − 3MC;
9, negative control; 10, pRSVcat positive control; and 11, CAT positive control.

B. 1, pMCcat + 3MC; 2, pMCcat − 3MC; 3, pMC'cat + 3MC; 4, pMC'cat − 3MC;
5, BstEII deletion from pMCcat + 3MC; 6, CAT positive control; 7, BstEII deletion
from pMCcat − 3MC; 8, MstII deletion from pMCcat + 3CM; and 9, MstII deletion
from pMCcat − 3MC.

The final proof of the argument that the binding protein is required for the induction of cytochrome P-450c will consist of demonstrating its effectiveness in stimulating the initiation of segments of the gene in in vitro nuclear assays. Should this be the case, the binding protein could be considered as a transcriptional factor which shuttles between the cytosolic compartment and the nucleus. These studies are currently underway in our laboratory.

# 8. References

BRESNICK, E. (1966), Mol. Pharmacol. **2**, 406—410.
BRESNICK, E., R. L. FOLDES, and R. N. HINES (1984), Pharmacol. Rev. **36**, 435—515.
COLLINS, S. and M. A. MARLETTA (1984), Mol. Pharmacol. **26**, 353—359.
CONNEY, A. H. (1967), Pharmacol. Rev. **19**, 317—366.
ELBRECHT, A., S. Y. TSAI, M.-J. TSAI, and B. W. O'MALLEY (1985), DNA 4, 233—240.
GILLETTE, J. R., D. C. DAVIS, and H. A. SASAME (1972), Ann. Rev. Pharmacol. **12**, 57—84.
GORMAN, C. M., L. F. MOFFAR, and B. H. HOWARD (1982), Mol. Cell Biol. **2**, 1044—1051.
GREENLEE, W. F. and A. POLAND (1979), J. Biol. Chem. **254**, 9814—9821.
GUENTHNER, T. M. and D. W. NEBERT (1977), J. Biol. Chem. **252**, 8981—8989.
HEINTZ, N. H., B. TIERNEY, E. BRESNICK, and W. I. SCHAEFFER (1981), Cancer Res. **41**, 1794—1802.
HINES, R. N., J. B. LEVY, R. D. CONRAD, P. L. IVERSEN, M.-L. SHEN, A. RENLI, and E. BRESNICK (1985), Arch. Biochem. Biophys. **237**, 465—476.
HOLDER, G. M., B. TIERNEY, and E. BRESNICK (1981), Cancer Res. **41**, 4408—4414.
HOUSER, W. H., R. N. HINES, and E. BRESNICK (1985), Biochemistry **244**, 7839—7845.
HOUSER, W. H., R. ZELINSKI, and E. BRESNICK (1986), Arch. Biochem. Biophys. **251**, 361—368.
IVERSEN, P. L., R. N. HINES, and E. BRESNICK (1986), Bioessays 4, 15—19.
JONES, P. B. C., D. R. GALEAZZI, J. M. FISHER, and J. P. WHITLOCK (1985), Science **227**, 1499—1502.
KARIN, M., A. HASLINGER, H. HOLTGREVE, R. I. RICHARDS, P. KRAUTER, H. M. WEST-PHAL, and M. BEATO (1984), Nature **308**, 513—519.
KUMURA, S., F. J. GONZALES, and D. W. NEBERT (1984), J. Biol. Chem. **259**, 10705 to 10713.
MATHIS, J. M., R. A. PROUGH, R. N. HINES, E. BRESNICK, and E. R. SIMPSON (1986), Arch. Biochem. Biophys. **246**, 439—448.
NEBERT, D. W., H. J. EISEN, and O. HANKINSON (1984), Biochem. Pharmacol. **33**, 917—924.
OKEY, A. B., G. P. BONDY, M. E. MASON, G. F. KAHL, H. J. EISEN, T. M. GUENTHNER, and D. W. NEBERT (1979), J. Biol. Chem. **254**, 11636—11648.
OKEY, A. B., G. P. BONDY, M. E. MASON, D. W. NEBERT, C. FORSTER-GIBSON, J. MUN-CAN, and M. J. DUFRESNE (1980), J. Biol. Chem. **255**, 11415—11422.
OKEY, A. B. and L. M. VELLA (1982), Eur. J. Biochem. **127**, 39—47.
OKEY, A. B., A. W. DUBE, and L. M. VELLA (1984), Cancer Res. 44, 1426—1432.
POLAND, A., E. GLOVER, and A. S. KENDE (1979), J. Biol. Chem. **251**, 4936—4946.
POLAND, A. and E. GLOVER (1980), Mol. Pharmacol. **17**, 86—94.
ROSENTHAL, N., M. KRESS, P. GRUSS, and G. KHOURY (1983), Science **222**, 749—755.
RYAN, D. E., P. E. THOMAS, D. KORZENIOWSKI, and W. LEVIN (1979), J. Biol. Chem. **254**, 1365—1379.

SCATCHARD, G. (1949), Ann. N.Y. Acad. Sci. **51**, 660—672.
SOGAWA, K., O. GOTOH, K. KAWAJIRI, and Y. FUJII-KURIYAMA (1984), Proc. Natl. Acad. Sci. **81**, 5066—5070.
SPELSBERG, T. C. (1983), Biochemistry **22**, 13—21.
TIERNEY, B., D. WEAVER, N. H. HEINTZ, W. I. SCHAEFFER, and E. BRESNICK (1980), Arch. Biochem. Biophys. **200**, 513—523.
TIERNEY, B., S. MUNZER, and E. BRESNICK (1983), Arch. Biochem. Biophys. **225**, 826—835.
TUKEY, R. H., R. R. HANNAH, M. NEGISHI, D. W. NEBERT, and H. J. EISEN (1982), Cell **31**, 275—284.
WATERMAN, M. R. and R. W. ESTABROOK (1983), Mol. Cell. Biochem. **53/54**, 267—278.
YABUSAKI, Y., M. SHIMIZU, H. MURAKAMI, K. NAKAMURA, K. CEDA, and H. OHKAWA (1984), Nucl. Acids Res. **12**, 2929—2938.
YOAKUM, G. H. (1984), Biotechniques **2**, 24—30.
ZYTKOVICZ, T. H. (1982), Cancer Res. **42**, 4387—4393.

# Chapter 4
# Sex-dependent Regulation of Cytochrome P-450 Expression

R. KATO and Y. YAMAZOE

# 1. Introduction

The actions of a variety of drugs are more pronounced and persist longer in female rats than in males (KATO, 1974). The main mechanism of such sex related differences is due to differences in the activity of hepatic microsomal drug-metabolizing enzymes (KATO et al., 1962a; KATO, 1974). Although the drug-metabolizing enzymes from rat liver microsomes oxidatively metabolize a variety of drugs, KATO and GILLETTE (1965a) showed that the magnitude of the differences between the sexes is dependent on the substrates and their metabolic pathways. On the other hand, there are only slight differences between sexes in the NADPH-cytochrome c reductase activity and cytochrome P-450 content of rat liver microsomes (KATO et al., 1968b). Such minor variations may explain the small sex-related differences observed in aniline hydroxylation but do not explain the large differences found in hexobarbital hydroxylation and aminopyrine N-demethylation. These results indicate that androgen may induce some special form(s) of cytochrome P-450 which are capable of oxidizing a variety of drugs more rapidly in male rats than in females.

These sex differences sometimes cause sex-related differences in the effects of certain chemicals and of certain pathological states on the activity of drug-metabolizing enzymes (KATO and GILLETTE, 1965a, 1965b; KATO, 1977). For example, administration of morphine or methylcholanthrene decreased hexobarbital hydroxylation and aminopyrine N-demethylation in male rats, but not in females (KATO and GILLETTE, 1965b; KATO and TAKAYANAGI, 1966). Similar results were found with the effects of adrenalectomy, starvation, alloxan diabetes and hyperthyroidism (KATO and GILLETTE, 1965a, 1965b; KATO and TAKAHASHI, 1968; KATO et al., 1970a, 1970b; KATO et al., 1971).

These results also indicate that some particular form(s) of cytochrome P-450 responsible for sex-related differences in rat liver microsomes may selectively be affected in some pathological states (KATO and TAKAHASHI, 1968; KATO, 1974).

On the other hand, KUNTZMAN et al. (1964, 1966) showed a close similarity between the microsomal drug-metabolizing enzymes and testosterone-hydroxylating enzymes in rat liver. Testosterone hydroxylases are induced by phenobarbital, and the activity is clearly higher in microsomes from male rats than females. KATO and co-workers, moreover, showed clear evidence that the testosterone hydroxylating activity in liver microsomes from male rats was markedly decreased by several treatments, such as hyperthyroidism, adrenalectomy or morphine administration, but the same treatments caused only marginal effects in female rats (KATO et al., 1970d, 1970e, 1970f). These results indicate that both testosterone-hydroxylating and drug metabolizing enzymes in rat liver microsomes are regulated by the same or very similar mechanism.

SCHENKMAN et al. (1967) found more substantial evidence of differences between male and female rats in the substrate-induced difference spectrum of liver microsomes. Moreover, LEVIN and KUNTZMAN (1969) showed the presence of two components with regard to the decay of hepatic microsomal P-450: fast phase ($t^{1/2} = 7-8$ hr) and slow phase ($t^{1/2} = 42-46$ hr). The ratio of the fast phase component to the slow phase component was 1.9:1 in microsomes from adult male rats, but the ratio was 3.4:1 in both the immature and adult female rats (LEVIN et al., 1975).

All these results suggest the existence of sex-related cytochromes P-450 responsible for the metabolism of drugs and steroid hormones. These cytochromes P-450 are regulated by androgen and estrogen and other unknown factors (KATO et al., 1968b; KATO and ONODA, 1970). However, a definitive conclusion can not be drawn until the purification of these cytochromes P-450 has been accomplished.

## 2.  Existence of sex-specific cytochrome P-450 in rat liver microsomes

All previous studies have indicated a possible presence of sex-specific cytochrome P-450 responsible for sex-related differences. CHUNG and CHAO (1981) attempted to demonstrate the presence of sex-specific cytochrome P-450 by means of solubilization and separation on sodium dodecyl sulfate-polyacrylamide gel electrophoresis. We also attempted to demonstrate the existence

**Table 1.** Different properties of cytochrome P-450-male and P-450-female

|  | P-450-male | P-450-female |
|---|---|---|
| 1) Peak in CO-binding spectrum (nm) | 451 | 449 |
| 2) Molecular weight (daltons) | 52,000 | 50,000 |
| 3) Peptide maps | male ╪ | female |
| 4) Benzphetamine N-demethylation | male > | female |
| Benzo(a)pyrene hydroxylation | male > | female |
| 5) Testosterone hydroxylation | | |
| 2α-OH | male (+++) | female (−) |
| 16α-OH | male (+++) | female (−) |
| 6) Microsomal content | | |
| male | +++ | − |
| female | − | +++ |

R. KATO; Y. YAMAZOE

of sex-specific cytochrome P-450 in rat liver through purification of cyto-chrome P-450 from male and female rats (KAMATAKI et al., 1981; KAMATAKI et al., 1982). We succeeded in the purification of sex-specific forms of cyto-chrome P-450 from male and female rat liver microsomes and designated these as P-450-male and P-450-female, respectively (KATO and KAMATAKI, 1982a; KATO and KAMATAKI, 1982b; KAMATAKI et al., 1983b). We developed an immunoquantitative method for quantification of P-450-male and P-450-female contents in adult rat liver microsomes, and we found that P-450-male is the major cytochrome P-450 (about one-third of the total cytochrome P-450) in the male liver and is undetectable in the female. P-450-female is the major cytochrome P-450 (about one-third of the total cytochrome P-450) in the female liver and is undetectable in the male.

P–450–male:

Met–Asp–Pro–Val–Leu–Val–Leu–Val–Leu–Thr–Leu–Ser–  X  –Leu–Leu–Leu–Leu–

P–450–female:

Met–Asp–Pro–Phe–Val–Val–Leu–Val–Leu–Ser–Leu–Ser–  X  –Leu–Leu–

Fig. 1. N-Terminal amino acid sequences of P-450-male and P-450-female (KAMATAKI et al., 1985a).

P-450-male and P-450-female are clearly different from each other in catalytic activity, molecular weight, absolute spectrum, CO-binding spectrum, peptide map and N-terminal amino acid sequence (KAMATAKI et al., 1983b; KAMATAKI et al., 1985a) (Table 1). It is noteworthy that many activities of drug oxidations are higher with purified and reconstituted P-450-male than with purified and reconstituted P-450-female, and the activities of testosterone $2\alpha$-, $16\alpha$-hydroxylase are found only with P-450-male. The N-terminal amino acid sequences of P-450-male and P-450-female are given in Figure 1.

On the other hand, GUENGERICH et al. (1982) purified a cytochrome P-450 from untreated male rat liver microsomes and designated it P-450 UT-A. Moreover, CHENG and SCHENKMAN (1982; 1983) also purified, from untreated male rats, a cytochrome P-450 designated as P-450 RLM$_5$ which is capable of hydroxylating testosterone at positions of $2\alpha$ and $16\alpha$.

Although these investigators did not report sex-specificity of the purified cytochromes P-450, P-450 UT-A and P-450 RLM$_5$ seem to be identical with or closely similar to P-450-male judging from the catalytic activity, molecular weight, absolute spectrum and $NH_2$-terminal amino acid sequence (WAXMAN, 1984; WAXMAN et al., 1985; DANNAN et al., 1986). After publication of our paper on sex-specific cytochrome P-450 (KAMATAKI et al., 1983b), other groups also reported sex-specific forms of cytochrome P-450 (RYAN et al., 1984a and b; WAXMAN, 1984; MORGAN et al., 1985b; MACGEOCH et al., 1984; MATSUMOTO et al., 1986). Judging from the $NH_2$-terminal amino acid sequences and other properties reported, P-450-male is probably identical with or closely similar to the forms designated as P-450h (RYAN et al., 1984b), P-450-2c (WAXMAN, 1984), $P-450_{16\alpha}$ (MORGAN et al., 1985b) P-450M-1 (MATSU-MOTO et al.1, 986), whereas P-450-female corresponds to the forms designated as P-450i (RYAN et al., 1984a) or $P-450_{15\beta}$ (MACGEOCH et al., 1984).

In addition to P-450-male and P-450-female, which are major forms of sex-specific cytochrome P-450, the existence of other forms of sex-specific cyto-chrome P-450 has been reported by several groups of investigators (KAMATAKI et al., 1982; CHENG and SCHENKMAN, 1983; WAXMAN et al., 1985; JANSSON et al., 1985; BANDIERA et al., 1986; GONZALEZ et al., 1986; MATSUMOTO et al., 1986; YAMAZOE et al., 1986). KAMATAKI et al. (1982) first reported the exist-ence of multiple sex-specific forms of cytochrome P-450. CHENG and SCHENK-MAN (1983) and JANSSON et al. (1985) purified two other forms of sex-specific cytochrome P-450 designated as P-450 $RLM_3$ and P-450 $RLM_2$ which show relatively high activities for testosterone $6\beta$-hydroxylation, and testosterone $15\alpha$- and $7\alpha$-hydroxylations, respectively.

MATSUMOTO et al. (1986) also purified two other forms of sex-specific cyto-chrome P-450 designated as P-450 (M-2) and P-450 (M-3) which may corre-spond to P-450 $RLM_2$ and P-450 $RLM_3$, respectively.

Another sex-specific form of cytochrome P-450, designated as P-450g, P-450 PCN/P-450p or P-450 PB-2a/PCN-E may be responsible for testosterone $6\beta$-hydroxylation (RYAN et al., 1984b; WAXMAN et al., 1985; WRIGHTON et al., 1985; BANDIERA et al., 1986).

Recently, two closely related cDNA clones of $P-450_{PCN}$ were isolated (GON-ZALEZ et al., 1986). As judged from northern blot analyses, one form of the cytochrome P-450 PCN-1, is an inducible form and the other form, P-450 PCN-2, is a constitutive and male-specific form, although P-450 PCN-2 has not yet been purified from rat livers.

We recently purified male specific cytochrome P-450 responsible for testo-sterone $6\beta$-hydroxylation and designated it as $P-450_{6\beta}$ (YAMAZOE et al., 1986b; YAMAZOE et al., 1988b). $P-450_{3\beta}$ probably corresponds to $P-450_{PCN-2}$, and it is suppressively regulated by growth hormone in male rats.

Another cytochrome P-450, P-450g, which also catalyses testosterone $6\beta$-hydroxylation, is reported to be male-specific in rats. The hepatic level is regulated developmentally similarly to that of P-450-male, but genetic poly-morphism has been observed among several strains of rats.

Moreover, the high spin form of PCB-inducible cytochrome P-450, designated P-448-H and corresponding to P-450d, is sex-dependent in intact rats. A higher level of P-448-H has been detected in liver microsomes from female rats (7 pmol/mg protein), than from male rats (KAMATAKI et al., 1983a; YAMAZOE et al., 1988c).

The hepatic level of P-450f is also two-fold higher in female rats than in male. The levels are developmentally regulated and appear at puberty similarly to P-450-male.

## 3.  Developmental and aging aspects of sex-specific forms of cytochrome P-450

Developmental aspects of the sex-dependent activity of microsomal drug-metabolizing enzymes in rat liver microsomes was first reported by KATO et al. (1962a, 1962b). Different developmental patterns of testosterone 6$\beta$-,

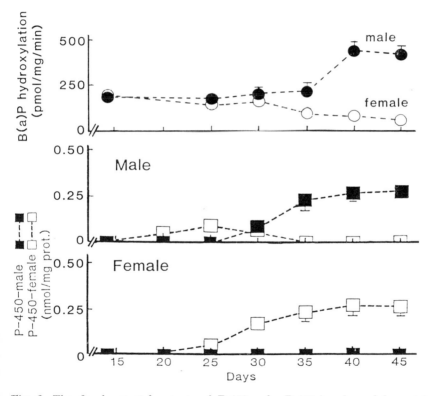

Fig. 2. The developmentel aspects of P-450-male, P-450-female and benzo(a)pyrene hydroxylase in male and female rats (MAEDA et al., 1984).

$7\alpha$- and $16\alpha$-hydroxylations were also reported (JACOBSON and KUNTZMAN, 1969). Sex-specific forms of P-450-male and P-450-female were not detected in liver microsomes from prepubertal male and female rats (KAMATAKI et al., 1983b). In male rats, a small amount of P-450-female was expressed transiently at the age of 20−25 days, but it was suppressed at the age of 30 days when P-450-male was initially expressed. P-450-male content increased to the adult level at the age of 45 days (Fig. 2) (MAEDA et al., 1984). On the other hand, in female rats, P-450-male was never expressed and P-450-female was expressed in rats more than 25 days old.

These results correlate with the developmental patterns of benzo(a)pyrene hydroxylase and aminopyrene N-demethylase in liver microsomes from male and female rats (MAEDA et al., 1984). Similar developmental patterns of

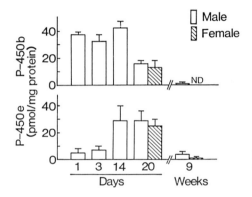

**Fig. 3.** Age-related changes in the contents of microsomal P-450b and P-450e (YAMAZOE et al., 1987).

$P-450_{2c/UT-A}$ and $P-450_{2d/UT-I}$, and $P-450_{16\alpha}$ have been reported by WAXMAN et al. (1985) and MORGAN et al. (1985b), respectively. The developmental patterns of testosterone $2\alpha$- and $16\alpha$-hydroxylase are closely correlated with that of $P-450_{2c/UT-A}$ (JACOBSON and KUNTZMAN, 1969; WAXMAN et al., 1985). YOSHIOKA et al. (1987) recently reported that P-450 (M-1) mRNA was expressed exclusively in the livers of mature male rats in a sex-specific manner.

Another interesting developmental pattern of sex-specific forms of cytochrome P-450, PB-2a/PCN-E, has been reported (WAXMAN et al., 1985). The level of PB-2a/PCN-E was high in immature male and female rats, but decreased in male rats with maturation, leaving a high level only in female rats.

Moreover, the phenobarbital inducible cytochrome P-450s, P-450b and P-450e, showed sex-related differences in their developmental pattern (YAMAZOE et al., 1987). P-450b was already higher at birth; the content decreased at 20 days of age and further decreased with maturation (Fig. 3). The content in adult female rats was not detectable and in the adult male rat was about

2% of 20 day old rats. On the other hand, P-450e was relatively low in new-born rats and increases to the maximum at 14—20 days of age. P-450e content decreased again, and the contents in male and female rats of 9 weeks of age were about 10 and 6%, respectively, of 20-day-old rats (YAMAZOE et al., 1987).

The levels of P-450-male and P-450-female remained high after the pubertal period in livers of male and female rats, respectively. However, the content of P-450-male was lower in aged male rats (Fig. 4). P-450-male disappeared

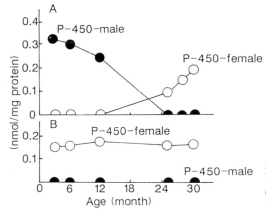

Fig. 4. The changes in P-450-male and P-450-female contents in liver micro-somes from aged male (A) and female (B) rats (KAMATAKI et al., 1985b).

in liver microsomes of 24—30 months old male rats and, interestingly, P-450-female was clearly expressed in these old male rats (KAMATAKI et al., 1985b). Microsomes from 30-months-old male rats showed an immunostained peptide pattern very similar to those of P-450-female and microsomes from young and old female rats, lending support to the notion that the real P-450-female is expressed in old male rats. On the other hand, in aged female rats, the content of P-450-female in liver microsomes was not significantly changed and P-450-male was not expressed at all.

The changes of activities of drug-metabolizing enzymes in liver microsomes were closely correlated with those of P-450-male and P-450-female contents (KAMATAKI et al., 1985b).

## 4.  Regulation of sex-specific forms of cytochrome P-450 by androgen and estrogen

The oxidations of a variety of drugs and steroid hormones are regulated by androgen and estrogen (KATO et al., 1962a; JACOBSON and KUNTZMAN, 1969; KATO and ONODA, 1970; KATO, 1974; COLBY, 1980).

Castration of adult male rats decreases the activities of drug-metabolizing enzymes showing a clear sex-difference, such as hexobarbital hydroxylation and aminopyrine N-demethylation, but those showing no clear sex-difference, such as aniline hydroxylation, are not affected, (KATO and GILLETTE, 1965a). Testosterone supplement completely restores the decreased activities in the castrated male rats. On the other hand, ovariectomy of adult female rats does not affect all activities of drug-metabolizing enzymes (KATO et al.,

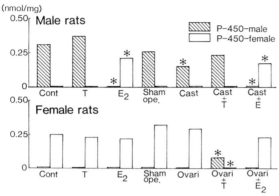

Fig. 5. Effect of testosterone (T) or estradiol ($E_2$) or gonadectomy on the expression of P-450-male and P-450-female (KATO and KAMATAKI, 1985). Sham ope.: sham operation; cast: castration; Ovari: ovariectomy; *: significant difference ($P < 0.01$) from control or sham-operated rats.

1962a; KATO and GILLETTE, 1965a). However, estradiol antagonizes the effect of testosterone to increase sex-dependent activities, such as hexobarbital and aminopyrine N-demethylation (KATO and ONODA, 1970).

Similar results are known about testosterone hydroxylating activities in liver microsomes of male and female rats (KATO et al., 1969; JACOBSON and KUNTZMAN, 1969).

Recently the effect of sex-steroid hormone on the level of the purified cytochrome P-450 has been investigated (KAMATAKI et al., 1983b).

Castration of adult male rats decreased the contents of the male specific form of cytochrome P-450, P-450-male and P-450$_{16\alpha}$, and testosterone supplement completely restored these contents (Fig. 5) (KAMATAKI et al., 1983b; MORGAN et al., 1985b). Moreover, castration plus estradiol treatment completely suppressed P-450-male expressions (KAMATAKI et al., 1983; KATO and KAMATAKI, 1985). Castration alone did not stimulate the expression of P-450-female, but treatment with estradiol caused almost full expression of P-450-female in adult male rats (KAMATAKI et al., 1983; KATO and KAMATAKI, 1985). Estradiol treatment of intact male rats also caused complete suppression of P-450-male and P-450$_{16\alpha}$, and full expression of P-450-female (KATO and KAMATAKI, 1985; MORGAN et al., 1985b).

R. KATO; Y. YAMAZOE

Ovariectomy of adult female rats did not have any significant effect on P-450-female contents and did not cause the expression of P-450-male. Ovariectomy plus testosterone treatment caused a partial expression of P-450-male and complete suppression of P-450-female (KAMATAKI et al., 1983b). In contrast, testosterone treatment of intact female rats neither suppressed P-450-female nor expressed P-450-male (KATO and KAMATAKI, 1985).

These results indicate that androgen stimulates the expression of P-450-male and suppresses the expression of P-450-female. On the other hand, estrogen is not necessary for the expression of P-450-female, but antagonizes the suppressive effect of androgen.

The content of P-450$_{6\beta}$ in hepatic microsomes of male rats was also decreased by castration, and testosterone supplement completely restored P-450$_{6\beta}$ content (YAMAZOE et al., 1988b).

A cytochrome P-450, designated as P-450a (THOMAS et al., 1981), with high testosterone 7$\alpha$-hydroxylase activity, has recently been purified from rat liver microsomes (NAGATA et al., 1987). P-450a was present at low levels in newborn rats and increased to a maximum level at 1 week of age in both males and females. At the age of 12 weeks, however, the P-450a level decreased in males but remained at an elevated level in females.

## 5. Regulation of sex-specific forms of cytochrome P-450 by hypothalamo-hypophyseal system

The regulation of activities of microsomal drug-metabolizing enzymes and steroid-hydroxylating enzymes by the hypothalamo-hypophyseal system has been well studied (KRAMMER et al., 1975; GUSTAFSSON and STENBERG, 1976; SKETT and GUSTAFSSON, 1979; COLBY, 1980; GUSTAFSSON et al., 1983).

WILSON (1970, 1973) first reported the modulation by growth hormone of drug metabolizing enzyme activity in rat liver microsomes. The activities in male rats were markedly decreased by injection of growth hormone or with an implanted growth hormone secreting pituitary tumor.

The effect of hypophysectomy on the hydroxylation of 4-androstene-3,17-dione was reported by GUSTAFSSON and STERNBERG (1974). Hypophysectomy decreased the activity of 16$\alpha$-hydroxylase, but not 6$\beta$-hydroxylase in male rats. In female rats, hypophysectomy increased the activity of 16$\alpha$- and 6$\beta$-hydroxylase.

The effect of neonatal imprinting on steroid hormone hydroxylation was first reported by EINARSSON et al. (1973). Castration in the neonatal period (0$-$5 days) completely suppressed male-type development of steroid hormone metabolism, and neonatally castrated rats did not respond or responded only slightly to adulthood testosterone treatment. However, neonatal testosterone treatment of neonatally-castrated male rats completely restored responsiveness to testosterone treatment at adulthood (SKETT and GUSTAFSSON, 1979).

These results suggest that the expression of sex-specific forms of cytochrome P-450 are regulated by the hypothalamo-hypophyseal system. Therefore, recent results along this line will be discussed in detail.

a) Effect of neonatal imprinting

Castration of male rats at birth completely suppressed the expression of P-450-male in adulthood and, interestingly, caused the expression of P-450-female (KAMATAKI et al., 1984). The treatment of neonatally castrated rats

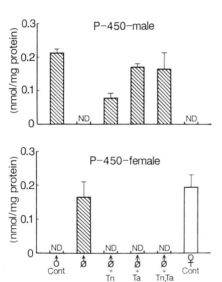

**Fig. 6.** Effect of neonatal castration and testosterone treatment on P-450-male and P-450-female contents of rat liver (SHIMADA et al., 1987). ♂/∅: neonatally castrated rats; Tn: neonatal testosterone treatment at 1, 3 and 5 days; Ta: adulthood testosterone treatment at 8 weeks old; Tn, Ta: neonatal and adulthood testosterone treatment.

with testosterone 1, 3, 5 days after birth partially restored the expression of P-450-male in adulthood (KAMATAKI et al., 1984). Similar results were reported on the regulation of P-450$_{16\alpha}$ by MORGAN et al. (1985a), and P-450$_{2c/UT-A}$ and P-450$_{2d/UT-I}$ by WAXMAN et al. (1985). However, if male rats were castrated at the age of 5 days or 14 days, no such effect was observed (KAMATAKI et al., 1984). These results indicate the existence of neonatal-androgen imprinting for the development of male-type expressions of sex-specific cytochrome P-450.

The changes in the activities of benzo(a)pyrene hydroxylase, 7-propoxy-coumarin O-depropylase, testosterone 2α- and 16α-hydroxylases have been closely correlated to those of the levels of P-450-male and P-450$_{2c/UT-A}$ (KAMATAKI et al., 1984; WAXMAN et al., 1985; SHIMADA et al., 1987). To our surprise, treating adult rats castrated at birth with testosterone caused full restoration of the expression of P-450 male (Fig. 6) (KAMATAKI et al., 1984; SHIMADA et al., 1987). The activities of $2\alpha$- and $16\alpha$-testosterone hydroxylases and the levels of ethylmorphine N-demethylation, propoxycoumarin O-depropylation and benzo(a)pyrene hydroxylation were increased in neonatally castrated rats to those of adult male rats with adult testosterone treatment (SHIMADA et al., 1987). Similar results have recently been reported on the full expression of P-450$_{2C/UT-A}$ and P-450$_{2a/PCN-E}$, in neonatally castrated rats with testosterone treatment in adulthood (DANNAN et al., 1986).

Treatment of neonatally castrated rats with testosterone at the age of 19 weeks did not cause a complete recovery of P-450-male content nor of the levels of benzo(a)pyrene hydroxylase and propoxycoumarin O-depropylation. These results indicated that a prolonged deficiency of androgen in neonatally

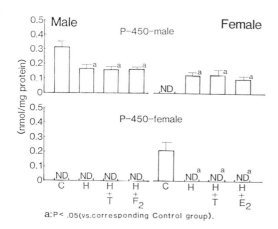

**Fig. 7.** Effect of hypophysectomy (H) and treatment with testosterone (T) or estradiol ($E_2$) on the expression of P-450-male and P-450-female. ND: not detectable; C: controls; H: hypophysectomy.

castrated rats caused them to lose a full response to androgen. Testosterone administration into neonatal female rats did not significantly alter the contents of sex-specific cytochrome P-450 nor drug and steroid hydroxylating activities. The additional testosterone administered in adulthood (8 weeks old) only slightly affected these parameters (SHIMADA et al., 1987). These results were in marked contrast to the effect of testosterone treatment in neonatally or prepubertally ovariectomized rats (KAMATAKI et al., 1983; DANNAN et al., 1986).

b) Effect of hypophysectomy

Hypophysectomy of male rats decreased P-450-male content, but P-450-female was not expressed (Fig. 7) (KAMATAKI et al., 1985a). Treatment with testosterone or estradiol did not cause any further effect on the contents of P-450-male and P-450-female. On the other hand, hypophysectomy of female rats caused the disappearance of P-450-female and expression of P-450-male

(KAMATAKI et al., 1985 b). The expression of P-450-male in hypophysectomized female rats is of special interest because androgen is deficient in these female rats. These results are especially interesting in comparison with the effect of castration. Testosterone or estradiol treatment also did not cause any signi-

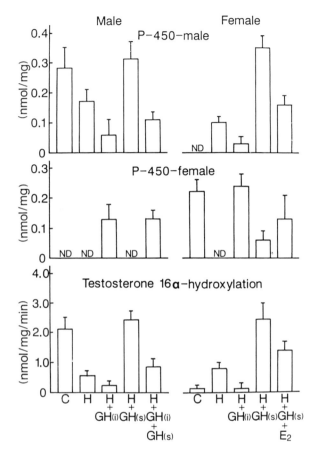

Fig. 8. The effects of different types of treatments with growth hormone on the contents of P-450-male and P-450-female and testosterone 16α-hydroxylation (YAMAZOE et al., 1986 a).

GH(i): growth hormone was infused with an osmotic mini-pump for 8 days;

GH(s): human growth hormone was subcutaneously injected twice a day for 7 days;

$E_2$: estradiol benzoate (500 μg/kg) was given every other day for 7 days.

ficant effect in hypophysectomized female rats, although testosterone treatment of ovariectomized female rats caused the expression of P-450-male and the complete suppression of P-450-female expression (Fig. 6) (KAMATAKI et al., 1983).

By hypophysectomy, the level of testosterone 2α- and 16α-hydroxylation was markedly decreased in male rats and the level was clearly increased in female rats in accordance with changes of P-450-male (Fig. 8) (YAMAZOE et al., 1986).

R. KATO; Y. YAMAZOE

Similar effects of hypophysectomy of male and female rats on hepatic content of P-450$_{16\alpha}$ and P-450$_{15\beta}$ have been reported by MORGAN et al. (1985 b) and MacGEOCH et al. (1984, 1985), respectively.

Moreover, hypophysectomy decreased the activities of benzo(a)pyrene hydroxylase and aminopyrine N-demethylase in male rats, but clearly increased the same activities in female rats.

These results indicate that the hypophysis factor(s) may regulate the expression of P-450-male and P-450-female and that the effects of androgen and estrogen are not produced in the absence of hypophysis. Thus, hypophysis may play a permissive role in the effect of androgen or estrogen on the expression of sex-specific forms of cytochrome P-450.

On the other hand, hypophysectomy elevates the content of P-450$_{6\beta}$ in male rat liver and expresses P-450$_{6\beta}$ in female rat liver (YAMAZOE et al., 1986 b; YAMAZOE et al., unpublished observation). Thus a hypophyseal factor, probably growth hormone, suppressively regulates P-450$_{6\beta}$ level, in contrast to the stimulatory regulation on P-450-male. Moreover, as mentioned later, P-450b and P-450e are major phenobarbital-inducible cytochromes P-450, and the level is very low in adult rats unless induced by hypophysectomy (YAMAZOE et al., 1987).

## 6. Modulation of sex-specific cytochrome P-450 by secretion pattern of growth hormone

As a pituitary factor responsible for sex-related differences in the oxidation of drugs and steroid hormones, growth hormone has been considered to be a most probable factor (RUMBAUGH and COLBY, 1980; COLBY, 1980; GUSTAFSSON et al., 1983). The secretion pattern of growth hormone is markedly different

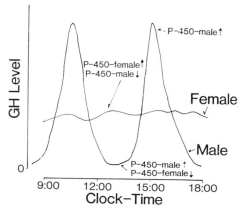

**Fig. 9.** Effect of serum pattern of growth hormone on the expression of P-450-male and P-450-female in male and female rats.

between male and female rats (EDEN, 1979). In male rats, the basal level of serum growth hormone is low, but there are surges of $6-8$ times for one day (Fig. 9). On the other hand, in female rats the basal level of serum growth hormone is relatively high without a clear surge.

To investigate the role of growth hormone on the regulation of sex-specific forms of cytochrome P-450, two types of administration methods were employed to mimic male-type and female-type secretions of growth hormone (MORGAN et al., 1985b; YAMAZOE et al., 1986a). Continuous infusion by an osmotic mini-pump implanted subcutaneously (GH(i)) in hypophysectomized rats mimics female-type secretion. Subcutaneous injection of growth hormone twice daily (GH(s)) in hypophysectomized rats mimics male-type secretion.

Subcutaneous injection of growth hormone twice daily into hypophysectomized male rats caused full expression of P-450-male and complete suppression of P-450-female (YAMAZOE et al., 1986a; KATO et al., 1986) (Fig. 10).

In contrast, continuous infusion of growth hormone into hypophysectomized male rats caused a further decrease in P-450-male and expression of P-450-female. It is of special interest that between hypophysectomized female and male rats similar results were obtained with two types of growth hormone treatments (YAMAZOE et al., 1986a).

Moreover, the anti-P-450-male IgG markedly inhibits testosterone $2\alpha$- and $16\alpha$-hydroxylases in liver microsomes from hypophysectomized rats treated with GH(s) as well as in microsomes from intact male rats. These results further support the conclusion that P-450-male expressed in liver microsomes in hypophysectomized female rats treated with GH(s) is the same P-450-male expressed in microsomes of intact male rats.

In accordance with the change in P-450-male content, similar changes in testosterone $2\alpha$- and $16\alpha$-hydroxylations, benzo(a)pyrene hydroxylation and aminopyrine N-demethylation were found in hypophysectomized rats treated with intermittent injection or continuous infusion of growth hormone (Figs. 7, 8) (YAMAZOE et al., 1986a).

Similar results were also reported with P-450$_{16\alpha}$ which is responsible for testosterone $16\alpha$-hydroxylation (MORGAN et al., 1985b) and with P-450$_{15\beta}$ responsible for androstan-$3\alpha$,$17\beta$-diol 3,17-disulfate $15\beta$-hydroxylation (MAC GEOCH et al., 1985).

These results indicate the importance of the secretion pattern of growth hormone on the regulation of P-450-male and P-450-female expression in rat liver microsomes.

The results of ectopic transplantation of pituitary gland further support the role of growth hormone (Fig. 10). The transplantation of male rat pituitary gland under the renal capsule, freeing it from hypothalamic inhibitory control, causes continuous release of growth hormone. Thus ectopic transplantation of pituitary gland under the renal capsule of male rats caused feminization of steroid hormone metabolism: increase in testosterone $5\alpha$-reduction and

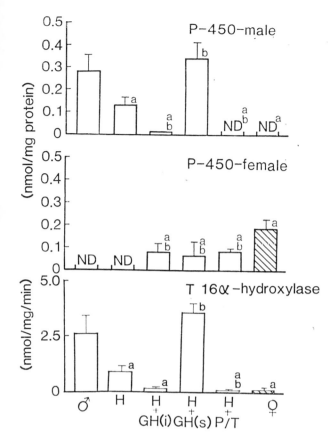

**Fig. 10.** Effect of growth hormone and pituitary transplantation under the renal capsule on P-450-male, P-450-female and testosterone 16α-hydroxylase in male rats (KATO et al., 1986).

H: hypophysectomy; GH(i): growth hormone infusion; GH(s): growth hormone subcutaneous injection; P/T: pituitary transplantation.
a: significant difference (P < 0.05) from control male rats;
b: significant difference (P < 0.05) from hypophysectomized male rats.

decrease in 5α-androstan 3,17-dione 16α-hydroxylation (DENF, 1974; GUSTAFS-SON and STENBERG, 1976; SKETT et al., 1978).

Ectopic transplantation of pituitary gland under the renal capsule of male rats completely suppressed the expression of P-450-male and stimulated the expression of P-450-female as observed in female rats or hypophysectomized rats treated with GH(i) (KATO et al., 1986). In accordance with the change of the sex-specific cytochrome P-450 contents, the levels of testosterone 2α-

and 16α-hydroxylases were almost completely suppressed in the pituitary gland transplanted in male rats. Similarly, the levels of 7-propoxycoumarin O-depropylase and benzo(a)pyrene hydroxylase were markedly decreased to the levels of intact female rats (KATO et al., 1986).

Ectopically transplanted pituitary gland in rats enhances not only secretion of growth hormone but also secretion of prolactin. Moreover, human growth hormone also has some activity identical to prolactin. These facts raised doubts that the effects of growth hormone injection or ectopic transplantation may be due to the action of prolactin. However, the administration of ovine prolactin to male rats failed to produce any significant effect on the levels of P-450-male and P-450-female or on drug-metabolizing enzymes and steroid-hydroxylating activities in liver microsomes (YAMAZOE et al., 1987).

Combined treatment with GH(i) and GH(s) of hypophysectomized rats caused a suppressive effect on GH(s)-induced increase of P-450-male and caused an antagonistic effect on the GH(s)-induced suppression of P-450-female (Figs. 8, 9) (YAMAZOE et al., 1986a). Moreover, the administration of GH(i) to intact male rats markedly depressed the level of P-450-male (KATO et al., 1986). All these results indicate that both high peak and low basal level of serum growth hormone are required for full expression of P-450-male (Fig. 10). On the other hand, a continuous intermediate level of serum growth hormone is required for full expression of P-450-female. The administration of testosterone to GH(s)-treated hypophysectomized rats did not cause any further increase in the level of P-450-male. However, the administration of estradiol to GH(s)-treated hypophysectomized rats caused a significant decrease in the level of P-450-male.

From these results and previous reports stating that testosterone or estrogen has no effect on the level of P-450-male and P-450-female in hypophysectomized rats, the following conclusion could be drawn: the major effect of androgen or estrogen is expressed through regulation of the secretion pattern of growth hormone, but estrogen also interferes with the action of growth hormone at the hepatic level.

## 7.    Species and tissue specificities of sex-specific cytochrome P-450

Although there are large sex-related differences in the oxidation of drugs and steroid hormone in rat livers, only small or no degrees of sex-related differences have been reported in the liver of other species of experimental animals (KATO, 1974; KATO, 1979).

For mouse livers, some strains show no sex-related differences in drug oxidations and other strains show slightly higher activities in female mice than in males (KATO, 1979; NOORDHOEK, 1972; BROWN et al., 1978). NOORD-HOEK (1972) observed slightly higher activities for hexobarbital metabolism and ethylmorphine N-demethylation in female mice than in males. However,

BROWN et al. (1978) clearly showed that ethylmorphine N-demethylase activity is greater in male mice than in females of the BALB/cJ strain, but the inverse male-female relationship exists for CRL:CD-1 mice. No consistent sex differences were present in DBA-2J, C3H/HeJ and C57BL/10J strains. However, cytochrome P-450 content in hepatic microsomes was slightly greater in male mice of all strains studied. Administration of testosterone caused a dose-related increase in ethylmorphine N-demethylase activity in female BALB/cJ mice, but not in female mice of other strains.

Moreover, SHAPIRO (1985) showed that testicular androgens normally repress the levels of hepatic hexobarbital hydroxylase activity in Cr1:CD-1(ICR) and that the ability of testosterone to suppress enzyme activity is dependent on the presence of the pituitary.

FORD et al. (1979) found clear sex-related differences in testosterone $16x$-hydroxylation in the 129/J strain of mouse. The testosterone $16x$-hydroxylase activity in liver microsomes was higher in male 129/J mice than in females. There are two forms of cytochrome P-450 responsible for testosterone $16x$-hydroxylases in the liver microsomes of 129/J mice (HARADA and NEGISHI, 1984b, NOSHIRO et al., 1986). One is a constitutive enzyme, designated as $C-P-450_{16a}$: the activity is about 3 times higher in males than in female 129/J mice (HARADA and NEGISHI, 1984b). Another is an inducible enzyme, designated as $I-P-450_{16a}$: the activity is higher in females but shows no detectable activity in male 129/J mice (DEVORE, 1985; NOSHIRO et al., 1986). The $C-P-450_{16a}$-dependent $16x$-hydroxylase and corresponding mRNA were higher in males than in female 129/J mice. In contrast, the $I-P-450_{16a}$-dependent $16x$-hydroxylase and $I-P-450-_{16a}$ mRNA were higher in females than in male 129/J mice (NOSHIRO et al., 1986).

The testosterone $16x$-hydroxylase activities due to $I-P-450_{16a}$ in hepatic microsomes from untreated female 129/J and BALB/cJ were 0.03 and 0.3 nmol/min/mg protein, respectively. No significant level of $I-P-450_{16a}$-dependent activity was detected in the microsomes from males of either mouse strain. Immunoblotting of microsomal protein, 54-KDa, with the antibody to $I-P-450_{16a}$ revealed approximately a 10 times greater activity in the microsomes from BALB/cJ than in those from 129/J female (0.03 and 0.26 nmol/mg protein/min, resp.). Moreover, the specific content of the hybridizable mRNA with cDNA probe was more than 10 times higher in BALB/cJ females than in males. The mRNA level in female 129/J mice was very similar to that of 129/J and BALB/cJ males. The repression of $I-P-450_{16}/$ in 129/J females was inherited as an autosomal recessive trait in 129/J and BALB/cJ pairs as indicated by the levels of mRNA in female $F_1$ offspring and the $I-P-450_{16}/$-dependent hydroxylase activity (NOSHIRO et al., 1986).

Female and male mice of eight more inbred strains (AKR/J, DBA2J, C57BL/6J, C$_3$H/HeJ, NZB/J, A/J, CBA/CaJ and P/J) were tested for levels of mRNA. The levels of mRNA were always 5- to 10-fold greater in the females than in the corresponding males, although there was some variation in the mRNA content in the males of different strains. NOSHIRO et al. (1986) concluded that 129/J females appear to be a genetic variant where the female-predominant expression of the mRNA is repressed.

On the other hand, the activity of testosterone $15\alpha$-hydroxylase was four times higher in 129/J female than in male mice. The level of cytochrome P-450 responsible for testosterone $15\alpha$-hydroxylase, designated as P-450$_{15\alpha}$,

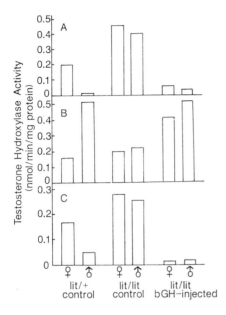

**Fig. 11.** Testosterone hydroxylase activities in liver microsomes from control and bGH-injected Little mice (NOSHIRO and NEGISHI, 1986). Panel A: I-P-450$_{16\alpha}$-dependent activity; Panel B: C-P-P450-$_{16\alpha}$-dependent activity; Panel C: P-450$_{15\alpha}$-dependent activity; bGH-injected: bovine growth hormone was injected once every 12 h for five consecutive days.

was higher in 129/J female than in male mice (HARADA and NEGISHI, 1984a; BURKHART et al., 1985). The level of P-450$_{15\alpha}$ mRNA was 6.6 times higher in 129/J female than in male mice (NOSHIRO and NEGISHI, 1986).

The effects of growth hormone on the expression of sex-dependent testosterone $16\alpha$- and $15\alpha$-hydroxylase at the activities as well as at the mRNA levels using Little (*lit/lit*) mice, a mutant strain raised from C57BL/6J which has autosomal recessive inheritance of growth, were studied (NOSHIRO and NEGISHI, 1986). The male isozyme of testosterone $16\alpha$-hydroxylase (C-P-450$_{16\alpha}$) was repressed in the liver of male *lit/lit* mice, and the injection of growth hormone (bovine) resulted in an increase of the isozyme at both activity and mRNA levels to those seen in control *lit/+* male mice (NOSHIRO and NEGISHI, 1986) (Fig. 11). The female isozymes of testosterone $16\alpha$-(I-P-450$_{16\alpha}$) and $15\alpha$-(P-450$_{15\alpha}$)hydroxylase, however, were increased in livers of both male and female *lit/lit* mice. The increased I-P-450$_{16\alpha}$ and P-450$_{15\alpha}$ in *lit/lit* mice were

R. KATO; Y. YAMAZOE

suppressed by growth hormone, but only when it was injected once every 12h (male type mimic injection). The results indicate that growth hormone acts as a masculinizing factor for testosterone hydroxylase activity by activating and inhibiting the expression of male and female isozymes of testosterone hydroxylases in mice.

On the other hand, when growth hormone was continuously infused (female-type mimic administration), it showed no significant effect on the expression of hydroxylases in *lit/lit* mice, suggesting that growth hormone may not be a feminizing factor for testosterone hydroxylases in female mice (NOSHIRO and NEGISHI, 1986). The changes of specific hydroxylase activities modulated by growth hormone in the mice correlated well with those amounts of hydroxylase mRNAs.

The roles of these sex-specific cytochrome P-450s in the oxidation of drugs in liver microsomes of male and female mice are not yet known, but contributions of each P-450 in drug oxidation should be clarified in relation to strain-dependent sex differences. Although marked sex-related differences in the oxidation of drugs and steroid hormones have been reported in rat liver, no clear sex-related differences have been reported in extrahepatic tissues (KATO, 1979; KATO and KAMATAKI, 1982). In accordance with these results, P-450-male, P-450-female and P-450 (M-1) mRNA were not expressed in other tissues such as kidney, lung, testis and ovary of matured rats (SHIMADA et al., 1985, unpublished observation; YOSHIOKA et al., 1987). CHHABARA and FOUTS (1974) found that there were no significant sex-related differences for rat lung and gut microsomes in oxidation of several drug substrates. However, rat kidney microsomes exhibited sex-related differences in rate of benzo(a)pyrene hydroxylation, although in contrast to the liver, the rate of hydroxylation was higher in kidneys from female rats than from males.

There was no clear sex-related metabolic oxidation of dimethylnitrosamine (N-demethylation) by mouse liver microsomes; only kidney microsomes from male C57BL/6J mice, but not females, were capable of N-demethylating dimethylnitrosamine (WEEKS and BRUSICK, 1975).

BAKSHI et al. (1977) conducted a series of studies using livers and kidneys from mice carrying the *tfm* allele from their normal litter mates. The *tfm* is an X-linked recessive mutation which confers androgen insensitivity (testicular feminization) on the hemizygous males (*tfm/y*) (BULLOCK et al., 1975). Normal male (*t/y*) litter males are not affected. The *tfm/y* mice, lacking androgen receptors, would be expected to have the same pattern of kidney metabolism as the non-androgen-stimulated normal females (+/+) or heterozygous carrier females (*tfm/+*). Liver microsomes from all animals were similar in their ability to activate markedly dimethylnitrosamine. In contrast, kidney microsomes from *tfm/y* mice activated dimethylnitrosamine to the same small extent as those from normal and carrier females and much less than those from male animals. Thus these results implicate testosterone as a regulator of dimethylnitrosamine N-demethylase in the kidney.

Moreover, castration reduced kidney microsomal dimethylnitrosamine activation in males. Testosterone treatment induced dimethylnitrosamine activation in normal females and castrated males and enhanced carcinogen metabolism in normal males (BAKSHI et al., 1977). In contrast, *tfm/y* mice did not respond; i.e., renal microsomes from both untreated and testosterone-treated animals had little ability to form mutagenic metabolites.

Similarly, in mice, only males are susceptible to chloroform nephrotoxicity, although no clear sex-related hepatotoxicity has been reported. There were clear sex-related differences in the level of renal cytochrome P-450 and of ethoxycoumarin O-deethylase activity (SMITH et al., 1984). The male levels were about five times as high as the female. The treatment of male and female mice with testosterone increased the renal cytochrome P-450 level and ethoxy-coumarin O-deethylase activity. Sex-related differences were no longer observed in testosterone-treated mice.

Testosterone hydroxylase activity in mouse renal microsomes was reportedly sex-dependent. Testosterone 15α- and 16α-hydroxylase activities were detected only in male AKR/J and BALB/cJ mice, not in females (HEIMS et al., 1984). The 15α-hydroxylase activity in male kidney was nearly as high or higher than that in male or female livers, depending on the strain and how activity was expressed. These results are in sharp contrast to the strain-specific female dominance shown in liver (BROWN et al., 1978; NOSHIRO et al., 1986). Castration of adult AKR/J males decreased 15α-hydroxylase activity to less than 1% of initial values, but cytochrome P-450 content decreased only about 50% in response to the operation. Treatment of castrated males with testosterone returned all values to sham-operated control levels, but treatment of female mice with testosterone had very little effect on 15α-hydroxylase activity and only a modest effect on cytochrome P-450 content. Moreover, HAWKE and WELCH (1985) reported interesting sex-related differences in the oxidation of drugs, steroid hormones and fatty acid by mouse renal microsomes.

Renal microsomes from C3H/HeJ mice were characterized by the following: 1) a 4- to 5-fold male predominance in cytochrome P-450 concentration; 2) absorption maxima of the reduced P-450-CO complex were 450 and 452 nm, respectively, for male and female mice; 3) a lack of sex difference in lauric acid 12-hydroxylase activity; 4) an 18-fold sex difference (M > F) in progesterone 16α-hydroxylase activity; 5) 8- to 10-fold sex difference (M > F) in progesterone 15α-hydroxylase, dimethylnitrosamine demethylase, and lauric acid 11-hydroxylase activities.

Treatment of female mice with testosterone selectively induced lauric acid 11-hydroxylase and dimethylnitrosamine demethylase activities 8- and 14-fold, respectively, but had no effect on progesterone 15α- and 16α-hydroxylase activites or on the high female rate of lauric acid 12-hydroxylation. Cytochrome P-450 content was increased 2.5 fold to 64% of the male level. The Soret peak maximum was shifted from 452 to 450 mm.

---

These results indicate that renal enzymes responsible for oxidation of drugs, steroid hormones and fatty acid may be regulated by different mechanisms or different sensitivity in response to androgen.

## 8. Sex-related difference in induction of cytochrome P-450 by xenobiotics

The effect of inducers of hepatic microsomal drug-metabolizing enzymes, such as phenobarbital, is higher in female rats than in male rats for sex-related metabolism, such as hexobarbital hydroxylation and aminopyrine N-demethylation, but not for sex-independent metabolism, such as aniline hydroxylation (KATO and TAKAYANAGI, 1966). The increase in testosterone 16$\alpha$-hydroxylase in male rats by treatment with phenobarbital was also reported (THOMAS et al., 1981). These results suggested that phenobarbital may induce its own cytochrome P-450, leaving intact or causing a decrease in sex-related cytochrome P-450 in male rats (KATO, 1974). Moreover, treatment with methylcholanthrene markedly decreased the activities of hexobarbital hydroxylase, aminopyrine N-demethylase and testosterone 16$\alpha$-hydroxylase in liver microsomes of male rats (KATO and TAKAYANAGI, 1966; KUNTZMAN et al., 1968; KREMERS et al., 1978).

Administration of phenobarbital, methylcholanthrene or polychlorinated biphenyls (Kanechlor 500) to male rats decreased the hepatic content of P-450-male. The content of P-450-female in female rats was decreased by the treatment with methylcholanthrene or polychlorinated biphenyls. Thus the percentages of P-450-male and P-450-female to the amount of total P-450 in liver microsomes of male and female rats were reduced to $20-40\%$ of control levels by treatment with phenobarbital, methylcholanthrene or polychlorinated biphenyl (KAMATAKI et al., 1986). In relation to the changes of P-450-male and P-450-female the ratios of inductions of testosterone 16$\alpha$-hydroxylase by phenobarbital and polychlorinated biphenyls were much more pronounced in female rats than in males. Similarly, the ratios of inductions of benzo(a)pyrene hydroxylase by phenobarbital, 3-methylcholanthrene and polychlorinated biphenyls were much more pronounced in female rats (KAMATAKI et al., 1986).

The administration of phenobarbital induced hepatic microsomal P-450b and P-450e to a greater extent in male rats than in female rats (Table 2) (YAMAZOE et al., 1987). The levels of P-450b and P-450e in phenobarbital-treated male rats were 922 and 365 pmol/mg protein, respectively, whereas those in phenobarbital-treated female rats were 221 and 86 pmol/mg protein, respectively. The levels of P-450b and P-450e were suppressively regulated by pituitary hormone(s) in adult rats and hypophysectomy clearly increased the levels of P-450b and P-450e. Moreover, there were no clear sex-related differences in the levels of P-450b and P-450e induced by phenobarbital in

**Table 2.** Microsomal contents of P-450b and P-450e in male and female rats (YAMAZOE et al., 1987)

| Treatment | Male | | Female | |
|---|---|---|---|---|
| | **P-450b** | **P-450e** | **P-450b** | **P-450e** |
| None | 1 ± 1 (1.0) | 4 ± 2 (1.0) | N.D.[a] | 2 ± 2 |
| Hypox[b] | 58 ± 10 (58.0) | 57 ± 12 (14.3) | 59 ± 11 (> 118) | 59 ± 7 (29.5) |
| PB[b] | 922 ± 78 (922.0) | 365 ± 88 (91.3) | 221 ± 43 (> 422) | 86 ± 15 (43.0) |
| Hypox + PB | 1,181 ± 114 (1,181) | 653 ± 65 (163.3) | 848 ± 92 (> 1,696) | 555 ± 38 (277.5) |

Data represent the mean ± standard deviation.

a: Not detected (less than 0.5 pmol/mg protein).

b: Hypox and PB indicate the treatment by hypophysectomy and with sodium phenobarbital, respectively.

Numbers in parentheses indicate the value relative to the respective nontreated control.

hypophysectomized rats. These results indicate that the inducing effect of phenobarbital on P-450b and P-450e is regulated by the serum level of growth hormone. In accordance with the changes of P-450b and P-450e levels, similar changes were observed in the activities of testosterone 16$\beta$-hydroxylation and O-pentylresorufin O-depentylation (YAMAZOE et al., 1987). Moreover, the hepatic level of translatable mRNA encoding a single peptide antigenically related to the phenobarbital inducible cytochrome P-450$_{PB-B}$ was higher in male rats than in females (GOZUKARA et al., 1984). Continuous infusion of growth hormone to male hypophysectomized male and female rats markedly decreased the levels of P-450b and P-450e (YAMAZOE et al., 1987). However, the suppression of P-450b level was much stronger than that of P-450e level. Ectopic transplantation of pituitary gland produced similar effects observed in the growth hormone-treated rats. The intermittent injection of growth hormone, however, only slightly decreased the level of P-450b and P-450e (YAMAZOE et al., 1987). These results clearly indicate that the regulation mechanisms of the expression of P-450-male and P-450-female are different from those of P-450b and P-450e. The latter are regulated by growth hormone, but some other factor may be necessary.

# 9.  Change of the level of sex-specific cytochrome P-450 in pathological states

Sex-related changes of drug-metabolizing enzymes and steroid-hydroxylating enzymes of rat liver microsomes under abnormal physiological or pathological states are well-known (KATO, 1974; KATO, 1977). For example, thyroxine treatment, morphine treatment, adrenalectomy, alloxan diabetes, or starvation markedly decreased hexobarbital hydroxylation, aminopyrine N-demethylation and testosterone hydroxylation in male rats but not in females (KATO and TAKAHASHI, 1968; KATO et al., 1970a, 1970b, 1970c, 1970d, 1970e, 1970f). These activities markedly decreased in tumor-bearing male rats, but decreased to a lesser extent in tumor-bearing female rats. The activity of aniline hydroxylation also decreased in tumor-bearing male and female rats (KATO et al., 1968a).

In the light of the existence of sex-specific forms of cytochrome P-450, it is reasonable to suppose that the level of P-450-male might decrease under the above mentioned abnormal states.

The administration of morphine or reserpine caused a marked decrease in the level of P-450-male in liver microsomes of male rats and the level of benzo(a)pyrene hydroxylation and propoxycoumarine O-depropylation (YAMA-ZOE et al., in preparation). The level of testosterone $2\alpha$- and $16\alpha$-hydroxylations but not $7\alpha$-hydroxylation markedly decreased.

The level of P-450-male was decreased in male rats with diabetes caused by alloxan or streptozotocin (YAMAZOE et al., in preparation). The level of P-450$_{6\beta}$ decreased concomitant with the level of testosterone $6\beta$-hydroxylation. Moreover, the level of P-450b was more markedly induced in male rats than in females by alloxan diabetes. It is of special interest that P-448-H, which is very low in liver microsomes of intact males, was markedly increased in alloxan-diabetic male rats (YAMAZOE et al., 1988c). The decreases in the levels of P-450$_{2C/UT-A}$ and P-450RLM5 in liver microsomes from diabetic male rats have recently been reported by ROUER et al. (1986) and FAVREAU and SCHENKMAN et al. (1987).

Administration of sodium glutamate in neonatal rats is known to cause a low serum growth hormone level in adult rats without strikingly affecting the levels of other pituitary hormones (MILLARD et al., 1982).

In neonatally glutamate-treated male rats, the level of P-450-male in liver microsomes was markedly decreased (YAMAZOE et al., 1988a). However, in contrast to the expression of P-450-male in hypophysectomized female rats, P-450-male was not expressed in neonatally glutamate-treated female rats.

These results also do not accord with the changes observed in hypophysectomized, aged or diabetic rats (Fig. 12). In liver microsomes from aged male rats, P-450-male had disappeared and P-450-female was fully expressed

---

(KAMATAKI et al., 1985a). Moreover, in aged female rats, P-450-female did not change and P-450-male was not expressed. In diabetic male rats, P-450-male was markedly decreased and P-450-female was clearly expressed (SHIMADA et al., 1987). These results suggest that multiple factors may be involved in the regulation of P-450-male and P-450-female.

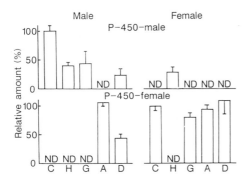

Fig. 12. The levels of P-450-male and P-450-female in liver microsomes of hypophysectomized, aged or diabetic male and female rats.
C: controls; H: hypophysectomized rats; A: aged rats; D: alloxan-diabetic rats.

The activities of benzo(a)pyrene hydroxylation and propoxycoumarin $O$-depropylation decreased in male rats, but not in female rats with neonatal glutamate-treatment. $O$-Ethylresorufin $O$-deethylation did not decrease in male rats treated neonatally with glutamate. Neonatal treatment with glutamate of male rats caused 40 and 30% decreases in testosterone $2\alpha$- and $16\alpha$-hydroxylations, but the decreases were not statistically significant. Moreover, the activity of testosterone $6\beta$-hydroxylase was markedly decreased in accordance with the decrease in the content of P-450$_{6\beta}$ (YAMAZOE et al., 1988a).

Hepatic cirrhosis caused an impaired regulation and activity of male specific androst-4-en-3,17-dione $16\alpha$-hydroxylase, P-450$_{UT-A}$ (MURRAY et al., 1987). The levels of six forms of cytochromes P-450 including the sex-specific forms of cytochromes P-450, P-450$_{UT-A}$, P-450$_{PCN-E}$, P-450$_{UT-I}$, were quantified, and the $16\alpha$-hydroxylase activity was determined in hepatic microsomes prepared from male rats with cirrhosis induced by the prolonged intake of a coline-deficient diet. The male-specific androst-4-en-3,17-dione $16\alpha$-hydroxylase decreased in cirrhotic liver to about 20% of controls. The content of P-450$_{UT-A}$ decreased concurrently from about 0.40 to less than 0.01 nmol/mg protein. The $6\beta$-hydroxylation decreased in cirrhotic liver to about 45% of the controls, despite a marked decrease in P-450$_{PCN-E}$ from 0.27 to less than 0.002 nmol/mg protein. Moreover, $7\alpha$-hydroxylation was less affected and $16\beta$-hydroxylation was unaffected in hepatic cirrhosis. Levels of three other forms of cytochrome P-450 — P-450$_{PB-C}$ (a constitutive form inducible by phenobarbital), P-450$_{ISF-G}$ (a major isosafrole-inducible form) and P-450$_{UT-I}$ — were apparently un-

R. KATO; Y. YAMAZOE

altered in cirrhotic livers. These results are consistent with the assertion that specific forms of cytochrome P-450 are subject to altered regulation in hepatic cirrhosis.

## 10. Role of sex-specific cytochrome P-450 in sex-related differences in drug metabolism

As shown in Figure 13, the hepatic levels of P-450-male in male and female rats subjected to various treatments were closely correlated with testosterone 2$\alpha$-hydroxylation and 16$\alpha$-hydroxylation (KATO et al., 1986). These results

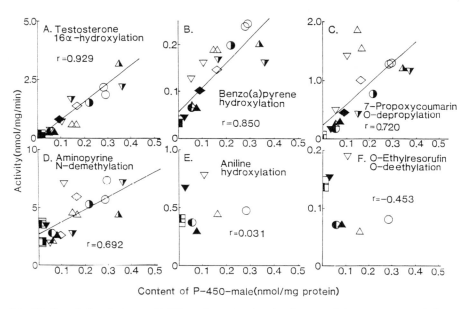

**Fig. 13.** Correlation between P-450-male content and microsomal testosterone and drug-oxidizing activities in male and female rats subjected to various treatments (YAMAZOE et al., 1986a).

indicate that in male rat liver microsomes testosterone 2$\alpha$- and 16$\alpha$-hydroxylations are almost exclusively catalyzed by P-450-male. However, the correlation between P-450-male level and testosterone 6$\beta$-hydroxylation and formation of androsten-3,16-dione is relatively low. Moreover, there was no correlation between P-450-male level and testosterone 7$\alpha$-hydroxylation. Anti-P-450-male IgG markedly inhibited the formation of 2$\alpha$-hydroxy- and 16$\alpha$-hydroxy testosterone and androsten-3,16-dione, but did not inhibit the formation of 6$\beta$-hydroxy and 7$\alpha$-hydroxytestosterone (Fig. 14) (KATO et al., 1986).

---

These results clearly indicate that testosterone 6β- and 7α-hydroxylations are catalyzed by cytochrome P-450 which are different from P-450-male. The correlation between the content of P-450-male and the activities of drug-metabolizing enzymes is given in Figure 13 (YAMAZOE et al., 1986a). The correlation between the content of P-450-male and benzo(a)pyrene hydroxylase activity was relatively high and the correlation coefficient was 0.850. The

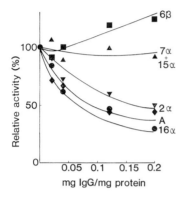

**Fig. 14.** Inhibition of testosterone hydroxylations in hepatic microsomes from male rats by anti-P-450-male IgG (KATO et al., 1986). Testosterone hydroxylase activities are given as percentages of the respective pre-immune IgG controls. A: androstenedione.

correlation coefficients for 7-propoxycoumarin O-depropylase and aminopyrine N-demethylase activities were 0.720 and 0.692, respectively (YAMAZOE et al., 1986a).

These activities are less correlated in comparison with testosterone 2α- or 16α-hydroxylation, suggesting that although the oxidations of these drugs are catalyzed by P-450-male, other forms of cytochrome P-450 may be involved in the oxidative processes.

As shown in Figure 13, there is a clear intercept at 0.057 pmol/mg protein/min in the benzo(a)pyrene hydroxylase activity, suggesting that about one-fourth of the activity in male rat liver microsomes is catalyzed by a cytochrome P-450 other than P-450-male. Similar high intercept values are also observed for 7-propoxycoumarin O-depropylase and aminopyrine N-demethylation.

On the other hand, there is no correlation between the content of P-450-male and aniline hydroxylase and O-ethylresorufin O-deethylase activities. Aniline hydroxylase activity is known as sex-independent and O-ethylresorufin O-deethylase activity is known to be higher in female rats than in male rats. Together with the inhibition study using a highly specific antibody against P-450-male, the results presented in Fig. 13 offer an excellent method for classification of P-450-male dependency of drug metabolism in rat liver microsomes (YAMAZOE et al., 1986a).

# 11.  References

BAKSHI, K., D. BRUSICK, L. BULLOCK, and C. W. BARDIN (1977), in: Origins of Human Cancer, HIATT, H. H., J. D. WATSON, J. A. WINSTEN (eds.), Cold Spring Harbor Laboratory, Cold Spring Harbor, 683—695.

BANDIERA, S., D. RYAN, W. LEVIN, and P. E. THOMAS (1986), Arch. Biochem. Biophys. 248, 658—676.

BROWN, T. R., C. W. BARDEN, and F. E. GREEN (1978), Pharmacology 16, 159—169.

BULLOCK, L., W. MAINWARING, and C. BARDIN (1975), Endocrinol. Res. 2, 25—32.

BURKHART, B., N. HARADA, and M. NEGISHI (1985), J. Biol. Chem. 260, 15357—15361.

CHENG, K.-C. and J. B. SCHENKMAN (1982), J. Biol. Chem. 257, 2378—2385.

CHENG, K. C. and J. B. SCHENKMAN (1983), J. Biol. Chem. 258, 11738—11744.

CHUNG, L. W. K. and H. CHAO (1981), J. Supramol. Struct 15, 194—204.

CHHABARA, R. S. and J. R. FOUTS (1974), Drug Metabol. Disp. 2, 376—379.

COLBY, H. D. (1980), in: Advances in Sex Hormone Research, (J. A. THOMAS, R. L. SINGHAL, eds.), Urban & Schwarzenberg, Munich, 27—71.

CONNEY, A. H., K. SCHNEIDMAN, M. JACOBSON, and R. KUNTZMAN (1983), Ann. Rev. Physiol. 45, 51—60.

DANNAN, G. A., F. P. GRUENGERICH, and D. J. WAXMAN (1986), J. Biol. Chem. 261, 10728—10735.

DENEF, C. (1974), Endocrinology 94, 1577—1582.

DEVORE, K., N. HARADA, and M. NEGISHI (1985), Biochemistry 24, 5632—5637.

EDEN, S. (1979), Endocrinology 105, 555—560.

EINARSSON, K., J.-Å. GUSTAFSSON, and STENBERG, A. (1973), J. Biol. Chem. 248, 4987—4997.

FAVREAU, L. V. and J. B. SCHENKMAN (1987), Biochem. Biophys. Res. Commun. 142, 623—630.

FORD, H. C., E. LEE, and C. C. ENGEL (1979), Endocrinology 104, 857—861.

GONZALEZ, F. J., B.-J. SONG, and J. P. HARDWICK (1986), Molec. Cell. Biol. 6, 2969 to 2976.

GOZUKARA, E. M., J. FAGAN, J. V. PASTEWKA, F. P. GUENGERICH, and H. V. GELBOIN (1984), Arch. Biochem. Biophys. 232, 660—669.

GUENGERICH, F. P., G. A. DANNAN, S. T. WRIGHT, M. V. MARTIN, and L. S. KAMINSKY (1982), Biochemistry 21, 6019—6030.

GUSTAFSSON, J.-Å., A. MODE, G. NORSTEDT, and P. SKETT (1983), Annu. Rev. Physiol. 45, 51—60.

GUSTAFSSON, J.-Å. and STENBERG, A. (1974), Endocrinology 95, 891—896.

GUSTAFSSON, J.-Å. and A. STENBERG (1976), Proc. Natl. Acad. Sci. U.S.A. 73, 1462 to 1465.

HARADA, M. and M. NEGISHI (1984a), J. Biol. Chem. 259, 1265—1271.

HARADA, N. and M. NEGISHI (1984b), J. Biol. Chem. 259, 12285—12290.

HAWKE, R. L. and R. M. WELCH (1985), Mol. Pharmacol. 27, 283—289.

HEIMS, A. H., R. L. HAWKE, L. RAYNOR, and S. GURMIT (1984), Biochem. Soc. Trans. 12, 45—48.

JACOBSON, M., and R. KUNTZMAN (1969), Steroids 13, 329—341.

JANSSON, J., J. MOLE, and J. B. SCHENKMAN (1985), J. Biol. Chem. 260, 7084—7093.

KAMATAKI, T., K. MAEDA, Y. YAMAZOE, T. NAGAI, and R. KATO (1981), Biochem. Biophys. Res. Commun. 103, 1—7.

KAMATAKI, T., K. MAEDA, Y. YAMAZOE, T. NAGAI, and R. KATO (1982), Life Sci. 31, 2603—2610.

KAMATAKI, T., K. MAEDA, N. MATSUDA, K. ISHII, Y. YAMAZOE, and R. KATO (1983a), Mol. Pharmacol. 24, 146—155.

KAMATAKI, T., K. MAEDA, Y. YAMAZOE, T. NAGAI, and R. KATO (1983b), Arch. Biochem. Biophys. **225**, 758—770.

KAMATAKI, T., K. MAEDA, M. SHIMADA, T. NAGAI, and R. KATO (1984), J. Biochem. **96**, 1939—1942.

KAMATAKI, T., M. SHIMADA, K. MAEDA, and R. KATO (1985a), Biochem. Biophys. Res. Commun. **130**, 1247—1253.

KAMATAKI, T., K. MAEDA, M. SHIMADA, K. KITANI, T. NAGAI, and R. KATO (1985b), J. Pharmacol. Exp. Ther. **233**, 222—228.

KAMATAKI, T., K. MAEDA, M. SHIMADA, and R. KATO (1986), J. Biochem. **99**, 841—845.

KATO, R. (1974), Drug Metab. Rev. **3**, 1—32.

KATO, R. (1977), Xenobiotica **7**, 25—92.

KATO, R. (1979), Pharmac. Ther. **6**, 41—98.

KATO, R., E. CHIESARA, and G. FRONTINO (1962a), Biochem. Pharmacol. **11**, 221—227.

KATO, R., E. CHIESARA, and P. VASSANELLI (1962b), Japan. J. Pharmacol. **12**, 26—33.

KATO, R., and J. R. GILLETTE (1965a), J. Pharmacol. Exp. Ther. **150**, 279—284.

KATO, R., and J. R. GILLETTE (1965b), J. Pharmacol. Exp. Ther. **150**, 285—291.

KATO, R., and T. KAMATAKI (1982a), Xenobiotica **12**, 787—800.

KATO, R., and T. KAMATAKI (1982b), in: Cytochrome P-450, Biochemistry, Biophysics and Environmental Implications, HIETANEN, E., M. LAITINEN, O. HANNINEN (eds.), Elsevier Biomedical Press B.V., Amsterdam, 421—428.

KATO, R., and K. ONODA (1970), Biochem. Pharmacol. **19**, 1649—1660.

KATO, R., K. ONODA, and M. SASAJIMA (1970a), Japan. J. Pharmacol. **20**, 194—209.

KATO, R., K. ONODA, and A. TAKANAKA (1970b), Japan. J. Pharmacol. **20**, 546—553.

KATO, R., K. ONODA, and A. TAKANAKA (1970c), Japan. J. Pharmacol. **20**, 554—561.

KATO, R., and A. TAKAHASHI (1938), Mol. Pharmacol. **4**, 109—120.

KATO, R., A. TAKAHASHI, and Y. OMORI (1969), Endocrinologia Japonica **16**, 653—663.

KATO, R., A. TAKAHASHI, and Y. OMORI (1970d), Biochem. Biophys. Acta **208**, 116 to 124.

KATO, R., A. TAKAHASHI, and Y. OMORI (1971), Biochem. Pharmacol. **20**, 447—458.

KATO, R., A. TAKAHASHI, K. ONODA, and Y. OMORI (1970e), Endocrinologia Japonica **17**, 207—213.

KATO, R., A. TAKAHASHI, T. OSHIMA, and E. HOSOYA (1970f), J. Pharmacol. Exp. Ther. **174**, 211—221.

KATO, R., A. TAKANAKA, and A. TAKAHASHI (1968a), Japan. J. Pharmacol. **18**, 224 to 244.

KATO, R., A. TAKANAKA, and M. TAKAYANAGI (1968b), Japan. J. Pharmacol. **18**, 482—489.

KATO, R., and M. TAKAYANAGI (1966), Japan. J. Pharmacol. **16**, 380—390.

KATO, R., Y. YAMAZOE, M. SHIMADA, N. MURAYAMA, and T. KAMATAKI (1986), J. Biochem. **100**, 895—902.

KRAMER, R. E., J. W. GREINER, and H. D. COLBY (1975), Life Sci. **17**, 779—786.

KREMERS, P., F. PASLEAN, and J. E. GIELSEN (1978), Biochem. Biophys. Res. Commun. **84**, 706—712.

KUNTZMAN, R., M. JACOBSON, K. SCHNIEDMAN, and A. H. CONNEY (1964), J. Pharmacol. Exp. Ther. **146**, 280—285.

KUNTZMAN, R., R. WELCH, and A. H. CONNEY (1966), Advan. Enzyme Regulation 4, 149—160.

KUNTZMAN, R., W. LEVIN, M. JACOBSON, and A. H. CONNEY (1968), Life Sci. **1**, 215 to 224.

LEVIN, W. and R. KUNTZMAN (1969), Mol. Pharmacol. **5**, 499—506.

LEVIN, W., D. RYAN, R. KUNTZMAN, and A. H. CONNEY (1975), Mol. Pharmacol. **11**, 190—200.

MacGeoch, C., E. T. Morgan, J. Halpert, and J.-Å. Gustafsson (1984), J. Biol. Chem. **259**, 15433—15439.

MacGeoch, C., E. T. Morgan, and J.-Å. Gustafsson (1985), Endocrinology **117**, 2085—2092.

Maeda, K., T. Kamataki, T. Nagai, and R. Kato (1984), Biochem. Pharmacol. **33**, 509—512.

Matsumoto, R., Y. Emi, S. Kawabata, and T. Omura (1986), J. Biochem. **100**, 1359 to 1371.

Millard, W. J., J. B. Martin, Jr., A. S. M. Sagar, and J. B. Martin (1982), Endocrinology **110**, 540—550.

Morgan, E. T., C. MacGeoch, and J.-Å. Gustafsson (1985a), Mol. Pharmacol. **27**, 471—479.

Morgan, E. T., C. MacGeoch, and J.-Å. Gustafsson (1985b), J. Biol. Chem. **260**, 11895—11898.

Murray, M., L. Zaluzny, G. A. Dannan, F. P. Guengerich, and G. C. Farrell (1987), Mol. Pharmacol. **31**, 117—121.

Nagata, K., T. Matsunaga, J. Gillette, H. V. Gelboin, and F. J. Gonzales (1987), J. Biol. Chem. **262**, 2787—2793.

Noordhoe. K. J. (1972), FEBS Lett. **24**, 255—259.

Noshiro, M. and N. Negishi (1986), J. Biol. Chem. **261**, 15923—15927.

Noshiro, M., C. J. Serabjit-Singh, J. R. Bend, and M. Negishi (1986), Arch. Biochem. Biophys. **244**, 857—864.

Ramperaud, A., D. J. Waxman, D. E. Ryan, W. Levin, and F. G. Walz, Jr. (1986), Arch. Biochem. Biophys. **243**, 174—183.

Rouer, E., Ph. Beauen, and J. P. Leroux (1986), Experientia **42**, 1162—1163.

Rumbaugh, R. C. and H. D. Colby (1980), Endocrinology **107**, 719—724.

Ryan, D. E., P. E. Thomas, D. Korzeniowki, and W. Levin (1979), J. Biol. Chem. **254**, 1365—1374.

Ryan, D. E., R. Dixon, R. H. Evans, L. Ramanathan, P. E. Thomas, A. W. Wood, and W. Levin (1984a), Arch. Biochem. Biophys. **233**, 633—642.

Ryan, D. E., S. Iida, A. W. Wood, P. E. Thomas, C. S. Lieber, and W. Levin (1984b), J. Biol. Chem. **259**, 1239—1250.

Schenkman, J. B., I. Frey, H. Remmer, and R. W. Estabrook (1967), Mol. Pharmacol. **3**, 516—525.

Shapiro, B. H. (1985), Life Sci. **36**, 1169—1174.

Shimada, M., N. Murayama, Y. Yamazoe, T. Kamataki, and R. Kato (1987), Japan. J. Pharmacol. **45**, 467—478.

Skett, P., P. Eneroth, and J.-Å. Gustafsson (1978), Biochem. Pharmacol. **27**, 1713—1716.

Skett, P. and J.-Å. Gustafsson (1979), in: Reviews in Biochemical Toxicology Vol. 1, (E. Hodgson, J. R. Bend, R. M. Philpot, eds.), Elsevier/North Holland Inc, Amsterdam, 27—52.

Smith, J. H., K. Maita, S. D. Sleight, and J. B. Hook (1984), Toxicology **30**, 305 to 316.

Thomas, P. E., L. M. Reik, D. Ryan, and W. Levin (1981), J. Biol. Chem. **256**, 1044—1052.

Waxman, D. J. (1984), J. Biol. Chem. **259**, 15481—15490.

Waxman, D. J., G. A. Dannan, and F. P. Guengerich (1985), Biochemistry **24**, 4409—4417.

Wilson, J. T. (1970), Nature **225**, 861—863.

Wilson, J. T. (1973), Biochem. Pharmacol. **22**, 1717—1728.

---

WRIGHTON, S. A., E. G. SCHUETZ, P. B. WATKINS, P. MAUREL, J. BARWICK, B. S. BAILEY, H. T. HARTLE, B. YOUNG, and P. GUZELIAN (1985), Mol. Pharmacol. **28**, 312 to 321.

YAMAZOE, Y., M. SHIMADA, T. KAMATAKI, and R. KATO (1986a), Japan. J. Pharmacol. **42**, 371—382.

YAMAZOE, Y., M. SHIMADA, N. MURAYAMA, S. KAWANO, and R. KATO (1986b), J. Biochem. **100**, 1095—1097.

YAMAZOE, Y., M. SHIMADA, N. MURAYAMA, and R. KATO (1987), J. Biol. Chem. **262**, 7423—7428.

YAMAZOE, Y., M. SHIMADA, N. MURAYAMA, K. YAMAUCHI, and R. KATO (1988a), Biochem. Pharmacol., **37**, 1687—1691.

YAMAZOE, Y., N. MURAYAMA, M. SHIMADA, K. YAMAUCHI, K. NAGATA, S. IMAOKA, Y. FUNAE, and R. KATO (1988b), J. Biochem. **104**, 785—790.

YAMAZOE, Y., M. ABU-ZEID, K. YAMAUCHI, N. MURAYAMA, M. SHIMADA, and R. KATO (1988c), Biochem. Pharmacol. **37**, 2503—2506.

YOSHIOKA, H., K. MOROHASHI, K. SOGAWA, T. MIYATA, K. KAWAJIRI, T. HIROSE, S. INAYAMA, Y. FUJII-KURIYAMA, and T. OMURA (1987), J. Biol. Chem. **262**, 1706 to 1711.

# Chapter 5

# Biotransformation
# of Xenobiotics During Ontogenetic Development

W. Klinger

## Abbreviations

| | |
|---|---|
| ADH | alcohol dehydrogenase |
| AHH | aryl hydrocarbon hydroxylase |
| ANF | alpha-naphthoflavone |
| BNF | beta-naphthoflavone |

| | |
|---|---|
| BP | benzo(a)pyrene |
| BSP | bromosulphophthaleine |
| $b_5$ | cytochrome $b_5$ |
| $b_5$ red | NADH-cytochrome $b_5$ reductase |
| c red | NADPH-cytochrome c reductase |
| GA | glucuronic acid |
| GSH | reduced glutathione |
| GSSG | oxidized glutathione |
| GT(s) | glucuronyl transferase(s) |
| ICG | indocyanine green |
| MC | 3-methylcholanthrene |
| OAP | o-aminophenol |
| P-450 | cytochrome P-450 |
| P-450 red | NADPH cytochrome P-450 reductase |
| PAB | p-aminobenzoic acid |
| PAH | polycyclic aromatic hydrocarbons |
| PB | phenobarbital |
| PNP | p-nitrophenol |
| TCDD | tetrachlorodibenzodioxin |
| Tm | maximum transport capacity |
| UDP | uridine diphosphate |
| UDPGA | uridine diphosphate glucuronic acid |
| UDPGA-DH | uridine diphosphate glucuronic acid dehydrogenase |

# 1.  Introduction

Since the early publications by KARUNAIRATNAM et al. (1949), VEST (1958), JONDORF, MAICKEL, and BRODIE (1959), and FOUTS and ADAMSON (1959) more than a thousand papers, including several reviews (KLINGER, 1982; SCHMUCKER, 1985), have been published concerning the time of appearance and the rates of development of various enzymes associated with the biotransformation of lipidsoluble xenobiotics.

The so-called lipid solubility is a term with a rather obscure background: the solubility of a given drug in various lipids and lipid solvents such as $CCl_4$, chloroform or heptane may be very different, so that water-octanol and water-oil partition coefficients differ and may serve as models only. These so-called lipid soluble foreign compounds which are not metabolized by the highly specific enzymes of carbohydrate, lipid or protein metabolism, would stay in the organism for weeks and months by continuous reabsorption in the gut and the kidneys if they were not metabolized to more hydrophilic metabolites which are more suitable for excretion and generally less toxic.

Xenobiotics are characterized by an extremely high structural variability; therefore, the enzymes responsible for their biotransformation are, in general, of low substrate specificity. These enzymes for the metabolism of foreign compounds were also investigated by the "physiological chemists" of the 19th century. The excretion of benzoic acid as conjugate with glycine was described by KELLER in 1842.

The insufficiency of glycine conjugation in the newborn was first published by VEST (VEST, 1959b), this "defect" in the newborn rat was described by BRANDT in 1964, more than 100 years after the detection of this reaction.

The glucuronidation of xenobiotics was published in 1879 by SCHMIEDEBERG and MEYER, and the insufficiency of newborn mice for glucuronidation was described by KARUNAIRATNAM et al. in 1949, 70 years after the first publication on this reaction. The age dependence of mixed function oxidation (mfO) reactions was detected as early as four years after AXELROD's et al. (1956) first description of this reaction by JONDORF et al. (1959) and FOUTS and ADAMSON (1959). LEVI et al. (1968) detected the organic anion-binding cytosolic proteins in rat liver and investigated the age course one year later (1969).

In 1970 they presented the data on the development in monkey liver to explain the "physiologic" jaundice in newborns.

Thus the developmental aspect of biotransformation has attracted increasing attention, especially after the accidents with chloramphenicol in newborns and by the Contergan tragedy. The developmental aspect of biotransformation became the first main field of "Developmental Pharmacology", the aims of which have been characterized as follows:

— to protect the newborn from deleterious effects of xenobiotics,

---

W. KLINGER

— to optimize drug therapy in the newborn and in prematures, and to include newborns and prematures into the progress of drug therapy,
— to bring about the prerequisits for a scientific drug therapy in aged humans (ANKERMANN, 1973).

It is indispensable to define the term "Developmental Pharmacology" in that the whole life span of an individuum is covered from the zygote to death. Toxicological aspects are involved or even preponderant, especially when early stages of development from the formation of the zygote till infancy are considered. Toxicological aspects are also a central problem in geriatric pharmacology, whatsoever the onset of senescence may be (ASTRACHANZEWA, 1977). These toxicological implications mainly arise from age dependent peculiarities in pharmacokinetics, and in this field biotransformation plays a central role.

This branch of developmental pharmacology was supported by the finding that these reaction types are also competent for biotransformation and elimination of endogenous compounds such as steroids and bilirubin.

It has become clear that the so-called biogenetic rule of the ontogeny being a short repetition of phylogeny, as postulated by Ernst HAECKEL (1866), has proven true with relation to some reactions.

## 2. Hepatic biotransformation

### 2.1. Morphological basis

Most biotransformation reactions with xenobiotics can be localized in the liver parenchymal cell, predominantly in both main forms of endoplasmic reticulum (ER), smooth (SER) and rough endoplasmic reticulum (RER). Except for some enzymes cytosol and, according to ER genesis, the nuclear membrane are active areas. Finally the whole cell is necessary for providing the necessary cofactors. Only few investigators attribute typical phase I reactions or the acetylation to liver mesenchymal cells.

In the rat, the liver is an important hemopoetic organ in fetal and early postnatal life until the 7th day. RER becomes visible at the 14th embryonic day as big vesicles, but free ribosomes are prevalent. The typical picture of the mature liver is reached at the 5th—7th or 10th postnatal day on the histological and submicroscopical level but the sublobular heterogeneity of the parenchymal cells has not reached maturity at the 20th day of postnatal life (KLINGER et al., 1986).

The growth pattern of rat hepatocytes during postnatal development with regard to ploidy, mitosis rate (hyperplasia) and/or hypertrophy was investigated in detail (DAVID, 1985).

Volume, surface and number of microvilli of liver cells as well as the volume of nuclei increase postnatally. Glycogen is visible at the 17th fetal day and heavy particles at the 19th day. Glycogen metabolism and its control in the pre- and perinatal period have been investigated in various laboratories. This might be of relevance for xenobiotic biotransformation, as a relationship between glycogen and xenobiotic metabolism has been shown (FOUTS, 1965).

In fetal liver cells nuclei are bigger, but much less mitochondria are found. In weanling rats, 2.6 times more but smaller mitochondria have been observed in comparison to newborn rats. In senescent rats cell volume increases, lysosomes are bigger, RER and SER are diminished, and the surface of the Golgi apparatus decreases. Only mitochondria do not alter or even have an enlarged surface (SCHMUCKER et al., 1974).

In the mouse the adult pattern can be observed in liver cells of newborn animals. The postnatal development of rabbit liver is similar to that of rat liver. In the guinea pig, which has a long gestation period, hepatic development occurs mainly before birth so that typical monooxygenase reactions and their inducibility can be detected prenatally. In man the early differentiation with the typical adult pattern by the 12th gestational week is remarkable. Regressive changes are observed after the 40th year of life.

## 2.2. Biochemical basis

Several more recent reviews exist on developmental biochemistry of the liver (WILSON et al., 1982); however only the data relevant for biotransformation reactions shall be reviewed. The ER is not only the site of various hydroxylation reactions, but it is also the main organelle for the synthesis of cellular structural components. During developmental growth faster synthesis and slower degradation of protein can be observed, whereas in senescent rats microsomal protein as well as protein synthesis (amino acid incorporation) decrease.

Only a few ER membrane proteins have so far been investigated for their biosynthesis and transport to the final position. Cyt $b_5$, for example, is synthesized on ribosomes and inserted directly into its final position at the cytoplasmic ER surface. Most hepatic microsomal proteins as well as protein synthesis in senescent rats (amino acid incorporation in vivo and in vitro) decrease due to a weakening of stimulating and a strengthening of inhibiting factors with age.

The lipids of endoplasmic membranes undergo important quantitative and qualitative changes during development, especially in their fatty acid pattern. They strongly influence the fluidity of membranes and by this means the diffusion velocity of xenobiotics and the electron transport from $NADPH_2$ to P-450, and they are essential for and decisively influence the binding of xenobiotics to cytochrome P-450 binding sites (BERLIN et al., 1984).

---

W. KLINGER

Microsomal lipids themselves bind to P-450 and inhibit monooxygenase reactions, depending on age.

Lipid peroxidation is evidently not only influenced by age-dependent differences in lipid composition but also by age-dependent inhibitory and enhancing factors (PLAYER and HORTON, 1978).

Variations in the fatty acid pattern of the maternal diet influence the fatty acid composition of microsomal phospholipids but do not affect enzyme activities in newborn animals and are therefore of minor importance (PALADE and PORTER, 1967). The phospholipid-protein-ratio in microsomes increases with increasing age.

Neither in rats nor in guinea pigs of any age are NADP/NADPH$_2$ and NAD/NADH$_2$ concentrations limiting for the capacity of the microsomal electron transport chain (SCHENKER and O'DONNEL, 1965). Microsomal glucose-6-phosphate dehydrogenase for the NADH$_2$ and NADPH$_2$ regenerating systems has highest activities before birth in mouse, rat and human liver (PELKONEN et al., 1971), and this activity declines postnatally. The activity depends on diet and, in the rat after weaning, on sex.

### 2.2.1. Sex differences in rats and mice

Until now only in rats and mice marked sex differences in drug disposition have been observed. After weaning and at the beginning of sexual maturation, male rats exhibit distinctly higher activities than females; in mice the opposite holds true (cp. KLINGER, 1982; FUJITA et al., 1985). Postnatal age course after sexual maturation must, therefore, be investigated separately for both sexes in these species, and these data are given in the chapter by KATO in this volume. In rats this sexual differentiation was found to be due to a neonatal androgen-induced imprinting in which the hypothalamic-hypophyseal axis plays an obligatory role.

Evidently the mechanism of sexual differentiation in mice has not been investigated (VAN DEN BERG et al., 1978).

### 2.3. Transport mechanisms

Xenobiotics enter hepatic parenchymal cells and are bound and stored within the cell after or without biotransformation. The specific systems for uptake, binding and storage as well as for excretion have their own developmental pattern, and it was demonstrated that the importance of transfer steps for the overall transport via hepatocytes changes with ageing (CAGEN and GIBSON, 1979). In the rat, bile secretion begins at about the same developmental stage as glucuronidation.

---

Bile pigment is detectable in the small intestine one day before birth, while bile canaliculi develop earlier (SANDSTRÖM, 1972). Especially in newborn rats, overall hepatic excretory function is low and the excretion of organic acids, including bile acids and neutral xenobiotics, develops postnatally to reach maximum capacity at an age of 30 days. Thereafter the excretion capacity declines (BARTH and KLINGER, 1979).

### 2.3.1. Hepatic uptake-and storage

Hepatic uptake rate is dose-dependent, saturable and evidently carrier-mediated. However, it could also be achieved through simple diffusion and subsequent binding to cytoplasmic acceptor proteins. Competition studies suggest at least four independent hepatic uptake systems responsible for the handling of organic anions, bile acids, organic cations and neutral compounds.

In adult rats the initial hepatic ICG uptake capacity is about ten times larger than the excretory capacity. Therefore hepatic uptake cannot be rate limiting for overall drug transport (KROCHMANN, 1974).

The hepatic uptake capacity for digoxin and digitoxin, eosin and ouabain is distinctly lower in newborn and in young animals a compared to adults (e.g. KLINGER, 1982).

In newborn rats the immature hepatic uptake leads to acumulation of some drugs in plasma with increased lethality (KLAASSEN, 1975) and of bilirubin with the consequence of non-hemolytic hyperbilirubinemia (HEIMANN et al., 1977). Inducers such as phenobarbital, spironolactone and PCN enhance the hepatic uptake of ouabain in neonates, whereas methylcholanthrene is ineffective. A stimulation of ouabain uptake by PCBs has been shown only in 15-day-old rats, not in adults (CAGEN et al., 1979).

Immediately after the first description of the organic anion-binding cytosolic proteins (ligandin), it was reported that the capacity of these acceptor or transport proteins develops in guinea pig and primates postnatally. This deficiency in newborns could be the cause or at least contribute to the icterus neonatorum. Newborn guinea pigs eliminate not only free bilirubin more slowly than adults, but also bilirubin glucuronide. This deficiency is even more pronounced in the fetus (SCHENKER et al., 1964).

### 2.3.2. Hepatic excretion

Similar to hepatic uptake, different transport systems of hepatic excretion are postulated for acids, bases, neutral compounds, bile acids and metals, which are transported actively into bile (bile/plasma concentration ratios are higher than 1). The maximal hepatic transport capacity (Tm) for organic anions such as eosine, BSP and ICG is lower in newborn animals than in adults.

Hepatic transport function in infantile as well as in very old animals is lower than in adults. As in adult and young, liver storage capacity is higher than excretory capacity, the overall hepatic transport into bile seems to be limited in all postnatal ages by the small rate of transport from liver into bile.

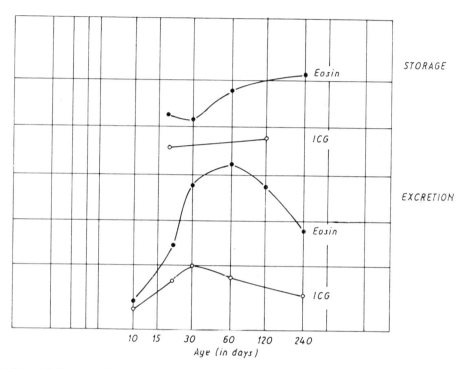

**Fig. 1.** Influence of age on hepatic storage and biliary excretion of the organic anions eosine and indocyanine green (ICG) in male rats. Data according by BARTH et al. (1986).

## 2.4. Biotransformation

Drug metabolism has been divided into two phases:

phase I:    hydroxylation, dealkylation, deamination, dehydrogenation, reduction, hydrolysis, or similar modifications of an administered molecule, producing a conjugable grouping

and

phase II: biosynthetic reactions whereby drugs with conjugable grouping or/ and after phase I reactions (to form these groupings) are conjugated with endogenous polar molecules (WILLIAMS, 1959). These conjugates are generally readily excreted.

### 2.4.1. Phase I biotransformation reactions

### 2.4.1.1. Oxidative reactions

### 2.4.1.1.1. Cytochrome P-450 dependent oxidation (monooxygenation, mixed function oxidation)

*NADPH-Cytochrome P-450 reductase*

According to our present knowledge about $10-15$ molecules of P-450 surround one molecule of reductase within the microsomal membrane and exhibit lateral mobility. Addition of purified, solubilized reductase to microsomes enhances monooxygenase activities (YANG et al., 1978).

Though cytochrome c reduction most often is used to determine NADPH-P-450 reductase activity and immunological identity seems to be proven (MASTERS et al., 1971), different age courses of cytochrome c and P-450 reduction give evidence for different age courses of activity of one enzyme with different substrates, possibly due to distinct age-dependent influences of microsomal constituents on availability or binding of these two substrates. Many papers are dedicated to the question of whether or not the reductase is rate limiting in the microsomal electron transport chain. HOLTZMAN (1979) summarizes that the overwhelming body of data indicate that the flow of electrons through the stimulated reductase is not the rate-limiting process in hydroxylation reactions. Basal reductase activity changes little with postnatal age and neither the stimulation by type I substrates nor induction with phenobarbital reveals age-dependent differences (MÜLLER et al., 1975).

*NADH-Cytochrome $b_5$ reductase*

It is widely accepted that the second electron is preferably introduced via NADH-cyt $b_5$, but NADPH may act as electron donor as well and the system also works depending on the substrate without $b_5$, e.g. in reconstituted systems containing P-450, NADPH-P-450 reductase and phospholipid. Until now the age course of NADH-cyt $b_5$ reductase activity has not been investigated. In vitro NADH enhances ethylmorphine N-demethylation activity of rat liver microsomes of all age groups by about $50\%$, in controls as well as after phenobarbital pretreatment (MÜLLER and KLINGER, 1977); but the influence of NADH differs with various substrates and is age dependent.

*Cytochrome P-450 concentration*

The total concentration of P-450 as a function of age was studied by a number of authors, for review cp. MÜLLER and KLINGER (1978) and KLINGER (1982). In some studies a distinct postnatal increase in P-450 concentration was observed in various species while others found small changes after the first week of life. Evidently the total P-450 concentration is responsible neither for the age dependence of drug oxidation nor for the age-dependent induction by inducers of the phenobarbital or methylcholanthrene-type: the postnatal increase in biotransformation activity is higher than that of P-450, and phenobarbital and methylcholanthrene enhance biotransformation activity to a greater extent than P-450 concentration. Now there exists ample evidence for multiple forms of P-450 with overlapping activities towards a variety of substrates (GUENGERICH et al., 1982). The main subspecies show different developmental patterns in the rabbit as well as in other species (ATLAS et al., 1977; KAHL et al., 1980).

The first step of monooxygenation is the binding of xenobiotics to the protein moiety of one or more P-450 subspecies, which is associated with spectral changes. For many substrates the spectral dissociation constants (Ks) are similar to the Km value of the corresponding biotransformation reaction, and the magnitude of spectral changes is generally correlated to the rate of biotransformation activity. Few investigations in animals of different ages demonstrate great differences in the magnitude of spectral changes between young and adult animals (COHEN and MANNERING, 1974).

The concentration of P-450 or of the active form of a relevant subspecies or of its binding sites need not be totally responsible for the age differences because changes in the microenvironment could also be of importance. Phospholipids and their fatty acid composition change with age (cp. "Biochemical Basis"), and thus changes in the properties of the microsomal membranes as well as in P-450 reducibility are to be expected. However, the age and induction by PB dependent spectral changes due to the binding of hexobarbital to P-450 after solubilization and partial purification of P-450 are the same as with intact microsomes; evidently the portion of "active P-450" is not influenced by the membrane structure.

Endogenous substrates which block P-450 by their binding could be responsible for age differences in the active portion of P-450, but after removal of endogenous compounds from solubilized P-450 by charcoal the hexobarbital-binding P-450 portion increases by a factor of two to three in young and adult control and phenobarbital-treated rats. However, the age dependence of hexobarbital binding is still evident. Thus, endogenous substrates do play a role in some aspects of drug metabolism, but they do not cause age differences in phase I biotransformation processes (KLINGER et al., 1975). For investigating P-448-dependent biotransformation reactions, the activity of aryl hydrocarbon hydroxylase (AHH) with benzo(a)pyrene (BP) as substrate was measured in dependence on age, the inducer methylcholanthrene and a relatively selective inhibitor, alpha-naphthoflavone = 7,8-benzoflavone (ANF).

A marked postnatal increase in AHH activity was found in the liver of male Wistar rats with maximum levels reached at about 60 days and a decline thereafter. Induction by postnatally administered methylcholanthrene was dose-dependent regarding the maximum as well as the duration of the effect (KLEEBERG et al., 1979).

We confirmed the finding of WIEBEL and GELBOIN (1975) that ANF has a striking stimulatory effect in neonates. The suggestion of the existence of a special kind of "neonatal" benzo(a)pyrene-hydroxylating P-450 seems to be possible. But recent findings on the developmental regulation of P-450 genes in the rat do not yet favour this assumption (GIACHELLI and OMIECINSKI, 1987):

The mRNA sizes for the two P-450 isozymes P-450c and P-450d do not appear to change with age, and PB responsiveness has been shown to be first evident in rats on day 21 of gestation (P-450b and P-450e). This finding is in agreement with enhanced benzphetamine N-demethylase and benzo(a)pyrene hydroxylase activities after transplacental induction in the last two days of pregnancy (CRESTEIL et al., 1986). Moreover it has been shown that no gross gene rearrangements are coincident with the onset of PB inducibility within the P-450b or P-450e genes.

But in contrast to the findings by many authors that MC also induces AHH prenatally in the rat (cp. KLINGER et al., 1979), both P-450c and P-450d mRNAs are detectable after inducer treatment only postnatally (GIACHELLI and OMIECINSKI, 1987). This contradiction needs further investigation.

In general there is increasing evidence that all P-450 isozymes are developmentally regulated, in rats additionally a sex-related expression can be demonstrated and these developmental patterns are different in different species and even in different strains (BANDIERA et al., 1986; GESTEIL, 1987; RAMPERSAUD et al., 1987; SONG et al., 1986; SUN et al., 1986; YASUMORIN et al., 1987).

*Hydroxylation of aliphatic compounds*
Both pento- and hexobarbital are nearly completely metabolized and the sleeping time is closely correlated to the side chain oxidation, as was directly demonstrated for various age classes. Similar to the hydroxylation of aromatic compounds, the main development of pento- and hexobarbital hydroxylation activities is observed postnatally and maximum values are reached around the 30th day of life in rats. Then we find a sexual differentiation: a decrease of activities after the 30th day of life in females, a further increase or constant values in males and finally a decrease, so that in very old rats sexual differences are no longer demonstrable (KLINGER, 1970; FUJITA et al., 1985). In mice, adult values of hexobarbital hydroxylation are reached even on the 21st day of life. After sexual maturation, the activities are lower in males and higher

in females. As in rats, sleeping time and metabolic breakdown are strictly correlated. Similar to rats and mice, hexobarbital hydroxylation capacity mainly develops postnatally in guinea pigs, rabbits and pigs.

*Epoxidation*
So far the aldrin epoxidation has been investigated in three and seven weeks old male and female SD rats (WOLFF and GUENGERICH, 1987). Whereas in both genders P-450 concentration was similar in both age groups, and rose by about 70 to 100%, aldrin epoxidase increased only in male rats (3-fold). In young adult female rats inducibility by PB, pregnenolone-16-a-carbonitrile or dexamethasone was much more pronounced than that in male ones.

*N-Dealkylation*
N- as well as O-dealkylation is a special case of hydroxylation: primarily instable N-hydroxy-alkyl-derivatives are formed which split into free amines and aldehydes. In rats with all substrates, maximum activities are observed at the beginning of sexual maturity and sexual differentiation of biotransformation activity, that is about the 30th day of life. In females the activities fall with the onset of sexual maturation after the 30th day of life; in males activity may show a further increase or constant values or only a small decrease up to an age of 60 days, so that in rats aged 60 — 120 days the sex differences are most pronounced (cp. KLINGER, 1982).

Despite the distinct age differences in activity, Vmax need not significantly change with age and the MICHAELIS-MENTEN-constant Km can decrease. This has been shown e.g. for indomethacin N-demethylation (CLOZEL et al., 1986). In guinea pigs a quick postnatal development of N-demethylation within the first three postnatal days is observed (KUENZIG et al., 1974). In mice, maximum activities are reached between the 14th and 20th day of life. With the onset of sexual maturation, females develop higher activities than males, contrary to rats (VAN DEN BERG et al., 1978). In the chicken, highest activities are observed as early as one day after hatching.

In man, evidence is given directly for small N-dealkylation activity in newborns, contrasting to the findings with human fetal liver. These observations are supported by the unanimous findings of prolonged elimination of drugs in the neonatal period.

*O-Dealkylation*
Substrates for this reaction are usually codeine, p-nitroanisole and ethoxycoumarine. In rats p-nitroanisole metabolism mainly develops postnatally with maximum values in 30-day-old rats of both sexes, thereafter the sexual differentiation occurs as with N-dealkylation. Contrariwise, ethoxycoumarine O-deethylation reaches maximum activity as early as the fifth day of life. In mice, adult values for p-nitroanisol O-demethylation are reached after

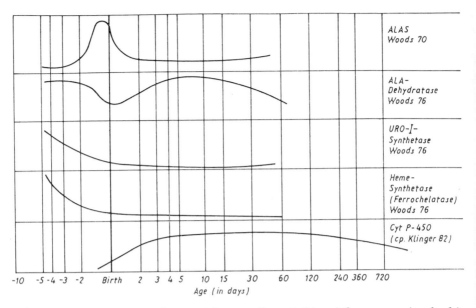

**Fig. 2.** Influence of pre- and postnatal age on the activities of the enzymes involved in hepatic heme synthesis and of cytochrome P-450 concentration in male rats.

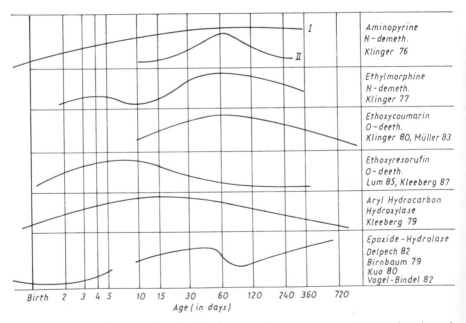

**Fig. 3.** Influence of pre- and postnatal age on various monooxygenase reactions in male rats.

three weeks of life. In rabbits about 2/3 of the adult activity of codeine O-demethylation are reached with in the first three weeks of life (cp. KLINGER, 1982). For an overview compare Figures 2 and 3.

*N-Hydroxylation*
UEHLEKE (1964) reviewed age differences in methemoglobin formation by arylamines, which are hydroxylated to form hydroxylamines. In rats N-hydroxylation activity develops postnatally to reach maximum values in 30-day-olds; thereafter the activities decline (UEHLEKE, 1973). In rabbits, too, maximum liver activities with N,N-demethylaniline as substrate are reached after 30 days of life (BEND et al., 1975). In mice a very steep increase in N,N-dimethylaniline N-oxidation is observed in the first three postnatal days, 50% of the adult value is reached. Thereafter the increase is not so rapid. Around the 20th day of life higher activities can be observed in females compared to males (WIRTH and THORGEIRSSON, 1978).

*S-Oxidation*
This reaction was investigated with chlorpromazine as substrate. In rabbits no activity was detectable in newborns, adult values were reached after the third week (FOUTS, 1961).

*Oxidative deamination*
Many papers deal with the influence of age on MAO activities in various organs and tissues, but only liver will be considered here. Depending on the substrate used or on the additional differentiation of MAO A and B by selective inhibitors, different age courses have been found, especially in the rat (BLATCH-FORD et al., 1976). In the domestic pig the different forms of hepatic MAO show different postnatal developments, in general an increase (BLATCHFORD et al., 1976).

In man, fetal hepatic MAO A has a low activity, but MAO B and benzyl-amine oxidase activity in fetal liver is the same as in adult ones (LEWINSOHN et al., 1980).

*Biotransformation of halogenated aliphates*
As early as 1912 WHIPPLE observed that pups are almost nonsusceptible to chloroform liver damage. The resistance of newborn rats to the hepatotoxic action of $CCl_4$ was observed by CAMERON and KARUNARATNE (1936) and DAWKINS (1963); but after the 3rd day of life $CCl_4$ destroys P-450 as in adults (SASAME et al., 1968). Later on, it was found that the toxicity of haloalkanes is closely related to their metabolism and especially to activation by the P-450 system and covalent bonding to proteins and lipids.

More recently, metabolic activation and covalent bonding with microsomes from developing rats have been investigated. The irreversible binding of halogenated aliphates to protein and lipids develops with age and with increasing P-450 concentrations (UEHLEKE and WERNER, 1975).

### 2.4.1.1.2. Dehydrogenation

*Alcohol dehydrogenation*
In the adult human liver about 75% of the alcohol degrading capacity is considered to be due to hepatic alcohol dehydrogenase and 25% to the microsomal ethanol oxidizing system (MEOS) (LIEBER and DeCARLI, 1970).

Only liver alcohol dehydrogenase (ADH) will be considered here. In all species investigated, ADH activity increases postnatally: in the mouse, in the rat, and in man (cp. KLINGER, 1982).

Although in the rat and at all developmental stages only one isoenzyme regarding pH optimum and Km values for NAD or ethanol was detectable, in man different isoenzymes with different pH optimum, Km values and electrophoretic pattern could be detected: in the fetus there is one, in newborns there are two and in adults four isoenzymes are detectable (PIKKARAINEN, 1971).

Rats of old age eliminate ethanol more slowly and have lower $LD_{50}$ values than young adults (HAHN and BURCH, 1983).

*Aldehyde dehydrogenation*
An increase of activity with age has been described and only one isoenzyme was electrophoretically detectable in rats (LINDAHL, 1977).

### 2.4.1.2. Reductive reactions
### 2.4.1.2.1. Ring reduction

The biotransformation of nicotine to form conitine comprises reduction of the pyridine ring. This activity is very low in fetal mouse liver, begins to rise before birth and then increases about 10-fold in the first four postnatal weeks to reach adult values (STALHANDSKE et al., 1969). Ring A reductase for the reduction of steroids, 7-dehydrocholesterol reductase and 5α-reductase mainly develop postnatally. Rat hepatic and renal folic acid reductase activity also increases during the first 10 postnatal days, thereafter a decrease has been observed (cp. KLINGER, 1982).

### 2.4.1.2.2. Aldehyde reduction

Sleeping time after administration of chloralhydrate increases in 1- to 24-month-old rats (MENDE and VIAMONTE, 1967). This could be due to a decreasing

---

W. KLINGER

reduction activity to the active metabolite trichloroethanol, but also to a delayed formation of the detoxification product urochloralic acid. No clear cut decision can be made.

### 2.4.1.2.3. Azo reduction

Even in newborn rabbits reduction of neoprontosil is detectable (Fouts, 1962). Activity in pig liver is low and increases threefold in the first six postnatal weeks to reach adult values (Short and Davis, 1970). Reduction of p-dimethyl-aminoazobenzol decreases by about $30-40\%$ in old aged rats (Kato and Tanaka, 1968; Fujita et al., 1982).

### 2.4.1.2.4. Nitroreduction

Mainly the age course of p-nitrobenzoic acid reduction has been investigated. In livers of newborn rats no activity was measurable, it can be detected after the 5th day of life, increases 4-fold from the 10th to the 60th day of life and thereafter decreases again (Müller et al., 1971). However, microsomal and cytoplasmic nitro-reduction must be distinguished. The typical age course must be referred to the P-450 dependent nitro-reduction only, as neither Km nor Vmax values of cytosolic nitro-reduction proved to be age dependent (Klinger et al., 1975).

### 2.4.1.3. Hydrolysis
### 2.4.1.3.1. Hydrolysis of esters

Reviews exist on the isoenzyme status of esterases and of their developmental pattern in different species. For differentiation of esterases which hydrolyse xenobiotics it seems to be suitable to characterize them by the substrates used. α-Naphthylacetate is hydrolysed by fetal rat liver supernatant with low activity only. Activity develops slowly, mainly postnatally; in the second week of life, $60\%$ of the adult activity is measurable. In the fourth week maximum values are reached, which are distinctly above adult ones.

Concomitant with the postnatal development of activity, the isoenzyme pattern changes (Schwark and Ecobichon, 1969). In guinea pigs much higher activities are observed prenatally, and postnatal development is faster than in rats. In pigs a steep increase is observed perinatally. Procaine is split by the liver of 5-day-old rats with $5\%$ activity of adult males, adult values are reached in 60-day-old rats (Yeary et al., 1973). Highest specific activities are observed in the microsomal fraction. In pig liver procaine hydrolysing activity

---

is low at birth, it increases 15-fold during the first four weeks, and thereafter a decrease is observed. Pethidin is eliminated with maximum velocity in 40-day-old rats. In order to investigate the age course of pethidin effects, the influence of age on analgesia threshold must be regarded (NICAK, 1971).

Measurements of pethidin hydrolysis in animals of different ages are not known at present.

Urethanes are also hydrolysed; 20-day-old rats eliminate urethane more quickly than 100-day-old ones, correspondingly the sleeping time is shorter (BRÄUNLICH, 1968).

Age courses of hydrolysis, elimination rate and duration of effect of cariso-prodol and meprobamate are nearly identical: very low hydrolysis activities in newborn rats increase to a maximum in 30-day-old ones, thereafter a decrease is observed (KATO, 1964).

### 2.4.1.3.2. Hydrolysis of amides

Isocarboxazide cleavage was not demonstrable with fetal rat liver; at birth very low activities increased considerably after the 5th day of life, constant values being observed from the 60th to the 100th day of life (SATOH and MOROI, 1971). Cyclophosphamide hydrolysis was not detectable in fetal mouse liver, adult values were measured after the 21st day of life (SHORT and GIBSON, 1971).

### 2.4.1.3.3. Hydrolysis of epoxides (epoxide hydrolase, epoxide hydrase, epoxide hydratase)

Even this important detoxifying enzyme for highly toxic, mutagenic, carcino-genic and teratogenic epoxides, especially PAH-epoxides, develops mainly postnatally in all mammals investigated although activity can be detected early in fetal life. This holds true for rats and mice (DELPECH et al., 1982) as well as for guinea pigs and rabbits (JAMES et al., 1977).

The conversion of thio- to oxophosphates shows the same activity increase in male and female rats up to an age of 30 days, thereafter the activity in females decreases. In males a doubling can be observed between the 30th and 60th day of life.

### 2.4.2. Phase II reactions

Phase II comprises one (sometimes more) biosynthetic reaction(s) whereby endogenous polar compounds are conjugated with the products of phase I

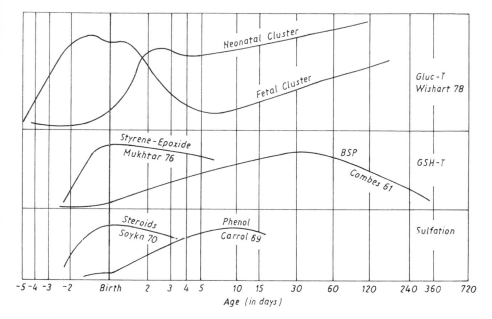

**Fig. 4.** Influence of age on the activities of various glucuronyl-, glutathione- and sulfate-transferases.

reactions or with the administered substance itself if it possesses a conjugatable grouping (WILLIAMS, 1959). In Figure 4 the postnatal development of the most important phase II reactions is demonstrated.

## 2.4.2.1. Glucuronidation

Reviews on the development of glucuronidation already exist (DUTTON, 1978; LUCIER et al., 1979).

We distinguish four types of glucuronidation:

1. Ether bonding between OH-groups and C-1 of GA, o-aminophenol (OAP) p-nitrophenol (PNP), phenolphthalein serve as substrates in most investigations,

2. N-glycoside formation between amino (or imino) groups and C-1 of GA, rather rare,

3. S-glycoside formation between SH-groups and C-1 of GA, very rare,

4. Ester bonding between a carboxyl group and C-1 of GA, for example bilirubin glucuronide formation.

---

Glucuronidation needs UDPGA and microsomal GTs. UDPGA formation, i.e. UDPGA-DH activity, and GT activities are rate limiting. In most adult animals GTs are rate limiting. Different GTs exist with different developmental patterns, different inducibility by phenobarbital and methylcholanthrene and different susceptibility to glucocorticoids with precocious prenatal development or to TCDD as inducer in fetal and early postnatal life, not only for these four types of glucuronidation but also for different substrates for the ether bonding type. The existence of different GTs seems to have been proven, even though they have not yet been solubilized, isolated and characterized.

The investigation of GT activity is complicated by physical or chemical influences on these membrane-bound enzymes. The activating effect of various detergents and other activating substances (digitonin, UDP-N-acetylglucosamine, diethylnitrosamine) varies considerably with age (BURCHELL and DUTTON, 1975); thus the age-dependent activity of the "native" enzyme is dependent to a high degree on the method used and comparisons are difficult.

### 2.4.2.1.1. Availability of UDPGA

In liver, UDPGA levels are low in fetal mammals. But compartmentational and turnover factors must be considered. UDPGA-DH activities, too, are low in fetal mammal liver. But as different substrates are glucuronidated with rather different velocities, UDPGA availability could be rate limiting only for the substrate with the highest glucuronidation velocity, providing availibility is the same for all GTs.

### 2.4.2.1.2. Activities of glucuronyl transferases

GT activities are given in most papers, unless the excretion of glucuronides in intact organisms is investigated. Among mammals, hepatic GTs in rats have been investigated most recently. WISHART (1978a) differentiated a late-fetal and a neonatal cluster of GTs, the first developing in the last five fetal days and reaching higher than adult activities before birth (with OAP, PNP, 1-naphthol, 4-methylumbelliferon, 5-hydroxytryptamine and 2-aminobenzoate as substrates), the second developing after birth (with bilirubin, testosterone, $\beta$-estradiol, morphine, phenolphthaleine and chloramphenicol as substrates) cp. Figure 4. These results confirm the findings of LUCIER and DANIEL (1977), who distinguished between non-steroid and steroid-type GTs. Whereas phenolphthalein behaved like a steroid, TCDD induced only non-steroid GTs. Glucocorticoids induced precocious intrauterine development only of the late fetal cluster GTs.

---

W. KLINGER

Although in the perinatal period sex differences are not detectable, adult male rats generally have distinctly higher GT activities than females. Neonatal treatment with sexual hormones alters sexual differentiation, the high male values are particularly depressed by estrogens (LAMARTINIERE et al., 1979).

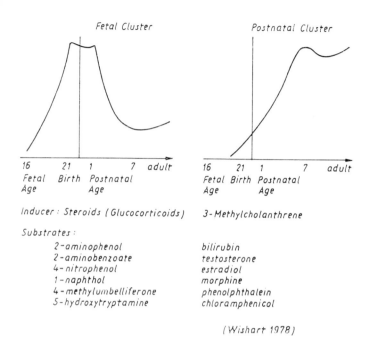

UDP-Glucuronosyltransferase Development

Fetal Cluster

Postnatal Cluster

| 16 | 21 | 1 | 7 | adult |
| Fetal | Birth | Postnatal | | |
| Age | | Age | | |

Inducer : Steroids (Glucocorticoids)    3-Methylcholanthrene

Substrates :

2-aminophenol
2-aminobenzoate
4-nitrophenol
1-naphthol
4-methylumbelliferone
5-hydroxytryptamine

bilirubin
testosterone
estradiol
morphine
phenolphthalein
chloramphenicol

(Wishart 1978)

**Fig. 5.** Characterization of the fetal and postnatal cluster of UDP-glucuronosyltransferases by pre- and postnatal development of activities, inducibilities and substrates in rats.

In mice it was observed for the first time that newborn animals cannot glucuronidize (KARUNAIRATNAM et al., 1949). In newborn mouse liver neither UDPGA nor GT activities with OAP and menthol as substrates were detectable. Even after addition of UDPGA, OAP-glucuronidation in newborn mouse liver homogenate was not measurable. In the third week of life 50% and in the fourth week 100% of adult activity are reached. STOREY (1965) observed adult activity for PNP-glucuronidation in newborn mouse liver. Glucuronidation of p-nitrothiophenol in mouse liver is detectable at the 16th day of gestation, ontogenetic development resembles that of PNP. According to this result PNP should belong to the late-fetal cluster as in the rat, but why not CAP?

This question needs reinvestigation, for OAP and estriol GTs were also different in the mouse and bilirubin-GT was inducible by PB immediately before birth, but not in earlier fetal life (BURCHELL and DUTTON, 1975).

Many investigations have been carried out with guinea pigs. UDPGA-DH activity is detectable at birth and increases to reach adult levels around the 15th—20th day of life. Accordingly, UDPGA concentration in fetal liver is low and increases postnatally about 20-fold (FLODGARD and BRODERSEN, 1967). But even with addition of UDPGA fetal and newborn guinea pig liver has no or low glucuronidation capacity; adult activities with bilirubin, OAP, PNP, phenol, phenolphthalein and menthol are reached three weeks after birth. In vivo the guinea pig fetus is able to glucuronidize bilirubin, but only small amounts are excreted in bile. At birth the excretion is already well developed (SCHENKER et al., 1964). Oxazepam glucuronidation is also well developed at birth. Remarkable activities before birth and fast postnatal development of α-naphthol-GT agree with the above mentioned characteristics (ECOBICHON et al., 1978).

Cats have low capacities for glucuronidation in all stages of development (SCHACHTER et al., 1959).

Development of OAP glucuronidation in the rabbit begins in the fetal period, bilirubin glucuronidation develops mainly postnatally (FLINT et al., 1963). As in rats, PNP-GT evidently belongs to the late-fetal cluster, as does salicylate (SCHACHTER et al., 1959).

In man elimination and glucuronidation of bilirubin have been investigated intensively. Until now it has not been clarified whether low activity in UDPGA formation and/or in bilirubin GT are the main factors in newborn hyperbilirubinemia (cp. SCHRÖTER, 1970). Late fetal and newborn liver has low activities for the glucuronidation of 4-methylumbelliferon, salicylamide, acetaminophen, acetanilide, estriol, oxazepam.

Development to adult levels occurs in the first 3—4 months of life, (cp. KLINGER, 1982).

It is of clinical significance that the low glucuronidation activity in newborns can be inhibited by pregnanediol, novobiocine, chloramphenicol and vitamin K (SALAND et al., 1974).


### 2.4.2.1.3. Glycosylations

Analogous to the formation of glucuronides by means of UDP-sugars (glucose, xylose), the biosynthesis of glycosides and UDP-glycosyltransferases play a minor role. Bilirubin glycosylation in rat liver also develops mainly postnatally, the pattern being different from bilirubin-GT development (VAISMAN et al., 1976). The UDP-glycosyltransferase activities in fetal liver including human are low or not detectable (BURCHELL, 1974).

---

### 2.4.2.2. Conjugation with glutathione

Glutathione (GSH) protects cells (mainly liver cells were investigated) against electrophilic chemicals or metabolites. This conjugation is catalysed by GSH-transferase (GSH-T). The conjugates are either excreted with bile or further metabolized to mercapturic acids, which are excreted in bile or urine.

The irreversible binding with electrophilic compounds is not covered here (for reversible binding, storage and transport see section 3).

#### 2.4.2.2.1. Development of GSH and GSSG concentration in liver

GSH concentration in rat liver is low before birth, develops postnatally to reach adult values after eight weeks in females; in males a further increase is observed (NEISH and KEY, 1968). GSSG concentration increases in rat liver more rapidly than GSH, so that the GSH-GSSG ratio decreases with age. In old age GSH levels in rat liver are lowered (KITAHARA et al., 1982) and can be enhanced by cyt c administration. In mouse fetal liver GSH concentrations are very low, at birth about $25-30\%$ of the adult level are found and adult values are reached on the 10th day of life (HART and TIMBRELL, 1979). Acetaminophen and paracetamol are less toxic in fetal and neonatal mice because GSH depletion by these compounds does not occur until about 10 days and adult proportion is reached at on age of 15 days; the development of the ability to detoxify the toxic reactive metabolite(s) precedes the development of the system producing it (them) (HART and TIMBRELL, 1979). In newborn guinea pigs, too, GSH concentration is much lower than in adults.

#### 2.4.2.2.2. Development of glutathione-transferases

GSH-Ts have now been isolated and purified showing overlapping specificities for different substrates, among them epoxides.

GSH-T B has been identified to be ligandin (HABIG et al., 1974). Little is known of the developmental pattern of these different GSH-T-isoenzymes. Newborn rat ligandin activity is only $20\%$ that of adult values which are reached after $4-5$ weeks of life. Within the different GSH-Ts ligandin (GSH-T B) is prevalent in newborn rat liver; however its proportion is $50\%$ of that in adult animals (HALES and NEIMS, 1976).

Thyroxine or cortisol administered at day three or 15 did not exert any influence; phenobarbital induced as early as on the 5th day of life. With styrene and methylcholanthrene epoxide as substrates, adult values were measured immediately before birth (MUKHTAR and BRESNICK, 1976).

With BSP as substrate a steep increase of conjugation activity immediately before birth has been observed; nevertheless the main development occurs

---

postnatally with a 13-fold enhancement to reach adult values after five weeks. No inhibitory factor could be demonstrated in investigations with mixtures of fetal and adult liver. But inhibitory factors are suggested to act in pregnancy as the activity in pregnant liver is lower than in the normal female one, very probably maternal gestagens act as inhibitors for GSH-Ts as for UDPGA-GTs (COMBES and STAKELUM, 1962). In guinea pigs BSP conjugation develops more quickly (SCHENKER et al., 1965).

For GSH-Ts with low activity after birth, precocious development can be induced by glucocorticoid administration to the newborn (the earlier, the more effectively). Corticoid administration to pregnant rats before birth did not induce the precocious development that could be achieved before birth for UDPGA-GTs (MUKHTAR et al., 1979). Sex differences have been observed in rats after sexual maturation, males having higher activities than females after a sex-dependent postnatal development (FUJITA et al., 1985 b).

### 2.4.2.2.3. Development of BSP-clearace

Because of the importance in human medicine as well as in experimental toxicology, postnatal development of BSP-clearance shall be handled separately. BSP-clearance was used as a liver function test in man, including children, long before BSP conjugation with GSH had been detected by GRODSKY et al. (1959, 1961). The specific systems for hepatic uptake, binding, conjugation and excretion have their own developmental pattern, cp. sections No 2.3.1. and 2.3.2. Postnatal development of BSP clearance is the sum of these processes and has been investigated carefully.

Premature babies have a much slower BSP elimination from blood than adults after administration of very low doses (less than 5 mg/kg) and, most evidently, after high doses (WICHMANN et al., 1968).

Maturation is indicated first by the appearance of a steep initial elimination phase. Linear elimination characteristics as in adults are observed after $6-8$ weeks. Age dependent changes in liver blood flow do not play any role, development of BSP-clearance only reflects maturation of liver cell function.

Mature newborns also have a delayed BSP elimination from blood (HERLITZ, 1926). Maximum values are observed in 6-month-old babies, thereafter the elimination velocity declines, especially in old age. BSP retention does not show an age dependence in old age in humans, but the storage capacity for BSP steadily decreases with the progression of the hepatocyte polyploidization in old age (NAKANISHI et al., 1980). The delayed elimination of BSP by prematures is not only due to a low conjugation capacity, but also to a low excretory capacity for free and conjugated BSP (VEST, 1962). The slow BSP elimination in guinea pigs, however, is mainly due to the low conjugation capacity, because BSP-conjugate is readily excreted even by newborns (SCHENKER et al., 1965). In rat, storage capacity plays a predominant role, which is not developed in newborns, and decreases again in old aged rats (LEEUW-ISRAEL et al., 1970). According to KANAI et al., (1985) storage capacity does not change up to an age of 30 months but biliary transport capacity decreases.

Thus BSP clearance develops differently in various species and the rate-limiting processes in the slow postnatal BSP elimination from blood differ from species to species.


### 2.4.2.3. Conjugation with sulfate

By sulfation, ionized half-esters of sulfuric acid are formed, predominantly with phenols, alcohols, and primary amines. ATP is needed for the formation of "active" sulfate: 3-phosphoadenosine-5-phosphosulfate (PAPS), sulfotrans-ferases then catalyze the conjugation reaction. All enzymes are cytosolic. PAPS formation for the biosynthesis of sulfated polysaccharides is evidently high even in the fetus.

The sulfation of some steroids also develops to high activities in fetal life. This has been considered to be a necessary protection against circulating maternal steroids (DUTTON, 1978).

The sulfation of foreign compounds mainly develops postnatally. Higher perinatal activity could be observed if steroid sulfotransferases have an over-lapping specificity for xenobiotics. On the other hand sulfation of some steroids develops mainly postnatally as well. Phenolsulfotransferase activity is very low in fetal rat liver, increase begins shortly before birth and adult levels are reached as early as about the 5th day of life (CARROLL, 1969).

On the contrary, in mice this activity seems to be fully developed at birth and shows a transient maximum in 30-day-old mice. But changes in Km values, probably due to changes of the enzyme protein during development, also indicate a postnatal development (YAFFE et al., 1968).

Sulfation seems to play a compensatory role in newborns: acetaminophen in the human neonate has a lower glucuronidation rate than in adults, but a higher sulfation rate (LEVY et al., 1975). In 7- to 10-year-old children, too, the percentage of salicylamide sulfation is higher than the glucuronidation of this drug (ALAM et al., 1977).


### 2.4.2.4. Conjugation with acetate

Acetylation occurs at aromatic amino or hydrazine groups of exogenous com-pounds. The rate-limiting enzyme, acetyl coenzyme A-arylamine N-acetyl-transferase, is found in many tissues. Using a new radioassay, a first peak of activity was found in the late fetal rat liver, followed by a second peak three days after birth, the fetal activities being higher than after two weeks of life (SONA-WANE and LUCIER, 1975). In rabbits activity increased steadily from fetal levels to adult ones by the second week after birth. Partially purified liver N-acetyltransferase from rapid-acetylating rabbits reaches adult levels at 3−4 weeks.

Premature infants acetylate sulfonamides more slowly than do mature newborns, both distinctly below adult values. The same was found with PAB as substrate, but the acetylation rate was higher than the conjugation rate with glycine (VEST, 1965).

### 2.4.2.5. Conjugation with glycine and other amino acids

These acylation reactions comprise conjugation of aromatic carboxyl groups with the alpha-amino group of amino acids, mainly glycine. They play a major role when glucuronidation capacity is either permanently low, as in the cat family, or temporarily low as in newborns (DUTTON, 1978). The exogenous compound is activated with ATP to react with coenzyme A as endogenous fatty acids. Thereafter the conjugation with the amino acid occurs.

In rats practically no activities in fetal and newborn liver were detectable, very low activities in 5-day-old rats increased to adult values within 30 days. The activity of N-acyltransferase paralleled the excretion of the conjugate (BRANDT, 1964).

The formation of hippuric acid in mice also develops mainly in the postnatal period with a 10-fold increase in the first 30 days of life (GORODISCHER et al., 1971).

Paraaminobenzoic acid (widely used in medicine for a liver function test) is conjugated in the human neonate with low activity, complete conjugation and excretion of the conjugate is observed after the 8th week of life, adult values for elimination are reached in the 6th month (VEST, 1959).

## 3. Extrahepatic biotransformation

### 3.1. Introduction

In adult mammals extrahepatic biotransformation capacity for xenobiotics is generally low (UEHLEKE, 1969; VAINIO and HIETANEN, 1980), but its importance increases with increasing age (SUN and STROBEL, 1986).

More recently biotransformation capacity of the intestine is believed to contribute to the first pass-effect after oral administration (CHHABRA, 1979), but to date investigations on the development of this intestinal "barrier" have been very rare as will be shown subsequently.

---

W. KLINGER

## 3.2. Phase I reactions

### 3.2.1. Oxidative reactions

#### 3.2.1.1. Cytochromes P-450 dependent oxidations
#### 3.2.1.1.1. Cytochrome P-450 concentrations

Cytochrome P-450 concentration in kidneys of adult rats is 15%, in lung 10% of that in liver (KATO, 1966).

In rabbit lung and small intestine and in pig lung, kidney and intestinal mucosa, P-450 concentration develops in parallel to liver P-450 concentration but on a lower level. $b_5$ concentration shows a barely detectable age course, if measurable at all, in rat kidney, rabbit lung and pig lung, kidney and intestinal mucosa.

NADPH-P-450 (cyt c)-reductase activity shows a distinct postnatal development in rabbit lung and small intestine (cp. KLINGER, 1982).

#### 3.2.1.1.2. Hydroxylations of aromatics and aliphatics

In general these activities develop postnatally in all organs investigated, partially in parallel to the development in liver, but on a far lower activity level. Nevertheless AHH especially plays an important role in the activation of PAH to ultimate carcinogens in organs other than liver, e.g. in lung (FOUTS and DEVEREUX, 1972) and skin (MUKHTAR and BICKERS, 1983).

#### 3.2.1.1.3. N- and O-dealkylations

Being a special kind of P-450 dependent hydroxylation, dealkylation develops in extrahepatic tissues, if measurable at all, mainly postnatally and on a lower level in comparison to liver (e.g. BEND et al., 1975).

#### 3.2.1.1.4. N-Oxidation

At optimum pH values rabbit lung N-oxidase activity was remarkably higher than in liver in all age groups, at physiological pH values in the same activity range (BEND et al., 1975).

#### 3.2.1.1.5. Oxidative deamination

The different MAO isoenzymes differentiated by various substrates and inhibitors have been investigated in various extrahepatic tissues and in distinct areas of brain. In all species and tissues investigated, the different MAO isoenzymes have different developmental patterns, which can be described but without the possibility of establishing general rules. These MAO activities are only similar in that they develop mainly postnatally.

### 3.2.1.2. Dehydrogenation

In contrast to the distinct age course of liver aldehyde dehydrogenase, this enzyme showed age dependence in the central nervous system of rats but was evidently correlated to the voluntary ethanol consumption in different rat strains (AMIR, 1978).

### 3.2.1.3. Reduction

Nitro- and azoreduction show the typical postnatal development in pig, the activities in all age groups being distinctly below those in liver (SHORT et al., 1972). The azo reduction activity in lung however comes up to about 50% of liver activity if related to fresh weight. If related to microsomal protein, lung is nearly as active as liver.

### 3.2.2. Hydrolysis

### 3.2.2.1. Hydrolysis of esters

In rat kidney hydrolysis is detectable as early as three days before birth, activity increases sharply one day before birth, but after birth the increase is not so intensive until the fourth week of age. Thereafter a distinct decline follows and adult values are below those of newborns (SCHWARK and ECOBICHON, 1969). In the intestinal mucosa the main development occurs after the 20th day of life (MASNEROVA et al., 1966).

In fetal guinea pigs the esterase activity in plasma, liver and kidney is extremely low, it increases rapidly in the perinatal period. About three days after birth, adult levels are reached (CHOW and ECOBICHON, 1974). In pig the hydrolysis of procaine in lung, kidney and intestine is low or not detectable, it increases postnatally in the first 6 – 10 weeks of life to reach adult values, distinctly below hepatic activities (SHORT et al., 1972).

In man plasma pseudocholinesterase and arylesterase increase during the first year of life, then adult values are almost reached (ECOBICHON and STEPHENS, 1973).

Increasing hydrolytic activity for procaine and paraoxon was observed in plasma of prematures, newborns, one-year-old children and adults. Nearly the same age course was observed for hydrolysis of acetylosalicylic acid (WINDORFER et al., 1974).

---

W. KLINGER

### 3.2.2.2. Hydrolysis of amides

Isocarboxacide hydrolysis was detectable neither in fetal liver nor kidney from rats. It was measurable in liver, but not in kidney of newborn rats. The increase in activity continued in liver up to the 60th day of life, it stopped at the 30th day of life in kidney (SATOH and MOROI, 1971).

### 3.2.2.3. Hydrolysis of epoxides

Epoxide hydrolase (EH) development has been investigated in guinea pig intestine (BEND et al., 1975) and in lung, kidney and intestinal mucosa of guinea pigs, rats and rabbits in comparison to liver. Despite significant differences, the same postnatal development pattern as for P-450-dependent monooxygenases can be observed with one difference: in guinea pigs weeks instead of three days are needed to reach adult levels. Rat testis also showed appreciable activity. Most active are the spermatogenetic cells with a very distinct age course (MUKHTAR et al., 1978).

## 3.3. Phase II reactions

### 3.3.1. Glucuronidation

In rats the glucuronidation capacity of the gut is remarkable, in 1-month-old rats o-aminophenol glucuronidation has higher activities in gut than in liver. LUCIER and DANIEL (1977) also measured rather high GTs for non-steroid (late-fetal) substrates in kidney and uterus, whereas steroid-GTs were barely detectable. The development patterns for the different GTs in these various organs and tissues remain to be investigated in detail, but the development in clusters is very probable. Glucocorticoids trigger late-fetal GT cluster not only in liver, but also in lung, kidney and small intestine (WISHART and DUTTON, 1977).

### 3.3.2. Conjugation with glutathione

Very few data for extrahepatic conjugation with GSH are available. In rat kidney BSP binding capacity is low and nearly no age course is detectable (MÜLLER and KLINGER, 1974). Low GSH concentration in lung might be rate limiting for epoxide conjugation which develops postnatally as in liver but reaches only 30% of the adult liver level.

In guinea pig kidney, lung, and small intestine, the adult values are about

10–30% that of adult liver, but these values are already fully developed in lung and small intestine at birth and only double postnatally in kidney (BEND et al., 1975).

### 3.3.3. Conjugation with sulfate

In rat and guinea pig the low phenolsulfotransferase activity in kidney and lung increases in the first two weeks of life more rapidly than that of steroid-sulfotransferase, adult values reaching about 1/7 that of liver (PULKKINEN, 1966). On the other hand the activity of sulfotransferase in small intestine of 13-day-old suckling rats is higher than in adults.

### 3.3.4. Conjugation with acetate

N-acetyltransferase activity increases in rabbit gut in parallel to liver activities from low fetal to adult values by the second week of life, whereas kidney does not reveal a development in the perinatal period up to the second week postnatally. In lung a decrease after birth and an increase from the 7th to the 14th day of life has been observed without reaching the high fetal activities (SONAWANE and LUCIER, 1975).

## 4. Influence of age on pharmacokinetics

Both biotransformation and renal excretion determine elimination half-life time from blood and from other organs, tissues or compartments. Since the fatal cases after treatment of prematures with chloramphenicol, the peculiarities of drug elimination in newborns have become evident to the medico-scientific public, and it was a pediatrician, H. DOST, who created the term pharmacokinetics and who provided the basis for this new discipline (DOST, 1957). In the early sixties the term "Developmental Pharmacology" was employed, most investigations were concerned with pharmacokinetics and the fundamental processes of it. Despite rather high activities of the human fetus for the biotransformation of foreign compounds, investigated mainly in the first trimenon (PELKONEN, 1980), the human neonate has prolonged half-life times in blood for most drugs, maturation occurs during the first year of life, maximum values for the elimination are reached in schoolchildren, aged about 12–16 years, and thereafter a tendency to prolongation of half-lives can be observed (MORSELLI, 1977; ASSAEL, 1982). For many drugs the postnatal development of elimination velocity reflects the development of biotransformation, but also changes in liver blood flow may be observed as has been shown for the hepatic clearance of propranolol (IWAMOTO et al., 1985).

# 5.   Regulation of development

In man and also in laboratory animals with long gestation periods (e.g. guinea pigs and rabbits) an early differentiation during pregnancy has been observed, whereas in the rat differentiation and development occur mainly after birth. Possible fetal and birth-correlated trigger and control mechanisms have been discussed by PELKONEN (1980).

The biogenetic rule that ontogenesis repeats briefly phylogenesis (Ernst HAECKEL, 1866), holds, in general, true for ontogentic development of biotransformation reactions (BRODIE and MAICKEL, 1961). The basic mechanisms which control the changes in old age are in principle unknown. Changes in hormone production, release and efficacy are considered to be part of ageing, but not the reason. Special interrelations between hormone and drug metabolism in the fetal and perinatal period directly influence development of drug metabolism in the fetus and the newborn; maternal progesterone inhibits the development of hydroxylation activity in the fetal and perinatal liver.

Glucocorticoids trigger the development not only of GTs, but also of P-450, and precocious development of P-450 in neonatal rat liver is possible by glucocorticoid treatment (LEAKEY and FOUTS, 1979). But then the question arises what is the mechanism to trigger ACTH and glucocorticoid release. But as there are pronounced differences in the trigger-capacity of different glucocorticoids (MÜLLER, personal communication) and the most effective glucocorticoid dexamethasone is used in extremely high doses, this so-called trigger-effect might be rather a pharmacological inducer-effect (CRESTEIL, 1987).

The postnatal decrease of genetic redundancy is at present a hypothesis to explain ageing, but the mechanism remains obscure. We interpreted postnatal development and ageing as a genetically controlled repression-phenomenon, consecutively the postnatal development with increasing activities to be a de-repression of the responsible gene loci and the loss of activities in old age to be an increasing repression. This hypothesis was formulated in connection with investigations on interrelations of development and enzyme induction. It was shown that low activities of various enzymes in infantile animals can be elevated by inducers up to the adult level. But after withdrawal of the inducer, these precocious adult values are not stable, they go back to the values which correspond to the age of the animals, and normal development continues (KLINGER et al., 1968).

In general, induction of various biotransformation reactions by different inducer types distinctly depends on the development stage of the animal or human being. But the age-dependent modification of enzyme induction is also dependent on the tissue with profound differences of sensitivity and development patterns (SONG et al., 1986; SUN et al., 1986; SUN and STROBEL, 1986). Moreover sexual hormones are not only endogenous substrates of phase I and phase II reactions, but also modify the age course and the induction response (DANNAN et al., 1986).

NEBERT postulates some form of temporal control, associated with the Ah locus (GUENTHNER and NEBERT, 1978; NEBERT, 1979; KAHL et al., 1980). Whether the Ah complex includes temporal genes, remains open. The temporal control may influence either the structural gene expression directly or the expression of regulatory genes indirectly. So this temporal control of biotransformation activity and capacity would be a special case, because temporal genes have already been characterized for other enzymes (cp. KAHL et al., 1980). But the biological clock, the time signal, the regulation, the control and mechanism of this temporal control remains an open field of investigation.

According to most recent investigations oncogenes could be essentially involved in regulation and control of developmental processes including ageing (GILDEN and RICE, 1983; NISHIMURA and SEKIYA, 1987):

— Both drug metabolism and tumor incidence rate are strongly influenced by the stage of postnatal development.

— Carcinogens such as MC also act as inducers.

— Enhanced formation of e.g. protein kinases and growth factors is an attribute of both carcinogenesis and enzyme induction, so it may be speculated that an activation of oncogenes is not only the initial step of the multi-step transformation in chemical carcinogenesis, but may also play a role in ontogenetic development of biotransformation, in its induction and in changes of expression of biotransformation enzymes by hormones and xenobiotics in distinct sensitive phases (so-called imprinting and booster-phenomena), so that the first contact with an effective xenobiotic can be compared with the initiation step in carcinogenesis.

Finally it must be emphasized that all these theories and hypotheses are supported by some experimental findings, but taken together this field is in the stage of speculation.

# 6. References

ALAM, S. N., R. J. ROBERTS, and L. J. FISCHER (1977), J. Pediat. 90, 130—135.
AMIR, S. (1978), Psychopharmacology 57, 97—102.
ANKERMANN, H. (ed.) (1973), Entwicklungspharmakologie. VEB Verlag Volk und Gesundheit Berlin.
ASSAEL, B. M. (1982), Pharmac. Ther. 18, 159—197.
ASTRACHANZEWA, L. Z. 1977 (ed., transl.: W. KLINGER), Geriatrische Pharmakologie, VEB Verlag Volk und Gesundheit Berlin.
ATLAS, S. A., A. R. BOOBIS, J. S. FELTON, S. S. THORGEIRSSON, and D. W. NEBERT (1977), J. biol. Chem. 252, 4712—4721.
AXELROD, J. (1956), J. Pharmacol. 117, 322—330.

BANDIERA, S., D. E. RYAN, W. LEVIN, and P. E. THOMAS (1986), Arch. Biochem. Biophys. **248**, 658—676.
BARTH, A. and W. KLINGER (1979), Wiss. Ztschr. Humboldt-Univ. (Berlin) **37**, 413—414.
BARTH, A., W. KLINGER, and H. HOPPE (1986), Arch. Int. Pharmacodyn. Ther. **283**, 16—29.
BARTH, A., W. KLINGER, and H. HOPPE (1987), Arch. int. Pharmacodyn. Ther. **283**, 16—29.
BEND, J. R., M. O. JAMES, T. R. DEVEREUX, and J. R. FOUTS (1975), in: Basic and Therapeutic Aspects of Perinatal Pharmacology (P. L. MORSELLI, S. GARATTINI, F. SERENI, eds.), Raven Press New York 229—243.
BERLIN, E., P. G. KLIMAN, M. A. KHAN, and G. R. HENDERSON (1984), Effect of age and sex on lipoprotein fluidity. In: First colloquium in biological sciences, SCOTT, W. N., F. L. STRAND (Eds.), New York Academy Sciences (New York) 116—118.
BIRNBAUM, L. S. and M. B. BAIRD (1979), Chem. Biol. Interact. **26**, 254—256.
BLATCHFORD, D., M. HOLZBAUER, D. G. GRAHAME-SMITH, and M. B. H. YOUDIM (1976), Br. J. Pharmac. **54**, 251—252.
BRÄUNLICH, H. (1968), Arch. int. Pharmacodyn. Ther. **164**, 387—396.
BRANDT, I. K. (1964), Developm. Biol. **10**, 202—215.
BRODIE, B. B., R. P. MAICKEL (1962), Proc. ist Internat. Pharmac. Meet. **6**, 299—324.
BURCHELL, B. (1974), J. Steroid Biochem. **5**, 261—267.
BURCHELL, B. and G. J. DUTTON (1975), Biol. Neonate **26**, 122—128.
CAGEN, S. Z. and J. E. GIBSON (1979), J. Pharmacol. exp. Ther. **210**, 15—21.
CAMERON, G. R. and W. A. E. KARUNARATNE (1936), J. Pathol. Bacteriol. **42**, 1—21.
CARROLL, J. (1969), Am. J. clin. Nutr. **22**, 978—985.
CHHABRA, R. S. (1979), Environm. Health. Perspect. **33**, 61—69.
CHOW, A. Y. K. and D. J. ECOBICHON (1974), Biol. Neonate **25**, 23—30.
CLOZEL, M., K. BEHARRY, and J. V. ARANDA (1986), Biol. Neonate **50**, 83—91.
COHEN, G. M. and G. J. MANNERING (1974), Drug Metab. Dispos. **2**, 285—292.
COMBES, B. and G. S. STAKELUM (1961), J. clin. Invest. **40**, 1030—1031.
COMBES, B. and G. S. STAKELUM (1962), J. clin. Invest. **41**, 750—757.
CRESTEIL, T., C. BEAUNE, C. CELIER, J. P. LEROUX, and F. P. GUENGERICH (1986), J. Pharmacol. exp. Ther. **236**, 269—276.
CRESTEIL, T. (1987), BioEssays **7**, 120—124.
DANNAN, G. A., D. J. PORUBEKM, S. D. NELSON, D. J. WAXMAN, and F. P. GUENGERICH (1986), Endocrinol. **118**, 1952—1960.
DAVID, H. (1985), The Hepatocyte, Development, Differentiation, and Ageing. VEB Gustav Fischer Verlag Jena.
DAWKINS, N. J. R. (1963), J. Pathol. Bacteriol. **85**, 189—196.
DELPECH, I., L. KIFFEL, J. MAGDALOU, J. C. ANDRE, and G. SIEST (1982), 8th Europ. Workshop on Drug Metabolism, Sart Tilman, Belgium, Sept. 5—9, Abstr. No 41, p. 73.
DOST, F. H. (1957), Der Blutspiegel, VEB Georg Thieme Leipzig.
DUTTON, G. J. (1978), A. Rev. Pharmac. Toxicol. **18**, 17—35.
ECOBICHON, D. J. and D. S. STEPHENS (1973), Clin. Pharmac. Ther. **14**, 41—47.
ECOBICHON, D. J., R. W. DYKEMAN, and M. M. HANSELL (1978), Can. J. Biochem. **56**, 738—745.
ESTABROOK, R. W. (1971), in: Handbook of Experimental Pharmacology XXVIII/2, (B. B. BRODIE, J. R. GILETTE, H. S. ACKERMAN, eds.), Springer Berlin—Heidelberg—New York 264—284.
FLODGARD, H. J. and R. BRODERSEN (1967), Scand. L. clin. Lab. Invest. **19**, 149—155.
FOUTS, J. R. and R. H. ADAMSON (1959), Science **129**, 897—898.
FOUTS, J. R. (1962), Proc. 1st Int. Pharmacol. Meet, Stockholm 1961 **6**, Pergamon Press Oxford 257—271.

FOUTS, J. R. and T. R. DEVEREUX (1972), J. Pharmac. exp. Ther. **183**, 458—468.

FUJITA, S., T. UESUGI, H. KITAGAWA, T. SUZUKI, and K. KITANI (1982), in: Liver and Ageing, (K. KITANI, ed.), 55—72.

FUJITA, S., H. KITAGAWA, M. CHIBS, T. SUZUKI, M. OHTS, and K. KITANI (1985a), Biochem. Pharmacol. **34**, 1861—1864.

FUJITA, S., H. KITAGAWA, H. ISHIZAWA, T. SUZUKI, and K. KITANI (1985b), Biochem. Pharmacol. **34**, 3891—3894.

GIADHELLI, C. M. and C. J. OMIECINSKI (1987), Molec. Pharmac. **31**, 477—784.

GILDEN, R. V. and N. R. RICE (1983), Carcinogenesis **4**, 791—794.

GRODSKY, G. M., J. V. CARBONE, and R. FRANSKA (1959), J. Clin. Invest. **38**, 1981 to 1988.

GRODSKY, G. M. (1961), Praxis **50**, 1299—1301.

GUENGERICH, F. P., G. A. DANNAN, S. T. WRIGHT, M. V. MARTIN, and L. S. KAMINSKY (1982), Xenobiotica **12**, 701—716.

GUENTHNER, T. M. and D. W. NEBERT (1978), Eur. J. Biochem. **91**, 449—456.

HABIG, W. H., M. J. PABST, G. FLEISCHNER, Z. GATMAITEN, I. M. ARIAS, and W. B. KAKOBY (1974), Proc. Natl. Acad. Sci. U.S.A. **71**, 3879—3882.

HAECKEL, E. (1866), Generelle Morphologie der Organismen, Bd. II: Allgemeine Entwicklungsgeschichte der Organismen, Berlin.

HAHN, H. K. and R. E. BURCH (1983), Alcohol. Clin. Exp. Res. **7**, 299—301.

HALES, S. F. and A. H. NEIMS (1976), Biochem. J. **160**, 231—236.

HART, J. G. and J. A. TIMBRELL (1979), Biochem. Pharmac. **28**, 3015—3017.

HEIMANN, G., B. ROTH, and E. GLADTKE (1977), Klin. Wschr. **55**, 451—456.

HERLITZ, C. W. (1926), Acta paed. **6**, 214—224.

HOLTZMAN, J. I. (1979), Pharmac. Ther. **4**, 601—627.

IWAMOTO, K., J. WATANABE, K. ARAKI, N. DEGUCHI, and H. SUGIYAMA (1985), J. Pharm. Pharmacol. **37**, 466—470.

JAMES, J., J. TAS, K. S. BOSCH, A. J. P. DEMSER, and H. C. SCHUYT (1979), Eur. J. Cell. Biol. **19**, 222—226.

JONDORF, O. R., R. P. MAICKEL, and S. B. BRODIE (1958), Biochem. Pharmacol. **1**, 352—354.

KAHL, G. F., D. E. FRIEDERICI, S. W. BIGELOW, A. B. OKEY, and D. W. NEBERT (1980), Devel. Pharmac. Ther. **1**, 137—162.

KANAI, S., K. KITANI, S. FUJITA, and H. KITAGAWA (1985), Arch. Gerontol. Geriatr. **4**, 73—85.

KARUNAIRATNAM, M. C., L. M. KERR, and G. A. LEVY (1949), Jap. J. Pharmacol. **11**, 31—36.

KATO, R. (1966), J. Biochem. **59**, 574—583.

KATO, R. and A. TAKANAKA (1968), Jap. J. Pharmac. **18**, 381—396.

KELLER, W. (1842), Liebigs Ann. **43**, 108.

KITAGAWA, H., S. FUJITA, T. SUZUKI, and K. KITANI (1985), Biochem. Pharmacol. **34**, 579—581.

KITAHARA, A., T. EBINA, T. ISHIKAWA, Y. SOMA, K. SATO, and S. KANAI (1982), in: Liver and Ageing, (K. KITANI, ed.), Elsevier Biomedical Press, New York 135—143.

KLAASSEN, C. D. (1975), J. Pharmacol. exp. Ther. **195**, 311—319.

KLEEBERG, U., G. GROHMANN, R. VOLKMANN, H. STEINERT, and W. KLINGER (1979), Pol. J. Pharmac. Pharm. **31**, 675—681.

KLEEBERG, U. (1987), Induktion Cytochrom P-450-abhängiger Biotransformationsreaktionen durch 3-Methylcholanthren. Mechanismen und postnatale Entwicklung. Thesis, University Jena.

KLINGER, W., T. KUSCH, A. NEUGEBAUER, F.-K. SPLINTER, and H. ANKERMANN (1968), Acta biol. med. germ. **21**, 257—269.

KLINGER, W. (1970), Arch. int. pharmacodyn. Ther. **184**, 5—18.
KLINGER, W., D. MÜLLER, F. REICHENBACH, U. KLEEBERG, H. LÜBBE, and H. REIN (1975), in: Basic and Therapeutic Aspects of Perinatal Pharmacology, (P. L. MORSELLI, S. GARATTINI, F. SERENI, eds.), Raven Press New York 255—275.
KLINGER, W. and D. MÜLLER (1976), Environm. Health Perspect. **18**, 13—23.
KLINGER, W. and D. MÜLLER (1977), Acta biol. med. germ. **36**, 1149—1159.
KLINGER, W. (1977), Development of Drug Metabolizing Enzymes, in: Drug Disposition During Development, (P. L. MORSELLI, ed.), Spectrum New York, pp. 71—88.
KLINGER, W., D. MÜLLER, and U. KLEEBERG (1979), in: The induction of drug metabolism, R. W. ESTABROOK and E. LINDENLAUB (Eds.), F. K. Schattauer Verlag Stuttgart—New York, 517—544.
KLINGER, W., F. JAHN, S. SPIEGLER, and K. D. SPIEGLER (1980), Zbl. Pharm. **119**, 1033—1038.
KLINGER, W. (1982), Pharmac. Ther. **16**, 377—429.
KLINGER, W., T. DEVEREUX, and J. R. FOUTS (1986), Arch. Toxicol. Suppl. **8**, 469—473.
KROCHMANN, E. (1974), Med. Diss. Univ. Hamburg.
KUENZIG, W., J. J. KAMM, M. BOUBLIK, F. JENKINS, and J. BURNS (1974), J. Pharmac. Exp. Ther. **191**, 32—44.
KUO, C.-H. and J. B. HOOK (1980), Life Sci. **27**, 2433—2438.
LAMARTINIERE, C. A., C. S. DIERINGER, E. KITA, and G. W. LUCIER (1979), Biochem. J. **180**, 313—318.
LEAKEY, J. and J. R. FOUTS (1979), Biochem. J. **182**, 233—235.
LEEUW-ISRAEL, F., C. F. HOLLANDER, and J. M. ARP-NEEFJES (1970), J. Gerontol. **24**, 140—142.
LEVI, A. J., Z. GATMAITAN, and I. M. ARIAS (1969), Lancet **II**, 139—140.
LEVI, A. J., Z. GATMAITAN, and I. M. ARIAS (1970), New. Engl. J. Med. **283**, 1136—1139.
LEVY, G., N. N. KHANNA, D. V. SODA, O. TSUZUKI, and L. STERN (1975), Pediatrics **55**, 818—825.
LEWINSOHN, R., V. GLOVER, and M. SANDLER (1980), Biochem. Pharmac. **29**, 1221—1230.
LIEBER, C. S. and L. M. DECARLI (1970), J. biol. Chem. **245**, 2505—2512.
LINDAHL, R. (1977), Biochem. J. **164**, 119—123.
LUCIER, G. W. and O. S. DANIEL (1977), J. Steroid Biochem. **8**, 867—872.
LUCIER, G. W., E. M. K. LUI, and C. A. LAMARTINIERE (1979), Environm. Hlth. Perspect. **29**, 7—16.
LUM, P. Y., S. WALKER, and C. IOANNIDES (1985), Toxicology **35**, 307—317.
MASNEROVA, M., O. KOLDOVSKY, and K. KUBAT (1966), Experientia **22**, 518—519.
MASTERS, B. S. S., J. BARON, W. E. TAYLOR, E. L. ISAACSON, and J. LoSPALUTTO (1971), J. biol. Chem. **246**, 4143—4150.
MENDE, T. J. and L. VISMONTE (1967), Gerontologie **13**, 165—172.
MORSELLI, P. L. (1977), Drug Disposition During Development, Spectrum, New York.
MUKHTAR, H. and E. BRESNICK (1976), Cancer Res. **36**, 937—940.
MUKHTAR, H., I. P. LES, G. L. FOUREMAN, and J. R. BEND (1978), Chem. biol. Interact. **22**, 153—165.
MUKHTAR, H., J. E. A. LEAKEY, T. H. ELMAMLOUK, J. R. FOUTS, and J. R. BEND (1979), Biochem. Pharmac. **28**, 1801—1803.
MUKHTAR, H. and D. R. BICKERS (1983), Drug Metab. Dispos. **11**, 562—567.
MÜLLER, D., F. REICHENBACH, and W. KLINGER (1971), Acta biol. med. germ. **27**, 605—609.
MÜLLER, D. and W. KLINGER (1974), Acta biol. med. germ. **32**, 211—218.
MÜLLER, D., H. LÜBBE, and W. KLINGER (1975), Acta biol. med. germ. **34**, 1333—1337.
MÜLLER, D. and W. KLINGER (1977), Acta biol. med. germ. **36**, 1161—1166.
MÜLLER, D. and W. KLINGER (1978), Pharmazie **33**, 398—400.

MÜLLER, D., K. GREILING, H. GREILING, and W. KLINGER (1983), Biomed. Biochim. Acta **42**, 981—987.

NAKANISHI, K., M. FUKUDA, and S. FUJITA (1980), Exp. Gerontol. **15**, 103—112.

NEBERT, D. W. (1979), Genetic Aspects of Enzyme Induction by Drugs and Chemical Carcinogens. in: The Induction of Drug Metabolism. (R. W. ESTABROOK, E. LINDEN-LAUB, eds.), Symposium Medica Hoechst 14, F. K. Schattauer Verlag Stuttgart 419 to 452.

NEISH, W. J. P. and L. KEY (1968), Biochem. Pharmac. **17**, 497—502.

NICAK, A. (1971), Exp. Gerontol. **6**, 111—114.

NISHIMURA, S. and T. SEKIYA (1987), Biochem. J. **243**, 313—327.

PALADE, G. and K. PORTER (1967), Science **156**, 106—110.

PELKONEN, O. (1980), Pharmac. Ther. **10**, 261—281.

PIKKARAINEN, P. (1971), Life Sci. **10**, 1359—1364.

PLAYER, T. J. and A. A. HORTON (1978), FEBS Lett. **89**, 103—106.

PULKKINEN, M. O. (1966), Acta physiol. scand. **66**, 115—119.

RAMPERSAUD, A., S. BANDIERA, D. E. RYAN, W. LEVIN, P. E. THOMAS, and F. G. WALZ (1987), Arch. Biochem. Biophys. **252**, 145—151.

SALAND, J., H. MCNAMARA, and M. I. COHEN (1974), J. Pediatrics **85**, 271—275.

SANDSTRÖM, B. (1974), Acta hepato-gastroenterol. **19**, 170—172.

SASAME, H. A., J. A. CASTRO, and J. R. GILLETTE (1968), Biochem. Pharmacol. **17**, 1759—1768.

SATOH, T. and K. MOROI (1971), Arch. int. Pharmacodyn. Ther. **192**, 128—134.

SCHACHTER, D., D. J. KASS, and T. J. LANNON (1959), J. biol. Chem. **234**, 201—205.

SCHENKER, S. and R. SCHMID (1964), Proc. Soc. Exp. Biol. Med. **115**, 446—448.

SCHENKER, S., N. H. DAWBER, and R. SCHMID (1964), J. Clin. Invest. **43**, 32—39.

SCHENKER, S. and J. F. O'DONNEL (1965), Amer. J. Physiol. **208**, 628—632.

SCHENKER, S., J. GOLDSTEIN, and B. COMBES (1965), Amerc. J. Physiol. **208**, 563—572.

SCHMIEDEBERG, O. and H. MEYER (1879), Hoppe-Seylers Z. Physiol. Chem. **3**, 422.

SCHMUCKER, D. L., A. L. JONES, and E. S. MILLS (1974), J. Gerontol. **29**, 506—513.

SCHMUCKER, D. L. (1985), Ageing and Drug Disposition: An Update. Pharmacol. Rev. **37**, 133—148.

SCHRÖTER, W. (1970), Ergebn. Inn. Med. u. Kinderheilk. **29**, 220—227.

SCHWARK, W. S. and D. J. ECOBICHON (1969), Biochem. Pharmac. **18**, 915—921.

SHORT, C. R. and L. E. DAVIS (1970), J. Pharmac. exp. Ther. **174**, 185—196.

SHORT, R. D. and J. E. GIBSON (1971), Toxicol. appl. Pharmac. **19**, 103—110.

SHORT, C. R., M. D. MAINES, and B. A. WESTFALL (1972), Biol. Neonate **21**, 54—68.

SONAWANE, B. R. and G. W. LUCIER (1975), Biochim. Biophys. Acta **411**, 97—105.

SONG, B.-J., H. V. GELBOIN, S.-S. PARK, O. S. YANG, and F. J. GONZALEZ (1986), J. biol. chem. **261**, 16689—16697.

SOYKA, L. F., L. GYERMEK, and P. CAMPBELL (1970), J. Pharmac. exp. Ther. **175**, 276—282.

STALHANDSKE, T., P. SLANINA, H. TJÄLVE, E. HANSSON, and C. G. SCHMITERLÖW (1969), Acta pharmac. toxicol. **27**, 363—380.

STOREY, I. D. E. (1965), Biochem. J. **95**, 209—214.

SUN, J. and H. W. STROBEL (1986), Exp. Gerontol. **21**, 523—534.

SUN, J., P. P. LAU, and H. W. STROBEL (1986), Exp. Gerontol. **21**, 65—73.

UEHLEKE, H. (1964), Proc. Europ. Soc. Stud. Drug Tox. **4**, 140—144.

UEHLEKE, H. (1969), Proc. Europ. Soc. Stud. Drug Tox. **10**, 94—100.

UEHLEKE, H. (1973), Drug Metab. Dispos. **1**, 299—313.

UEHLEKE, H. and T. WERNER (1975), in: Basic and Therapeutic Aspects of Perinatal Pharmacology, (P. L. MORSELLI, S. GARATTINI, F. SERENI, eds.), Raven Press New York, 277—287.

---

VAINIO, H. and E. HIETANEN (1980), in: Concepts in Drug Metabolism, (P. JENNER, B. TESTA, eds.), Marcel Dekker New York—Basel, 251—284.
VAISMAN, S. L., K. S. LEE, and L. M. GARTNER (1976), Pediatr. Res. 10, 967—971.
VAN DEN BERG, A. P., J. NORDHOEK, E. M. SAVENIJE-CHAPEL, and E. KOOPMAN-KOOL (1978), Biochem. Pharmacol. 27, 627—633.
VEST, M. (1958), Arch. Dis Childh. 33, 473—476.
VEST, M. (1959), Suppl. Ann. paed. 69.
VEST, M. (1962), J. clin. Invest. 41, 1013—1020.
VEST, M. (1965), Biol. Neonate 8, 258—266.
VOGEL-BINDEL, U., P. BENTLEY, and F. OESCH (1982), Eur. Biochem. 126, 425—431.
WHIPPLE, G. H. (1912), J. Exp. Med. 15, 259—269.
WICHMANN, H. M., H. RIND, and E. GLADTKE (1969), Ztschr. Kinderheilk. 103, 262 to 276.
WIEBEL, F. J. and H. V. GELBOIN (1975), Biochem. Pharmacol. 24, 1511—1515.
WILLIAMS, R. T. (1959), Detoxication Mechanisms 2nd ed., Chapman and Hall London.
WILSON, P. D., R. WATSON, and D. L. KNOOK (1982), Gerontol. 28, 32—43.
WINDORFER, A., W. KUENZER, and R. URBANEK (1974), Z. Kinderheilk. 113, 33—12.
WIRTH, P. J. and S. S. THORGEIRSSON (1978), Biochem. Pharmac. 27, 601—603.
WISHART, G. T. and G. J. DUTTON (1977), Biochem. J. 168, 507—511.
WISHART, G. J. (1978), Biochem. J. 114, 485—489.
WOLFF, T. and F. P. GUENGERICH (1987), Biochem. Pharmac. 36, 2581—2588.
WOODS, J. S. and R. L. DIXON (1970), Biochem. Pharmacol. 19, 1951—1954.
WOODS, J. S. (1976), Biochem. Pharmacol. 25, 2147—2152.
YAFFE, S. J., J. KRASNER, and C. S. CATZ (1968), Ann. N. Y. Acad. Sci. 151, 887—899.
YANG, C. S., F. S. STRICKHART, and L. P. KICHA (1978), Biochem. Biophys. Acta 509, 326—337.
YASUMORIN, T., S. KAWANO, K. NAGATA, M. SHIMADA, Y. YAMAZOE, and R. KATO (1987), J. Biochem. 102, 1075—1082.
YEARY, R. A., D. GERKEN, and D. R. DAVIS (1973), Biol. Neonate 23, 371—380.

# Chapter 6

# Postoxidation Enzymes

A. LANGNER, H.-H. BORCHERT, and S. PFEIFER

**Abbreviations**

| | |
|---|---|
| AHAT | Arylhydroxamic acid-N,O-acetyltransferase |
| AT | Acetyltransferase |
| COMT | Catecholamine-O-methyltransferase |
| HIOMT | Hydroxyindole-O-methyltransferase |
| HMT | Histamine-N-methyltransferase |
| MT | Methyltransferase |
| PAP | 3'-Phosphoadenosine-5'-phosphate |
| PAPS | 3'-Phosphoadenosine-5'-sulfatophosphate |
| PNMT | Phenolethanolamine-N-methyltransferase |
| SAH | S-Adenosyl-L-homocysteine |
| SAM | S-Adenosyl-L-methionine |
| ST | Sulfotransferase |
| UDPGA | Uridine-5'-diphospho-D-glucuronic acid |

# 1. Introductory remarks

The aim of this chapter is to introduce aspects of conjugation reactions and to provide the reader with important data about structure-function relationships of conjugating enzymes not included in other chapters of this volume.

At the same time that the 'Chemical Defence Hypothesis' of C. P. SHERWIN was in its hey-day, QUICK (1927) wrote 'the idea is no longer tenable that these conjugations are more or less unimportant mechanisms concerned solely with the detoxication powers of the organism. If these synthetic processes are looked on as normal reactions applied to a foreign substance, it is possible to perceive how the study of conjugation may help solve various problems'. It is now beaming clear how true these words are, and it is to be anticipated that the next few years will see substantial insights in these directions.

# 2. Xenobiotic and endobiotic viewpoints of conjugation reations

The conjugation reactions of drug metabolism are a group of synthetic reactions in which a foreign compound, or a metabolite thereof, is covalently linked with an endogenous moiety to give a product known as a conjugate.

Since the discovery of glycine conjugation (URE, 1841; KELLER, 1842), sulfation (BAUMANN and MUNK, 1876), glucuronidation (SCHMIEDEBERG and MEYER, 1879), methylation (HIS, 1887) and acetylation (COHN, 1893), further conjugation reactions, their biochemical mechanisms and their importance in biotransformation of foreign compounds have been discovered.

**Table 1.** The principal conjugation reactions of drug metabolism

| Reaction | Substrates | Activated intermediate |
|---|---|---|
| *Reactions involving activated endogenous agents* | | |
| Glucuronidation | alcohols, phenols, carboxylic acids, amines, hydroxylamines, thiols, C—H-acidic compounds | UDP-glucuronic acid |
| Sulfation | alcohols, phenols, amines, hydroxylamines | PAPS |
| Methylation | phenols, amines, thiols, carboxylic acids | ϱ-adenosylmethionine |
| Acetylation | primary amines, hydrazides, hydrazines, amides | acetyl Co A |
| ϱ-transfer to cyanide | cyanides | thiosulfate ion |
| *Reactions involving activation of the xenobiotic* | | |
| Glutathione conjugation | metabolites of compounds with aromatic or aliphatic double bonds, halogenated aliphatic or cyclic hydrocarbons, α,β-unsaturated carbonyl compounds | frequently electrophilic metabolites such as epoxides |
| Conjugation with amino acids or amines | aliphatic and aromatic carboxylic acids | xenobiotic acyl Co A |

The principal conjugation reactions of drug metabolism and the corresponding substrate types are listed in Table 1. Among these are reactions involving activated endogenous conjugating agents and reactions involving activation of the xenobiotic. There are also a larger number of 'novel' reactions about which comparatively little information is available. The novel reactions can be divided into two types (EADSFORTH and HUTSON, 1984), either (i) new examples of old reactions, that is, new substrate types for any of the principal reactions, such as the N-oxide drug minoxidil (1), which gives rise to a novel type of sulfate and glucuronide (2), tripelennamine which can be glucuronidated to a quaternary N-glucuronide (3) and tocainide (4), which forms a N-carbamoylglucuronide (5), or (ii) novel reactions, in which a previously unknown endogenous conjugating agent is used (see PFEIFER and BORCHERT, 1975—1983; PFEIFER et al., 1988).

Examples of these latter are found with carbohydrates in the cases of ribose, galactose, glucose, and xylose conjugates, although glycosides with pentoses and hexoses are seldom in drug metabolism. To a certain degree the formation of ribose conjugates of antimetabolites such as azathioprine, fluorouracil or mercaptopurine (6), the formation of N-β-glucopyranoside with, for example, amobarbital or phenobarbital (7) and the formation of glucose esters with pranoprofen (8) may be of interest.

In the case of acylations, comparatively novel pathways are the formation of formyl and long chain fatty acyl conjugates.

Formylations for example are involved to a small extent in the biotransformation of hydralazine (9) to the cyclic compound (10) or in the formation of the metabolite (12) from aminophenazone (11).

Examples of conjugates with long chain fatty acids are the palmitic and stearic esters (14 and 15, respectively) of 11-hydroxy-$\Delta^8$-or-$\Delta^9$-tetrahydrocanabinole (13).

Among the amino acid conjugations, there are novel instances of the use of histidine, alanine and glutamic acid as well as the more important conjugations with glycine. Furthermore, it has been reported that the endogenous amine taurine, also forms conjugates with foreign compounds such as the analgesic fenclofenac (**16**). The formation of taurine conjugates with endogenous bile acids is reasonably well documented.

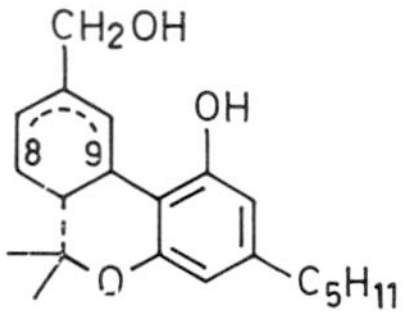


13

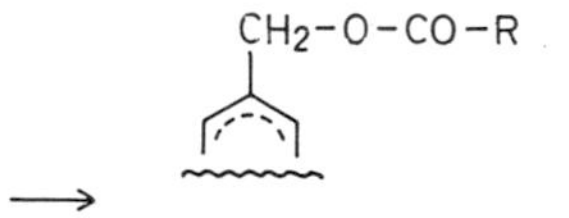


14 : $R = (CH_2)_{14}-CH_3$
15 : $R = (CH_2)_{16}-CH_3$

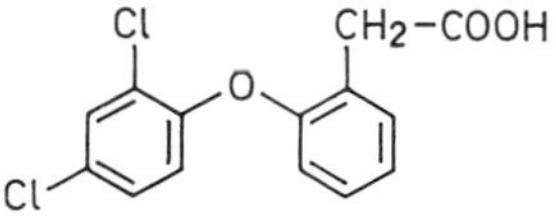


16

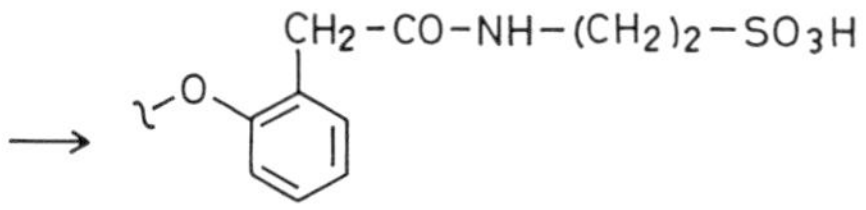


17

It is now well established that glutathione conjugates are transformed into cysteine adducts by $\gamma$-glutamyltransferase and cysteinyl-glycine dipeptidase which can be acetylated to mercapturic acids. In addition to this pathway, evidence has recently been presented that the extensive catabolism of glutathione conjugates is a source of other new types of xenobiotic conjugates (Bakke, 1986).

From the viewpoint of didactics, it is helpful to distinguish between two phases of drug metabolism, phase I in which chemical modifications of the parent drugs take place, such as oxidations, reductions or hydrolysis and phase II in which the inserted polar groups of phase I metabolites are conjugated with endogenous agents. Occasionaly the phase II reactions are also designated as postoxidation reactions because the functional group, which is conjugated, is inserted in the parent drug by oxidative biotransformation. However, both concepts of classification have failings as conjugations can also occur directly with the parent drug, if the corresponding functional groups are present. Furthermore reactions other than oxidations can occur in phase I or reactions other than conjugations can take place postoxidatively, such as the widespread hydrolysis of epoxides.

Although in phase II highly water-soluble metabolites of decreased pharmacological activity are formed which can be easily excreted, there are some exceptions. For example, $N^4$-acetylated metabolites of some sulfanilamides are less water-soluble and have a longer biological half-life than the parent compounds.

There are also examples showing that phase II-reactions lead to more toxic products. Thus, the ultimate mutagens or carcinogens, in the case of aromatic amines and azo-compounds such as 2-acetylaminofluorene (**18**) and dimethyl-4-aminobenzene (**21**), respectively, are sulfate conjugates (sulf) of the corresponding N-hydroxylated metabolites of these compounds (MILLER and MILLER, 1976).

$$\underline{18} \qquad \underline{19} \qquad \underline{20}$$

$$\underline{21} \qquad \underline{22} \qquad \underline{23} \qquad \underline{24}$$

Ac = acetate    Sulf = sulfate

Glutathione transferases are included in important detoxification mechanisms. Highly reactive metabolites are also able to bind covalently to glutathione transferase. In this case the enzyme becomes inactivated. Glutathione transferases can also act as storage proteins for hydrophobic substances, such as bilirubin, estrogens, hemin, chloramphenicol, penicillin and contrast media (JAKOBY, 1978). On occasions, conjugation with glutathione can also lead to more toxic products. Thus for instance, the weak mutagenic 1,2-dichloroethane (**25**) is activated to a strong mutagen (**26**) by a nucleophilic displacement of a chlorine atom by the SH-group of glutathione (GSH) catalysed by glutathione transferases (MULDER, 1979).

$$\underline{25} \qquad \underline{26}$$

Another aspect of phase II-reactions which is frequently overlooked is that the xenobiotic conjugation mechanisms have a potential impact upon the endogenous biochemical processes of the body (CALDWELL, 1980). In addition to the xenobiotic substrate, each of the conjugation reactions involve the participation of an endogenous conjugating agent as cosubstrate. The mobilization of the endogenous agent for this purpose may result in its depletion

A. LANGNER; H.-H. BORCHERT; S. PFEIFER

and thus modify its utilization in endogenous processes. Some examples of the endogenous use of the principal conjugating agents from the same pools used for drug metabolism are, in the case of glucuronic acid and sulfate, the conjugations of bilirubin, steroids, and catecholamines or the biosynthesis of heparin and chondroitin; in the case of methyl groups the methylation of catecholamines and nicotinamide or their biosynthetic role as one carbon pool; and in the case of amino acids, especially glycine, the conjugation of unusual acids in e.g. amino acidurias or the synthesis of proteins.

It is well known that the capacities of the principal conjugation reactions are partly limited, due to restrictions upon the supply of the endogenous conjugating agent. In addition, it is apparent that certain adverse effects of toxic chemicals can arise from interference with the normal function of endogenous conjugating agents due to limitations upon their supply. Some examples are the teratogenic effect in mice due to reduced sulfation of connective tissue arising from depletion of 3'-phosphoadenosine-5'-phosphosulfate (PAPS) by salicylamide (ROE, 1982), the growth impairment and reduced heme biosynthesis due to diversion of glycine into hippuric acid synthesis by benzoic acid (CALDWELL et al., 1986), cataract in rabbits, due to glutathione S-transferase depletion in lens by bromobenzene or naphthalene (CALDWELL, 1980), and a topical ataxic neuropathy, in part due to depletion of sulfane sulfur pools by cyanide (WESTLEY, 1980).

Further evidence for the close relationship between endogenous conjugating agents for phase II-metabolism and their role in endogenous metabolism is provided by certain inborn defects in metabolism of conjugating agents. The brachymorphic mouse is congenitally dwarfed due to its inability to sulfate chondroitin and other connective tissue macromolecules. This inability arises from the deficiency in the synthesis of PAPS. In addition to reduced sulfation of endogenous macromolecules, these mice exhibit a marked inability for xenobiotic sulfation (LYMAN and POLAND, 1983). Another example is the so-called Leber's optic atrophy, a rare genetic disorder marked by a sensitivity to cyanide in sources such as cigarette smoke and cider, exposure to which causes blindness (WESTLEY, 1980). Cyanide is normally metabolized by conversion to thiocyanate, catalyzed by the enzyme rhodanide synthetase. The sulfur atom is transferred from thiosulfate or another member of the sulfane sulfur pool of the body. The defect in Leber's optic atrophy is due to a deficiency of sulfane sulfur for the detoxication of cyanide. Its impact on the metabolism of endogenous compounds is seen in reduced synthesis of the amino acid cysteine and disordered function of vitamin $B_{12}$.

In conclusion it must not be overlooked that the conjugation reactions can be regarded as interfaces between biotransformation of drugs and metabolism of endogenous compounds.

# 3. Selected conjugating enzymes

## 3.1. Sulfotransferases

### 3.1.1. Introduction

In 1876 BAUMANN (1876) discovered an enzyme that catalyzes the sulfate conjugation of xenobiotic compounds. He fed phenol to dogs and was able to show that some of it was excreted in the urine as a sulfate conjugate.

During the past decade there has been increased interest both in the enzymes that catalyze sulfate conjugations and in the possible role of this reaction in the metabolism of neurotransmitters, steroids or xenobiotic compounds (SANDLER, 1981). The formation of sulfate esters in mammalian tissues is a major pathway of metabolism for substrates bearing a hydroxy group, and serves as a means for preparing lipophilic endogenous compounds and xenobiotics for excretion in bile and urine (YOUNG et al., 1984; WEINSHILBOUM, 1986). Recently several studies have indicated that sulfation may also lead to very unstable highly reactive compounds, particularly the sulfate conjugates of some N-hydroxy-arylamines (SHIRAI and KING, 1982). Furthermore sulfate conjugates may act as important biosynthetic intermediates in the metabolism of steroids and catecholamines or act as storage forms of various steroid hormones and amine neurotransmitters that can be released by enzymatic desulfation by sulfatases (HOBKIRK, 1985).

Sulfotransferases (STs) catalyze sulfation of diverse substrates such as alkyl and allyl alcohols and amines, phenols and phenolic steroids, carbohydrates and glycolipids (SANDLER et al., 1981; RAMLI and WHELDRAKE, 1981; IWASAKI et al., 1986; FOYE and KULAPADITHAROM, 1985). The various forms and the precise number of different STs in species ranging from primitive single-cell-organisms to man is unknown (SANDLER, 1981). Since sulfate conjugation seems essential in every animal tissue, the sulfate conjugation system is expected to be present in the whole body (SANDLER, 1981; SINGER, 1985).

Various STs seem to be involved in different physiological and pathological processes in the organism. However, their role has not been clarified completely. Phenol STs catalyze the sulfate conjugation of endogenous catecholamines and their metabolites (WEINSHILBOUM, 1986). It was hypothesized (VAN KEMPEN et al., 1982; PICOTTI et al., 1981a, 1981b; VAN KEMPEN and PENNINGS, 1981) that phenol ST is involved in the termination of the action of catecholamines and other amine neurotransmitters and in the biosynthesis of storage or transport forms of catecholamines and amine neurotransmitters. The plasma concentration of conjugated amine neurotransmitters and their unknown relationship to hypertension was discussed (WEINSHILBOUM, 1986; KUCHEL et al., 1984).

Sulfate conjugation of bile acids seems to prevent the biosynthesis of mono-hydroxylated bile acids showing toxic activities in man (SINGER, 1985).

A. LANGNER; H.-H. BORCHERT; S. PFEIFER

Several steroid STs act as regulatory enzymes (GREEN and SINGER, 1983; BOUTHILLIER et al., 1984a). STs which act upon estrogen in reproductive tissues, such as uterus are of particularly high affinity, appear to be under some biological control and may exert important effects upon estrogen action.

The high activities of glucocorticoid STs were determined in the liver suggesting that glucocorticoid sulfates are involved in control of corticosteroid metabolism, hypertension and cancer (SINGER and MOSHTAGHIE, 1981; SINGER et al., 1980).

The N-O-sulfation by hepatic STs is an important step in the carcinogenesis by some N-hydroxyarylamines (MEERMAN et al., 1981).

STs play an important role in drug metabolism (BONHAM CARTER et al., 1983). For example, paracetamol, a widely used analgesic (REITER and WEINSHILBOUM, 1982), the antihypertensive agent methyldopa (CAMPBELL et al., 1984, 1985) and the inhibitor of thrombocyte aggregation and the coronary agent trapidil (PFEIFER and BORCHERT, 1975–1983) are substrates of ST.

### 3.1.2. Localization and multiplicity of sulfotransferases in mammals

Sulfotransferases (STs) are found both in the cytoplasm and in the endoplasmic reticulum. It seems that the smaller molecules are degraded by the former group and large molecules are metabolized by the latter (SANDLER, 1981).

Sulfotransferases with different substrate specifity have been separated indicating that different enzyme forms exist (WEINSHILBOUM et al., 1981). These are widespread in the mammalian body, with highest specific activities in platelets, erythrocytes, in the jejunum, in different regions of the brain (the highest activities have been found in the cerebral cortex), in the zona reticularis of the suprarenal glands, in the liver and in the placenta (SANDLER, 1981; HOBKIRK, 1985; SINGER, 1984; YOUNG et al., 1985). Within the different tissues and organs, the enzymes are distributed heterogenously. In the human liver, for example, the activity of ST is higher in the periportal than in the pericentral sublobular regions (MOUELHI and KAUFFMAN, 1986).

In many tissues, such as platelets, brain, liver, small intestine and lung, there are multiple forms of STs (YOUNG et al., 1985; BARANCZYK-KUZMA and SZYMCZYK, 1987; WONG and YEO, 1982; WONG, 1982).

The most frequently studied source of phenol ST in man has been the blood platelet. The human platelet contains at least two independently regulated forms of phenol ST with different substrate specificities, different sensitivities to inhibitors, and different physico-chemical properties (REITER et al., 1983; VAN LOON and WEINSHILBOUM, 1984; ANDERSON and JACKSON, 1984). The two forms of platelet phenol ST have been separated by ion exchange chromatography. Both forms are soluble enzymes. One form is relatively thermolabile and catalyzes the sulfate conjugation of monoamines. This form has been referred to as the "TL" (thermolabile) or the "M" (monoamine metabolizing) form of phenol ST. The other form of platelet phenol ST is thermostable and catalyzes the sulfate conjugation of phenol and p-nitrophenol. It has been referred to as the "TS" (thermostable) or the "P" (phenol metabolizing) form. The terms thermolabile and thermostable resulted from experiments measuring enzyme activities at temperatures from 37 °C to 50°C.

The human brain also contains at least two forms of phenol ST, forms that are similar to the platelet TS and TL phenol ST with respect to substrate specificity, apparent Km constants, thermal stability, and sensitivity to inhibitors (REIN et al., 1984; YU et al., 1985; VAN KEMPEN and PENNINGS, 1982). There is a highly significant correlation between the activities of the TS form of phenol ST in cerebral cortex and platelets in man, but there is no significant correlation between activities of the TL forms in the two tissues (YOUNG et al., 1985). The mammalian liver appears to contain at least nine different phenol and steroid STs that catalyze phenol, glucocorticoid, androgen, estrogen, bile acid, mineralocorticoid and hydroxylamine sulfation. Two phenol STs (PS 1 and PS 2) are observed in rat liver (SINGER et al., 1985), PS 2 is present in largest amounts in male rats and PS 1 is present in roughly equal amounts in rats of both sexes. PS 1 and PS 2 were eluted from DEAE-Sephadex A 50. An estradiol specific ST was isolated by SINGER et al. (1928). This enzyme is present at similar levels in both sexes and does not catalyze dehydroepiandrosterone, cortisol, testosterone, estrone or corticosterone sulfation. ST I, II and III were separated by chromatographic, endocrinological and immunological methods. ST III is the glucocorticoid-preferring ST of the rat liver. ST I and II appear to be dehydroepiandrosterone-preferring steroid STs (SINGER et al., 1984; SINGER, 1984). A mineralocorticoid (deoxycorticosterone) ST has been identified in livers of female rats by LEWIS et al. (1981). Besides these ST a bile acid ST and a hydroxyarylamine ST have been isolated (SINGER et al., 1982). JAKOBY et al. (1980) and SEKURA and JAKOBY (1979) reported on the presence of three phenol STs I, II and IV. These enzymes sulfate catechols, p-nitrophenol and 2-naphthol. The phenol ST IV catalyzing the sulfation of a variety of substituted phenols, including catecholamines, tyrosine esters, and peptides containing $NH_2$ terminal tyrosine residues has also been identified by DUFFEL et al. (1981) from male rat liver. The relationship between the TS and TL forms and the aryl ST I, II and IV is not yet clear. MIZUMA et al. (1983) proposed that the aryl ST IV is the TL form and the aryl STs I and II could be TS forms. A 3β-hydroxysteroid ST has been discovered in rat liver (SINGER, 1985). This enzyme sulfates dehydroepiandrosterone and appears to be identicalwith the glucocorticoid ST II. JAKOBY et al. (1981) isolated homogenous preparations of three hydroxy steroid STs. It is not

A. LANGNER; H.-H. BORCHERT; S. PFEIFER

clear whether these enzymes differ from other STs. Overlapping substrate specificities have been observed suggesting that some enzymes are identical with others.

3β-Hydroxysteroid STs have been detected also in uterine tissue and in epididymis (BOUTHILLIER et al., 1985; BOUTHILLIER et al., 1984b; BROOKS et al., 1982). Estrogen STs have been identified in endocrine organs in various species. However, only the adrenal and placental estrogen ST have been purified and characterized. Human placenta contains a substantial amount of estrogen ST (TSENG et al., 1985; HOBKIRK and CORDY, 1985).

BARANCZYK-KUZMA et al. (1985) have purified a phenol ST from rat kidney and stomach. The isolated kidney enzyme appeared to have properties very similar to those of the liver enzyme and had the p-nitrophenol sulfating activity intermediate between those of rat liver and brain. The properties of the stomach enzyme were different from those of the rat liver and kidney enzymes. The phenol ST from stomach mucosa had the p-nitrophenol sulfating activity comparable to that of rat brain.

In some tissues of rat and guinea pig, arylamine STs could be detected. These enzymes sulfate primary and secondary amino groups (WONG and YEO, 1982; BARANCZYK-KUZMA, 1987). The sulfamates formed by these enzymes were detected in the urine. IWASAHI et al. (1986) described the N-sulfoconjugation of alicyclic alkyl- and arylamines.

In Table 2 a classification of STs is given.

**Table 2.** Classification of some important sulfotransferases (according to SINGER, 1985)

| Enzyme | Substrates | Source/Species | Reference |
|--------|-----------|----------------|-----------|
| *Phenol sulfotransferases* | | | |
| TL, M | monoamines | platelets of rat, human, dog | TOTH et al., 1986; REITER et al., 1983; SANDLER et al., 1981 |
| TS, P | phenol | platelets of human | BONHAM CARTER et al., 1983; YOUNG et al., 1985 |
| TL, M | monoamines | brain of rat, human | YOUNG et al., 1985 |
| TS, P | phenol, p-nitro-phenol | brain of rat, human | WEINSHILBOUM, 1986; YOUNG et al., 1984 |
| 1 and 2 | p-nitrophenol | liver of rat | SINGER, 1984 |

**Table 2.** (continued)

| Enzyme | Substrates | Source/Species | Reference |
|---|---|---|---|
| I, II and IV | 2-naphthol | liver of rat | SEKURA et al., 1981a |
| Phenol ST | p-nitrophenol | kidney and stomach of rat | BARANCZYK-KUZMA et al., 1985 |

*Hydroxylarylamine sulfotransferases*

| Enzyme | Substrates | Source/Species | Reference |
|---|---|---|---|
| N-Hydroxy-arylamine ST | N-hydroxy-2-acetylamino-fluoren | liver of rat | SHIRAI and KING, 1982; MEERMAN et al., 1981 |

*Steroid sulfotransferases*

| Enzyme | Substrates | Source/Species | Reference |
|---|---|---|---|
| Estrone ST | estrone | adrenal glands, placenta of bovine, human | HOBKIRK et al., 1985; TSENG et al., 1985; BROOKS et al., 1982 |
| Estradiol ST | estradiol | liver of rat | GREEN and SINGER, 1982 |
| 3β-Hydroxy-steroid ST | dehydroepiandro-sterone | liver of rat, human, adrenal glands of human, uterus, epididymis of rat | BOUTHILLIER et al., 1984a, 1984b, 1985 |
| Glucocorticoid ST I, II, III | cortisol | liver of rat, guinea pig | SINGER and BRILL, 1982; FEDERSPEIL and SINGER, 1981; SINGER et al., 1980 |
| Mineralo-corticoid ST | deoxycortico-sterone | liver of rat | LEWIS et al., 1981 |
| Bile acid ST | taurolithocholic acid | liver of rat, guinea pig, human, kidney of rat | KIRKPATRICK and BELSAAS, 1985; SINGER et al., 1982 |

*Alcohol sulfotransferases*

| Enzyme | Substrates | Source/Species | Reference |
|---|---|---|---|
| Alcohol-hydroxy-steroid-ST | butanol, dehydro-epiandrosterone | liver of rat | JAKOBY et al., 1981 |

     A. LANGNER; H.-H. BORCHERT; S. PFEIFER

### 3.1.3. Physico-chemical properties

The isolation and purification procedures of STs have been described and form the basis for studies of the relationship between the chemical structure and physico-chemical properties of STs and its mechanism of catalysis. These methods include anion exchange chromatography on DEAE-cellulose, absorption chromatography on hydroxyapatite or on ATP-agarose and other chromatographic and immunological procedures. Some characteristics of highly purified forms of STs are listed in Table 3.

### 3.1.4. Biochemical characteristics

#### 3.1.4.1. Cosubstrate of sulfotransferases

An essential requirement for sulfation to occur is the activation of inorganic sulfate to its group-donating form, adenosine 3'-phosphate 5'-sulfatophosphate (PAPS), which occurs in a two-step reaction, catalyzed by ATP-sulfurylase and APS-phosphokinase, respectively (SANDLER, 1981):

$$\text{ATP} + \text{SO}_4{}^{2-} \xrightarrow{\text{ATP-sulfurylase}} \text{APS} + \text{PP} \tag{I}$$

$$\text{ATP} + \text{APS} \xrightarrow{\text{APS-phosphokinase}} \text{PAPS} + \text{ADP} \tag{II}$$

Since step I is thermodynamically unfavourable, the reaction will proceed only when the products of the reaction are removed by subsequent reactions. In a second step (II) APS is converted to PAPS (**27**) (WEINSHILBOUM, 1986; PENNINGS and VAN KEMPEN, 1982).

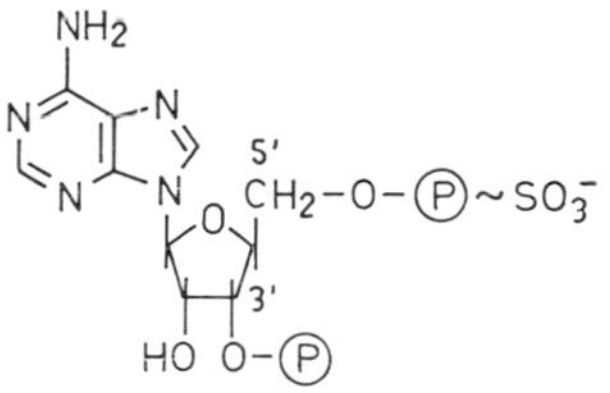

27

Presumably, the concentration of PAPS in a certain tissue is the result of its continuous synthesis and break-down. The enzymes 3'- and 5'-nucleotidase can hydrolyse the ribose-phosphate bonds in PAPS. Furthermore, sulfohydrolases in the liver hydrolyse the phosphate-sulfate bond (HOBKIRK, 1985). The unfavourable equilibrium of the first step, in addition to the fact that PAPS

---

**Table 3.** Physico-chemical properties of some important sulfotransferases

| Enzyme | Source | Molecular weight | Iso-electric point | Optimum pH value | References |
|---|---|---|---|---|---|
| Phenol ST | platelets of human | 65,000 | | 5.9 (dopamine)<br>6.8 (dopamine)<br>6.3 (phenol)<br>7.0—8.6 (5-hydroxytryptamin)<br>7.4 (methylumbelliferon)<br>7.8 (tyramine) | ABENHEIM et al., 1981;<br>BUTLER et al., 1983;<br>PENNINGS et al., 1981;<br>WONG et al., 1984;<br>VAN KEMPEN, 1981 |
| Phenol ST | brain of human | 62,000 | 5.8 | 7.0 (dopamine)<br>7.5 (phenol)<br>7.8—8.0 (tyramine) | REIN et al., 1984;<br>WHITTEMORE, 1985; YU et al., 1985 |
| Phenol ST 1 and 2 | liver of rat | 68,000<br>65,000<br>69,000—70,000<br>64,000 | 1: 8.1<br>2: 6.9 | 5.5—6.1 (4-nitrophenol, 2-naphthol) | BARANCZYK-KUZMA et al., 1981; SINGER, 1985; SECURA et al., 1981 b;<br>BORCHARDT and SCHASTEEN, 1982 |
| Aryl (phenol) ST IV | liver of rat | 61,000 | 5.8 | 5.5 (4-nitrophenol, 2-naphthol) | SECURA et al., 1981 b; SINGER, 1985 |
| Phenol ST | brain of rat | 68,000 | | 5.6 (4-nitrophenol) | BARANCZYK-KUZMA et al., 1981 |
| Phenol ST | kidney of rat | 68,000<br>69,000 | | 5.4—6.6 (4-nitrophenol) | BARANCZYK-KUZMA et al., 1981, 1985 |
| Phenol ST | stomach of rat | 32,000 | | 6.4 (4-nitrophenol) | BARANCZYK-KUZMA et al., 1985 |
| Phenol ST | liver of dog | 60,000 | | 5.5—6.5 | WHITTEMORE et al., 1986 |
| Glucocorticoid ST | liver of rat | 156,000<br>160,000 | 6.5 | 6.0 (cortisol) | HOBKIRK, 1985; SINGER, 1984, 1985 |

A. LANGNER; H.-H. BORCHERT; S. PFEIFER

**Table 3.** (continued)

| Enzyme | Source | Molecular weight | Iso-electric point | Optimum pH value | References |
|---|---|---|---|---|---|
| Glucocorti-coid ST III | liver of rat | 66,000 68,300 | | 6.0 (cortisol) | SINGER, 1985; HOBKIRK, 1985 |
| Glucocorti-coid ST II | liver of rat | 68,000 | | 5.0 | SINGER, 1985 |
| 3β-Hydroxy-steroid ST | epididymis of hamster | 160,000 | | 8.7; 10.0 (dehydro-epiandrosterone) | BOUTHILLIER et al., 1981, 1984 b |
| Estradiol-specific ST | liver of rat | 54,500 | | 7.75 | HOBKIRK, 1985 |
| Estrone ST | adrenal gland, placenta of bovine | 74,000 | 5.8 | 8.0 | SINGER, 1985 |
| Estrogen ST | uterus of guinea pig | 70,000 | 5.8 | | HOBKIRK, 1985 |
| Estrogen | placenta of human | 68,000 | 5.8 | | TSENG et al., 1984 |
| Bile acid ST | liver of rat | 130,000 | 5.3 | 6.5 | SINGER, 1985 |
| Bile acid ST | liver of human | 67,000 | 5.2; 5.5 | | CHEN and SEGEL, 1985 |
| Bile acid ST | liver of guinea pig | 76,000 | 5.6 | 6.8 | SINGER, 1985 |
| Bile acid ST | kidney of rat | 80,000 | 5.8 | | SINGER, 1985 |
| Alcohol ST I | liver of rat | 180,000 | 5.0 | 6.0 | JAKOBY et al., 1981 |
| Alcohol ST II | liver of rat | 290,000 | 7.9 | 5.5 | JAKOBY et al., 1981 |
| Alcohol ST III | liver of rat | 120,000 | 6.1 | 5.5 | JAKOBY et al., 1981 |
| N-Hydroxy-arylamine ST | liver of rat | 68,000 | 5.7 | 6.3 (N-hydroxy-2-acetylamino-fluoren) | SINGER, 1985 |

is a metabolically rather labile substance, may explain the low tissue concentration of PAPS. In the liver the concentration is about 30 μmol/kg.

The important role of the availability of inorganic sulfate for sulfation has been demonstrated in several studies in which the sulfation of a substrate was stimulated by the concomitant administration of inorganic sulfate or one of its precursors (KRIJGSHELD et al., 1981). Inorganic sulfate can be generated from cysteine by sulfoxidation of the thiol group in this amino acid (DAWSON et al., 1983). The availability of inorganic sulfur is a rate-limiting factor in the sulfation of xenobiotics and endogenous compounds in mammals.

### 3.1.4.2. Substrate specificity of sulfotransferases

Concerning the substrate specificity, various ST preparations have been distinguished.

Several endogenous phenolic amines, including catecholamines, tyramine and 5-hydroxytryptamine are substrates for phenol STs which catalyze the transfer of sulfate from PAPS to amines, their deaminated metabolites and a variety of phenolic drugs. Platelets have high phenol ST activity, which is mainly localized in the soluble fraction of the cytoplasm. The literature on the kinetics of platelet phenol ST gives rather poor information about the affinity of the amine substrates for the enzyme. The apparent Km values, as determined for human platelet phenol ST with several amine substrates are given in Table 4.

**Table 4.** Apparent Km values of human platelet phenol sulfotransferase for different amine substrates (according to PICOTTI et al., 1981 b)

| Substrate | Apparent Km ($\cdot$ $10^{-6}$ M) |
|---|---|
| 3-Methoxytyramine | 0.3 |
| 5-Hydroxytryptamine | 1.0 |
| Tyramine | 1.5 |
| Normetanephrine | 1.6 |
| Metanephrine | 1.6 |
| Dopamine | 1.9 |

At the substrate concentrations used, dopamine is the most rapidly esterified substrate, followed by tyramine and then by the catecholamine-methoxy derivatives. This indicates that, at least in the case of tyramine and catecholamine-methoxy derivatives, platelet phenol ST efficiently catalyzes the transfer of sulfate to OH-groups in the para-position.

A. LANGNER; H.-H. BORCHERT; S. PFEIFER

Human platelet phenol ST exists in two functional forms, M and P. Dopamine, tyramine, noradrenaline, adrenaline, 5-hydroxytryptamine, $\beta$-hydroxyamphetamine, isoprenaline, salbutamol and 1-naphthol are all specific substrates for the M form. Paracetamol is also predominantly metabolized by the M form. Salicylamide at low concentrations is a substrate for the P form but becomes an M substrate at higher concentrations (BONHAM CARTER et al., 1983). Phenol is a specific substrate for P at 10 µM. At 1 µM it also becomes a substrate for the M form. Similar observations have been made while investigating the two forms of brain ST. Kinetic parameters for human brain ST M and P are given in Table 5.

**Table 5.** Apparent Km and Vmax values of phenol sulfotransferases M and P from human brain for some substrates (according to YU et al., 1985)

| Substrate | Apparent Km ($\cdot\,10^{-6}$ M) | Vmax (pmol $\cdot$ min$^{-1}$ $\cdot$ mg$^{-1}$) |
|---|---|---|
| *Phenol sulfotransferase M* | | |
| Dopamine | 2 | 101 |
| m-Tyramine | 9 | 143 |
| p-Tyramine | 55 | 97 |
| Noradrenaline | 18 | 40 |
| Serotonin | 330 | 25 |
| *Phenol sulfotransferase P* | | |
| Phenol | 5 | 88 |
| p-Nitrophenol | 0.6 | 167 |
| m-Nitrophenol | 0.4 | 152 |

The phenol ST M exhibits an extremely high affinity to dopamine and m-tyramine based on the low Km values and is moderately active toward noradrenaline and p-tyramine. p-Tyramine and m-tyramine are readily conjugated in the presence of PAPS and M. The m-isomer exhibits a significantly higher affinity than that of the p-isomer. Phenol ST P exhibits a very high affinity to phenol and nitrophenols but is inactive toward the amine acceptors at the concentrations tested.

Sulfate conjugation in the liver and other peripheral tissues is also characterized by a multiplicity of phenol ST isoenzymes important for substrate specificity. SEKURA and JAKOBY (1981) investigated some kinetic parameters of the aryl ST IV from rat liver (Table 6). The sulfation of substrates catalyzed by aryl ST IV seems to be dependent on the pH value concerning the substrate specificity. At pH 5.5 IV sulfates phenols. At higher pH values the affinity for substrates other than phenols increases.

In Table 7 apparent Km values of STs for some steroids are given.

**Table 6.** Kinetic parameters of aryl sulfotransferase IV from rat liver for some substrates (according to SEKURA and JAKOBY, 1981)

| Substrate | Km ($\cdot\ 10^{-3}$ M) | Vmax (nmol $\cdot$ min$^{-1}$ $\cdot$ mg$^{-1}$) |
|---|---|---|
| $\beta$-Naphthol | 0.1 | 690 |
| Phenol | 0.92 | 27 |
| m-Chlorophenol | 0.32 | 450 |
| p-Chlorophenol | 0.34 | 290 |
| m-Methylphenol | 2.80 | 180 |
| p-Methylphenol | 1.20 | 18 |
| m-Nitrophenol | 0.44 | 800 |
| p-Nitrophenol | 0.17 | 450 |
| p-Methoxyphenol | 1.40 | 450 |
| Epinephrine | 0.43 | 120 |
| Tyramine | 0.46 | 72 |
| Dopamine | 0.16 | 68 |

**Table 7.** Apparent Km values of steroid sulfotransferases

| Enzyme | Source | Substrate | Km ($\cdot\ 10^{-6}$ M) | Reference |
|---|---|---|---|---|
| Glucocorticoid ST I and III | liver of rat | cortisol | 7.0 | SINGER, 1985 |
| Glucocorticoid ST | liver of guinea pig | cortisol | 150 | SINGER and BRILL, 1982 |
| Estradiol-specific ST | liver of rat | estradiol | 76 | SINGER, 1985 |
| Estrogen ST | placenta of human | estradiol<br>estrone | 4<br>20 | TSENG et al., 1985 |
| Bile acid ST | liver of rat | taurolithocholic acid | 50 | SINGER, 1985 |
| | kidney of rat | taurolithocholic acid | 40 | SINGER, 1985 |
| | liver of guinea pig | taurolithocholic acid | 77 | CHEN, 1982<br>SINGER, 1985 |
| | liver of human | glucolithocholic acid | 3.3 | SINGER, 1985 |
| $3\beta$-Hydroxy-steroid ST | liver of rat | dehydroepiandro-sterone | 6 | SINGER, 1985 |

A. LANGNER; H.-H. BORCHERT; S. PFEIFER

**Table 8.** Apparent Km values of alcohol-/hydroxysteroid sulfotransferases 1, 2 and 3 (according to JAKOBY et al., 1981)

| Substrate | Apparent Km ($\cdot 10^{-6}$ M) | | |
|---|---|---|---|
| | ST 1 | ST 2 | ST 3 |
| Dehydroepiandrosterone | 12 | 24 | 22 |
| Testosterone | 70 | 11 | 29 |
| Estradiol | 35 | 15 | 27 |
| Cortisol | 290 | 84 | 70 |
| Ethanol | 42,000 | 40,000 | 52,000 |
| Butan-1-ol | 3,000 | 3,300 | 3,000 |
| Isoamylalcohol | 1,200 | 2,000 | 1,500 |
| Amylalcohol | 1,700 | — | 1,200 |

For the alcohol-/hydroxysteroid STs 1, 2 and 3 isolated by JAKOBY et al. (1981), the following Km values have been described (Table 8).

It is apparent that the estrogen STs of certain reproductive tissues differ markedly from these liver enzymes in terms of specificity, affinity and several other characteristics (TSENG et al., 1985). The estradiol-specific ST of rat liver sulfates estradiol at all concentrations between 10 and 180 μM. It does not sulfate estrone, testosterone, dehydroepiandrosterone or cortisol (GREEN and SINGER, 1982). Estradiol is also the preferred substrate for estrogen ST from human placenta. But estrone, estriol and dehydroepiandrosterone are also sulfated by this form with lower rate of sulfuration (TSENG et al., 1985).

Glucocorticoid STs have been differentiated concerning their substrate specifity (SINGER, 1984). Rat liver glucocorticoid ST III prefers glucocorticoid substrates, while dehydroepiandrosterone is the best substrate for glucocorticoid ST I and II. Thus, endocrine-mediated differences of the amounts of the three enzymes in rat liver may indicate modified abilities to produce glucocorticoid sulfates.

### 3.1.4.3. Mechanism of sulfate conjugation

From the results of the kinetic investigations of the rat brain phenol ST a sequential ordered Bi Bi reaction mechanism is supposed (PENNINGS and VAN KAMPEN, 1982). The sulfate donor PAPS is the first substrate that adds to the enzyme and the sulfate acceptor is the second substrate. The sulfated product is the first product and PAP is the second product that leaves the enzyme. Substrate inhibition results from binding of the second substrate to the enzyme-PAP complex (Fig. 1).

An investigation of the mechanism of the aryl ST IV was carried out using 2-chloro-4-nitrophenol as a model substrate. In addition to the first reaction (Fig. 1), DUFFEL and JAKOBY (1981) described the transfer of sulfate from 2-chloro-4-nitrophenylsulfate in the absence of PAPS, but with requirement for PAP as an "exchange reaction" (Fig. 2). Kinetic, inhibition and binding studies with aryl ST IV are all consistent with a random rapid equilibrium Bi Bi kinetic mechanism with two dead end product inhibitor complexes.

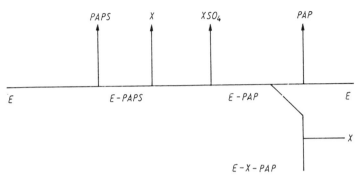

**Fig. 1.** Sequential ordered Bi Bi reaction mechanism of sulfation (according to PENNINGS and VAN KEMPEN, 1982). E = enzyme; X = second substrate.

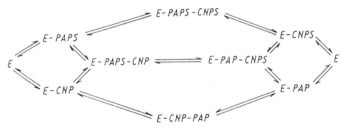

**Fig. 2.** Rapid equilibrium random mechanism for aryl ST IV (according to DUFFEL and JAKOBY, 1981). E = enzyme; CNP = 2-chloro-4-nitrophenol; CNPS = 2-chloro-4-nitro-phenyl sulfate.

A similar mechanism of sulfate transfer was proposed for an aryl ST from an anaerobic bacterium of human intestine as a model enzyme by KIM et al. (1986) (Fig. 3). A donor substrate, p-nitrophenylsulfate, combines a histidine residue of the enzyme active site with concomitant release of a phenolic compound, p-nitrophenol. The sulfate group of the histidine residue transfers to a neighbouring tyrosine group of the enzyme, and then to an acceptor with the binding of another donor to the histidine residue.

A sequential ordered Bi Bi reaction mechanism is proposed for phenol ST of human platelets similar to that of rat brain (PENNINGS et al., 1981).

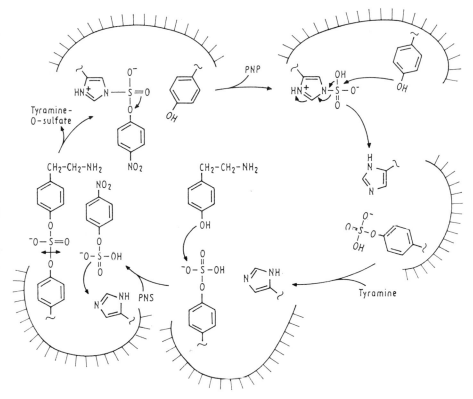

**Fig. 3.** Reaction mechanism of sulfate transfer by aryl ST (according to KIM et al., 1986). PNP = p-nitro-phenol; PNS = p-nitro-phenylsulfate.

The sulfation of cortisol by glucocorticoid ST III may occur according to a sequential ordered reaction mechanism, where the steroid is the first substrate and PAP is the first product (SINGER, 1985).

The glucocorticoid ST I from rat liver catalyzes the sulfation of cortisol by a Theorell-Chance-mechanism. This reaction mechanism is a special form of a sequential ordered mechanism characterized by a low concentration of the complex between enzyme and substrates owing to a high isomerisation velocity (SINGER, 1985).

### 3.1.5. Regulation of sulfotransferase activity

Several different mechanisms could play a role in the regulation of the sulfate conjugation. These mechanisms include age, humoral factors and inheritance.

---

The generation of PAPS requires both ATP and inorganic sulfate. The possibility exists that, under certain conditions, the concentration of inorganic sulfate might be limiting (WEINSHILBOUM, 1986). This limitation results in the decreased availability of PAPS necessary for sulfate conjugation. There is also the possibility that individual differences in phenol ST might be another factor responsible for variations in sulfate conjugations in various species (WEINSHILBOUM, 1986). Several attempts have been made to study the possible contribution of inheritance to individual differences in the regulation of phenol ST activity. REVELEY et al. (1982) found that the hereditability for platelet TS and TL form of phenol ST activities is quite high. The inheritance is responsible for much of the variation among individuals. Inherited differences in the thermal stability of the phenol ST may result from structural gene polymorphisms. The frequency distribution of TS form of phenol ST, thermal stability measured as heated/control ratios in platelet samples from humans, is bimodal. The bimodal frequency distribution raised the possibility that variations in platelet TS form of phenol TS might be under genetic control.

Results of experiments investigating regulation of phenol ST indicated that striking changes in rat liver and brain TS activity occur during growth and development (PEARSON et al., 1981).

Humoral factors affect the ST activity in a different manner (SINGER et al., 1982). For example, administration of dexamethasone results in large increases in rat kidney phenol ST activity (PEARSON et al., 1981). Many studies indicate that sex differences in rat liver ST activities appear to be primarily regulated by androgenic, estrogenic and progestational hormones (SINGER, 1984; WATANABE and MATSUI, 1984). The response of the enzymes to sex hormones differs between multiple forms of hepatic STs (SINGER and MOSHTAGHIE, 1981).

The adrenals are the main determinant of adult glucocorticoid levels since after adrenalectomy or hypophysectomy the enzyme activity decreases. Cortisol administration prevents the effect of adrenalectomy suggesting glucocorticoids are responsible for the adrenal effects (SINGER, 1984).

Regulation of sulfate conjugation can also be affected by the interaction between sulfation and glucuronidation. Usually the glucuronidation is a competing reaction (KOSTER et al., 1981; MULDER, 1986). The available data suggest that UDP-glucuronosyltransferase has a higher Km for the same substrate than the ST, so that at a low concentration of the substrate, sulfation predominates. The glucuronides have a preference for biliary excretion, whereas the sulfate conjugates usually are predominantly excreted in urine (Fig. 4).

Another factor influencing competition between glucuronidation and sulfation is the sublobular distribution of transferases in the liver (MOUELHI and KAUFFMAN, 1986). Glucuronosyltransferase is localized predominantly in pericentral regions. In contrast, ST activity is greater in periportal than peri-

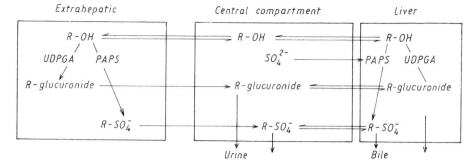

**Fig. 4.** Pharmacokinetic scheme for sulfation in vivo and its competition with glucuronidation (according to MULDER, 1986). R—OH = substrate; PAPS = cosubstrate for sulfation; UDPGA = cosubstrate for glucuronidation.

central regions. Various observations have shown that sulfation activity is localized anterior to glucuronidation activity in the liver flow path from periportal to centrilobular regions.

UDP-glucuronosyltransferase is localized in microsomes whereas STs are found predominantly in the soluble fraction. This phenomenon may be the cause for the differences in sulfation or glucuronidation of lipophilic compounds.

### 3.1.6. Inhibition and induction of sulfotransferases

REIN et al. (1981, 1982) observed that dichlorophenol is a selective inhibitor of phenol sulfation in platelets suggesting that phenol, on the one hand, and the monoamines, on the other hand, are metabolized in the human platelet by two separate enzyme activities. Pentachlorophenol is also an inhibitor of phenol ST (MULDER, 1986), but the duration of action of this compound is shorter than of dichloronitrophenol.

Whereas the inhibition of phenol ST from kidney, stomach and liver by dichloronitrophenol is competitive, the phenol ST of rat brain is inhibited in a noncompetitive manner (BARANCZYK-KUZMA et al., 1985). Dichloronitrophenol and pentachlorophenol also supress the activity of estradiol-specific ST suggesting similarities of these enzymes with phenol ST (SINGER, 1985). Some phenols with chloro- or nitro-substituents effectively inhibit the sulfation but to a lesser extent than dichloronitrophenol (KOSTER et al., 1979).

The inhibition of estrogen sulfoconjugation by other substrate analogous compounds, by some 2- and 4-substituted estra-1,3,5(10)-trien-17β-ols, was reported by HOSWITZ et al. (1986). These compounds were characterized to be specific inhibitors and important for a new approach in the treatment of hormone-dependent breast cancer.

Chemically modifying agents for amino acid residues, such as phenylglyoxal and N-ethylmaleimide, produce rapid inactivation of a rat liver phenol ST. Ribonucleotide dialdehydes appear to modify the active site of phenol ST and are irreversible inactivators (BORCHARDT et al., 1982b).

The effects of drugs on hepatic glucocorticoid ST activity have been tested by SINGER et al. (1984). Most of the drugs, such as spironolactone, pentachlorophenol, metyrapone, propranolol, aminoglutethimide, theophylline, acetylsalicylic acid and alloxan inactivate differently glucucorticoid ST I, II and III in vitro. However, no highly specific inhibitor of any one of the three enzymes was identified. Contrary to the in vitro effects, some of the above drugs caused significant elevation of the ST levels in vivo after multiple administration.

The rabbit liver ST is inhibited by substituted salicylic acids and related compounds, some hydroxamic acids and other agents capable of metal ion complexation (FOYE and KULAPADITHARON, 1984). Vanillin oxime, salicylhydroxamic acid and other salicylic acid derivatives of weaker acid strength than salicylic acid have been observed to be the most effective inhibitors.

Little data are available about induction of ST. In isolated rat hepatocyte subpopulations, the influence of phenobarbital and 3-methylcholanthrene has been studied (TONDA and HIRATA, 1983). The treatment of hepatocytes with phenobarbital did not enhance the sulfation of p-nitrophenol, while 3-methylcholanthrene enhanced the sulfation 2-fold. Since 3-methylcholanthrene alters the cell structure, it may induce particular functions to facilitate substrate-enzyme encounters or to elevate the level of PAPS. The rat hepatic ST activities toward androsterone and 4-nitrophenol after multiple administration of 3-methylcholanthrene, phenobarbital and sex steroids were determined by WATANABE and MATSUI (1984). ST activity toward androsterone increased slightly by pretreatment with phenobarbital and was significantly increased by estradiol and progesterone pretreatment in male rats. ST activity toward 4-nitrophenol was increased by administration of testosterone and progesterone in females. The exact mechanisms of these effects have not yet been clarified.

### 3.1.7. Concluding remarks

In general sulfate conjugation of xenobiotics results in the formation of less toxic metabolites, although important exceptions exist. For instance, sulfation of glucolithocholic acid was shown to be toxic since it induces a reversible cholestasis. The sulfation of N-hydroxyarylamines may be involved in hepatocarcinogenesis.

Sulfate conjugation of endogenous and exogenous phenolic compounds steroids and amines in most cases inactivates pharmacologically active substances and facilitates the elimination of these compounds in bile and urine.

A. LANGNER; H.-H. BORCHERT; S. PFEIFER

Numerous drugs are sulfated and, in general, detoxified thereby. An important consequence of sulfate conjugation is an increase in water-solubility of the substrates. Because the substrates usually are very lipid soluble, they are often slowly excreted in the unconjugated form. Therefore, conjugation results in an increased rate of their elimination in bile and urine.

Sulfate conjugation seems essential in every animal tissue. The sulfotransferases and the sulfate-activating system are expected to be present throughout the whole body.

The pharmacological intervention in ST activity may be a new approach for drugs influencing hormonal regulation, hormone-dependent pathological processes, cancerogenesis of N-hydroxyarylamines and biotransformation and excretion of drugs sulfated by STs.

## 3.2. Methyltransferases

### 3.2.1. Introduction

For several years the mechanism and regulation of biological alkyl transfer reactions have been studied in order to use this information to investigate physiological and pathological pathways, biotransformation reactions and to design potent and specific multisubstrate adduct inhibitors of these reactions. Methyltransferases (MTs) belong to a group of transferases catalyzing the alkyl transfer from the co-substrate serving as the 'activated alkyl group' to the substrate (COWARD, 1981). Biological N-, S- and O-methylation reactions are catalyzed by S-adenosyl-L-methionine dependent MTs. These N-, O- and S-MTs are involved in the biosynthesis or metabolism of small molecules, such as catecholamines, histamine, serotonin, melatonin, various thiols, and in modulating the activities of macromolecules, as proteins and nucleic acids (BORCHARDT, 1980).

The MTs are of great physiological and pathophysiological significance. The phenolethanolamine N-MT (PNMT) catalyzing the conversion of norepinephrine is only present in cells of the adrenal medulla and certain cell groups in the brainstem. The physiological role served by these neurons remains unknown although they are located near centers that regulate blood pressure and mediate the baroreflex. The PNMT activity is greater in these cell groups in spontaneously hypertensive rats (LOVENBERG, 1981). The catecholamines can be O-methylated in the position 3 by catechol O-MT (COMT). The O-methylated derivatives have reduced biological activity and the enzymatic O-methylation is one of the mechanisms by which the action of released catecholamines is terminated. Attempts have been carried out to correlate COMT activity in erythrocytes with mental diseases.

Another putative neurotransmitter, histamine, is modified by a MT reaction.

The enzyme histamine N-methyltransferase (HMT) appears to serve a role in terminating the action of histamine.

The successive N-methylation of phosphatidylethanolamine by N-MTs from liver microsomes constitutes a major route for the biosynthesis of phosphatidylcholine (MARIO et al., 1986). The phospholipid methylation is of great importance in participation in $Ca^{2+}$ influx and histamine release from mast cells (ISHIZAKA and ISHIZAKA, 1984).

The S-MTs are involved in detoxification of hydrogen sulfide formed by anaerobic bacteria in the intestinal tract (WEISIGER and JAKOBY, 1980).

Methylation, although a minor pathway of metabolism, is of importance for many endogenous substances and xenobiotics. Methylation may occur in different tissues, such as liver, brain, kidney, skin, blood cells, glands, nerve fibres and lung.

N-MTs have been found in livers of various species metabolizing drugs, such as nicotinamide, phenylethanolamine, $\beta$-phenetylamine, deoxyepinephrine and other derivatives of dopamine (YOUDE et al., 1984; SHIBATA, 1986; FONG and HWANG, 1983). COMT from pig liver has been detected to methylate various $\beta$-adrenoceptor agents with different substrate specificity (RAXWORTHY et al., 1986).

MTs seem to be involved in methylation of steroids. RAXWORTHY and GULLIVER (1982) reported the methylation of 2-hydroxyethinylestradiol by COMT.

The participation of S-MTs in thiol detoxification has been recognized (WEISIGER and JAKOBY, 1980). They are involved in the metabolism of the dialkyldithiocarbamates, e.g., disulfiram, and of the thiosubstituted pyridines and purines that include the antithyroid drugs, e.g., 2-thiouracil, 6-thiopurine and 6-propyl-2-thiouracil (WOODSON et al., 1983; KEITH et al., 1983 b).

The methylation reactions do not lead to more water soluble compounds in each case. The methylation may also form more lipid soluble metabolites with delayed excretion (PFEIFER et al., 1988). CALDWELL (1982a) reported on the methylation of some carboxyl acids which led to more lipophilic metabolites in comparison to the parent compounds. The methylation may produce drastic changes in physico-chemical properties. The N-methylation of some substances led to a quaternary structure of the compounds with change in polarity and solubility influencing the distribution and excretion of these substances. Some quaternary ammonium compounds have been found to possess high pharmacological activity or toxicity (CALDWELL, 1982b).

Some other compounds may also be activated in their biological activity or toxicity by MTs. Methylation of some polycyclic aromatic hydrocarbons is responsible for the induction of their significant carcinogenic activity (SILVERMAN and LOWE, 1982).

### 3.3.2. Localization and multiplicity of methyltransferases in mammals

Although the PNMT is only present in adrenal medulla and in neurons in brainstem (LOVENBERG, 1981), other MTs are widely distributed in mammalian tissues.

The highest level of COMT activity is generally found in rats and in human liver. But some activities of this enzyme have been reported to be present in rat and human brain, human placenta, breast tissue, erythrocytes and lung, in rat kidney and heart (LOVENBERG, 1981). In these tissues, the majority of the enzyme activity is found in the soluble fraction. Membrane-bound enzyme activity has also been found in red blood cells and in liver and brain microsomes (NISSINEN, 1984). The membrane-bound COMT has an affinity of at least one order of magnitude greater than the soluble COMT. The results of a study od NISSINEN (1984) demonstrate that the meta/para O-methylation ratios produced by the membrane-bound COMT differ significantly from meta/para ratios produced by the soluble enzyme (Table 9).

The meta/para ratios obtained with various tissues also show differences which may be due to multiple forms of membrane-bound COMT. It has been reported that soluble COMT is present in two forms designated A and B (MARZULLO and FRIEHOFF, 1981). COMT B possesses a higher meta/para O-methylation ratio than COMT A. COMT I and II are probably identical with form A and B, respectively (BORCHARDT, 1980).

**Table 9.** Methylation ratio of meta and para hydroxyls by membrane bound (MB) and soluble (SOL) COMT from rat (according to NISSINEN, 1984)

| Tissue | | Ratio of meta and para methylation | |
|---|---|---|---|
| | | Dopamine | Dihydroxybenzoic acid |
| Brain | MB | 61.0 | 23.7 |
| | SOL | 4.7 | 5.1 |
| Liver | MB | 21.5 | 11.7 |
| | SOL | 3.7 | 5.7 |
| Kidney | MB | 25.7 | 7.8 |
| | SOL | 3.7 | 5.6 |
| Lung | MB | 39.0 | 12.8 |
| | SOL | 4.4 | 5.7 |
| Heart | MB | n.d. | 24.6 |
| | SOL | 4.2 | 5.5 |

A hydroxyindole O-methyltransferase (HIOMT) was first characterized and purified from pineal tissue by AXELROD and WEISSBACH (1960). In mammals this enzyme is largely limited to the pineal gland, retina and Harderian gland, and it exhibits a high substrate specificity for N-acetylserotonin.

HMT is widely distributed in mammalian tissues. This activity has been detected in mouse and guinea pig liver, lung, kidney, heart, muscle, lymph nodes, spleen, in gastric mucosa of man, pig, dog and cow, in guinea pig, rat, mouse, and monkey brain, and in human erythrocytes. In brain, the majority of the activity occurs in the soluble supernatant fraction (BORCHARDT, 1980).

A non-specific N-MT in rabbit lung was described by AXELROD (1962). This enzyme has been detected with lesser amounts in adrenal gland, kidney, spleen, and heart. It was later designated as indolethylamine N-MT.

A non-specific N-MT was found in dog liver (FONG and HWANG, 1983). This enzyme is different from other N-methylating systems, especially in terms of substrate and species specificity.

DAMANI et al. (1986) reported on two amine N-MTs obtained from rabbit liver cytosol. Both enzymes possess overlapping substrate specificity methylating azaheterocycles.

Microsomal thiol S-MT activity has been observed in the livers of rat, mouse, rabbit and sheep with each at about the same specific activity using O-methylmercaptoethanol as the thiol substrate (WEISIGER and JAKOBY, 1980). Studies on intracellular distribution have shown that the enzyme is present in the particle fraction of the cell and is not in the cytosol. Tissue distribution of the enzyme revealed high concentrations in rat liver, kidney, and lung and relatively low in testis, spleen and intestine (WEISIGER et al., 1980). The microsomal thiol S-MT has also been found in rat brain. It was found to be unevenly distributed amongst various brain regions, with highest activities in the medulla oblongata and the hippocampus (HIEMKE and GHRAF, 1983). KEITH et al. (1983a) described that human erythrocytes appear to contain two forms of membrane-bound thiol S-MT distinguished by affinity to sulfhydryl substrates such as 2-mercaptoethanol. In the liver of mouse, two forms of thiol S-MT have been detected by OTTERNESS et al. (1986). Both forms differ in terms of affinity to 2-mercaptoethanol.

A thiopurine MT catalyzing the S-methylation of thiopurines and thiopyrimidines has been characterized by WOODSON and WEINSHILBOUM (1983). Classification of some important MTs is shown in Table 10.

### 3.2.3. Physico-chemical properties

In Table 11 some physicochemical properties, such as molecular weight, isoelectric point, and pH optimum of some methyltransferase activities are shown. These methyltransferases seem to occur as monomeric enzymes. For

**Table 10.** Classification of some important methyltransferases

| Enzyme | Source | Substrates | References |
|---|---|---|---|
| *O-Methyltransferases* | | | |
| Catechol O-MT (COMT) (form A or I and B or II) | brain, placenta, erythrocytes, lung, kidney, heart, lymphocytes | catecholamines, hydroxyestradiol, hydroxyestrone | RAXWORTHY et al., 1986; GORDONSMITH et al., 1982 |
| Hydroxyindole O-MT (HIOMT) | pineal tissue | serotonin | AXELROD and WEISSBACH, 1960 |
| *N-methyltransferases* | | | |
| Phenolethanolamine N-MT (PNMT) | adrenal medulla, brainstem | norepinephrine | LOVENBERG, 1981 |
| Histamine N-MT (HMT) | brain, liver, lung, kidney, heart, muscle, lymph-nodes, spleen, intestine, erythrocyte | histamine | BORCHARDT, 1980 |
| Indolethylamine N-MT | lung, adrenal gland, kidney spleen, heart, liver, pineal gland, erythrocytes, brain, stomach, intestine | serotonin, tryptamine, tyramine, nor-epinephrine, dopamine, amphetamine, normorphine | AXELROD, 1962 |
| Nicotinamide N-MT | liver | nicotinamide | SHIBATA, 1986; |
| N-MT | liver | trichlorotetra-chloroisoquino-lines | FONG and HWANG, 1983 |
| Amine N-MT | liver | azahetcrocycles | DAMANI et al., 1986 |
| *S-Methyltransferases* | | | |
| Thiol S-MT (alkylthiol S-MT) | erythrocytes, liver, kidney lung, brain, intestine | aliphatic sulf-hydryl compounds | WEISIGER and JAKOBY, 1980 |
| Thiopurine MT (arylthiol MT) | erythrocyte, kidney | thiopurines, thiopyrimidines | WOODSON and WEINSHILBOUM, 1983 |

**Table 11.** Physico-chemical properties of some methyltransferases

| Enzyme | Source | Molecular weight | Iso-electric point | Optimum pH value | References |
|---|---|---|---|---|---|
| COMT (soluble) | liver of rat | 23,000 | 5.1; 5.2; 5.3 | 7.3—8.2 | BORCHARDT, 1980, 1981 |
| COMT (membrane bound) | liver of rat | 26,000 | 6.2 | | GROSSMANN et al., 1985 |
| COMT (soluble) Form I or A Form II or B | liver of rat | 23,000 45,000 47,000 | 4.9 4.8 | 7.5 8.0; 9.5 | MARZULLO and FRIEDHOFF, 1979; BORCHARDT, 1980 |
| COMT (soluble) | brain of rat | 23,000 | 5.2 | | HEYDORN et al., 1986 |
| COMT (soluble) | brain of human | 27,500 | 5.0 | | JEFFERY and ROTH, 1985 |
| HIOMT | pineal gland of rat | 78,000 —80,000 | | | AXELROD and WEISSBACH, 1961 |
| HMT | brain of guinea pig | 100,000 | | 7.2—7.4 | THITHAPANDHA and COHN, 1978 |
| HMT | brain of rat, mouse | 29,000 | | | SELLINGER et al., 1978 |
| PNMT | adrenal medulla of bovine | 38,500 | | | CONNETT and KIRSHNER, 1970 |
| Indolethylamine MT | lung of rabbit | | | 8.0—8.5 | AXELROD, 1962 |
| Arylamine N-MT | liver of rat | 27,000 | 4.8 | 7.5 | LYON and JAKOBY, 1981 |
| N-MT | liver of dog | | | 8.0 | FONG and HWANG, 1983 |
| Thiol S-MT | liver of rat | 28,000 | | 7.5 | WEISIGER and JAKOBY, 1980, 1981 |
| Thiol S-MT | brain of rat | | | 7.0 | HIEMKE and GHRAF, 1983 |
| Thiopurine S-MT | lymphocytes of human | | | 6.6 | VAN LOON and WEINSHILBOUM, 1982 |

A. LANGNER; H.-H. BORCHERT; S. PFEIFER

the optimal activity of COMT the presence of $Mg^{2+}$ is necessary. Sulfhydryl groups play an important role for enzyme activity. Agents that modify SH-groups, such as N-ethylmaleimide and p-chloromercuribenzoat, inhibit the activity of methyltransferases (WALKER et al., 1981; WEISIGER and JAKOBY, 1981).

### 3.2.4. Biochemical characteristics

### 3.2.4.1. Cosubstrate of methyltransferases

The cosubstrate of the methylconjugations is adenosyl-L-methionine (SAM, 28). It is generated from ATP and the essential amino acid methionine. The transfer of the adenosyl group to the S-atom of L-methionine is catalyzed by methionine adenosyltransferase (ATP-L-methionine-S-adenosyltransferase; S-adenosylmethioninesynthetase, GUCHAIT, 1979).

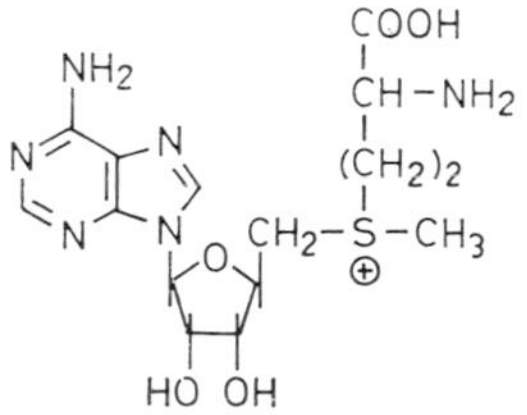

28

The MTs exhibit a strict requirement for SAM as the methyl donor. If there is a high requirement for the donor, the reserve of methionine may be exhausted in the organism. Xenobiotics, such as pyrogallol and L-dopa, strongly decrease the SAM level in brain and liver. However, injection of methionine leads to the elevation of SAM (MULDER, 1982). BORCHARDT (1980) investigated the interactions between HMT and COMT and the methyl donor. The enzyme exhibited high specificity for SAM. The S-configuration at the asymmetric sulfonium pole of SAM was necessary for optimal enzymatic binding and methyl donation in this enzyme-catalyzed reaction. The corresponding R-isomer was inactive as a methyl donor, but it exhibited potent inhibitory activity for the enzyme. Furthermore, a series of structural analogues of SAM with modifications in the amino acid, sugar or base portions of the molecule have been synthesized and evaluated as substrates. The kinetic parameters are listed in Table 12.

From these results, the following conclusions could be drawn by BORCHARDT (1980) concerning the interaction of SAM with its enzyme binding site:

i)  For the amino acid portion of SAM, the terminal carboxyl group, the terminal amino group, the configurations of the amino acid asymmetric

**Table 12.** Kinetic parameters for methyl donor substrates for COMT and HMT (according to BORCHARDT, 1980)

| Substrate | COMT | | HMT | |
| --- | --- | --- | --- | --- |
| | Km (µM) | Vmax* | Km (µM) | Vmax* |
| S-Adenosyl-L-methionine | 9.66 | 1 | 2.54 | 1 |
| S-Adenosyl-D-methionine | — | — | 93 | 0.34 |
| S-3'-Deoxyadenosyl-L-methionine | 377 | 2.7 | 220 | 0.62 |
| S-Aristeromycinyl-L-methionine | 125 | 0.08 | 81 | 0.36 |
| S-Tubercidinyl-L-methionine | 135 | 1.34 | 43.2 | 0.20 |
| S-8-Azaadenosyl-L-methionine | 1,170 | 9.30 | 143.5 | 0.91 |
| S-N⁶-Methyladenosyl-L-methionine | 348 | 2.24 | 179.5 | 0.80 |
| S-3-Deazaadenosyl-L-methionine | 635 | 4.76 | 25.8 | 0.97 |
| S-N⁶-Methyl-3-deazaadenosyl-L-methionine | — | — | 633 | 2.37 |

* Maximal velocities are given as the ratio of the value for the corresponding substrate to that obtained for SAM.

    carbon and the sulfonium centre, or the distance between the sulfonium centre and the asymmetric amino acid carbon are requirements for a maximum potential as a methyl donor.

ii) For the ribose portion of SAM, the 2'- and 3'-hydroxyl groups are requirements, whereas the 1',5'-oxygen bridge of the ribose portion appears less crucial for binding but essential for methyl donation.

iii) For the base portion of SAM, the 6-amino group and the 3-, 7-, and 8-positions of the purine ring are important for maximal binding.

    With purified rat thiol MT in a standard assay system, the Km for SAM was found to be temperature dependent: 1 µM at 37 °C, 0.6 µM at 30 °C, 42 nM at 15 °C.

    CALDWELL (1982b) reported on another methyl donor. In rat brain some primary and secondary amines, such as amphetamine, mescalin, and desipramine, are methylated by a cytosolic enzyme using 5-methyltetrahydrofolic acid as methyl donor.

### 3.2.4.2. Substrate specificity of methyltransferases

The COMT exhibits high specificity for the catechol functionality of the methyl acceptor substrate, but exhibits a broad specificity with regard to other substituents on the aromatic nucleus (BORCHARDT, 1980). In Tables 13 and 14 kinetic parameters for representative catechol substrates are listed.

**Table 13.** Kinetic parameters for methyl acceptor substrates for COMT (according to BORCHARDT, 1980)

| Substrate | Km (mM) | Vmax* |
|---|---|---|
| Dopamine | 0.78 | 0.90 |
| α-Methyldopamine | 0.67 | 0.69 |
| Norepinephrine | 0.26 | 0.66 |
| Epinephrine | 0.40 | 0.28 |
| Isoproterenol | 0.04 | 0.12 |
| N-Acetyldopamine | 0.52 | 0.70 |
| N-Acetylnorepinephrine | 0.26 | 1.11 |
| 3,4-Dihydroxyphenylethanol | 0.27 | 1.58 |
| 3,4-Dihydroxyphenylglycol | 0.40 | 2.16 |
| 3,4-Dihydroxyphenylacetic acid | 0.31 | 1.16 |
| 3,4-Dihydroxymandelic acid | 0.94 | 1.16 |
| 3,4-Dihydroxybenzoic acid | 0.25 | 1.04 |
| 3,4-Dihydroxyacetophenone | 0.02 | 0.70 |
| 3,4-Dihydroxypropiophenone | 0.16 | 0.95 |
| 3,4-Dihydroxybromobenzene | 0.70 | 1.53 |
| 3,4-Dihydroxytoluene | 0.76 | 1.44 |

\* Maximal velocities are given as the ratio for the corresponding substrate to that obtained with 3,4-dihydroxybenzoic acid.

**Table 14.** Kinetic parameters of the COMT from pig liver for some substrates (according to GORDONSMITH et al., 1982)

| Subtrate | Km value (mM) | Vmax (mU · mg$^{-1}$) |
|---|---|---|
| L-Dopa | 1.7 | 291.9 |
| D-Dopa | 2.05 | 194.6 |
| DL-Dopa | 1.86 | 252.5 |
| 3,4-Dihydroxyphenylacetic acid | 0.68 | 778.3 |
| (−)-Adrenalin | 0.51 | 590.9 |
| DL-α-Methyldopa | 3.35 | 206.2 |
| Dopamine | 0.75 | 157.5 |
| α-Methyldopamine | 0.56 | 727.3 |
| (−)-Noradrenaline | 0.52 | 342.8 |
| (+)-α-Methylnoradrenaline | 1.09 | 639.3 |
| (+)-Isoprenalin | 0.37 | 474.8 |
| (+)-α-Ethylisoprenalin | 1.50 | 608.0 |

These data demonstrate that the enzyme catalyzes the methylation of catecholamines and catechols as well as endogenous catecholamine metabolites. This enzyme also methylates catechol drugs and numerous nonphysiological catechols (GORDONSMITH et al., 1982).

O-Methylation of catecholamines and related physiological substrates leads to the formation of the meta O-methylated products. Nonphysiological catechols are methylated to both the meta and para O-methylated products (BORCHARDT, 1980). The meta and para methylated product ratio is dependent on the nature of the aromatic substituent. For substrates containing highly polar substituents, the meta methylated products predominate, whereas for substrates with nonpolar substituents the ratio of the meta to para product is close to unity. Investigations from RAXWORTHY (1986) indicated that stereochemical and steric determinants are important in the interaction of COMT with physiologically and clinically important $\beta$-adrenoceptor agents. O-methylation of isoprenaline and noradrenaline enantiomers was found to be stereo-selective. COMT shows selectivity toward the $(-)$-isomer with respect to the $(+)$-form or racemic mixture.

The catechol O-methylation is strongly dependent on pH (RAXWORTHY and GULLIVER, 1986). The Km for 3,4-dihydroxyphenylacetic acid, for example, is decreased with increasing pH using COMT from pig liver.

RAXWORTHY and GULLIVER (1982) have found estrogens as substrates for COMT from pig liver (Table 15). They assumed that the O-methylation is a major route of ethinylestradiol metabolism which implies that COMT via SAM neutralizes the impaired bile secretion by ethinylestradiol in female rats.

The HMT exhibits a high specificity for histamine. The Km values for histamine vary from 8 to 43 µM. Imidazoles, as histidine, 1-methyl-imidazole-acetic acid and imidazoleacetic acid are not methylated (BORCHARDT, 1980).

The indolethylamine N-MT from rabbit lung methylates a variety of endogenous compounds, such as serotonin, tryptamine, tyramine, norepinephrine and dopamine, and drugs, such as desipramine, amphetamine and normorphine. The N-MT of the rat liver exhibits a broad substrate specificity for primary,

**Table 15.** Kinetic parameters of COMT from pig liver for estrogens (according to RAX-WORTHY and GULLIVER, 1982)

| Substrate | Km value (µM) | Vmax value (mU · mg$^{-1}$) |
|---|---|---|
| 2-Hydroxyethinylestradiol | 11.0 | 521.2 |
| 2-Hydroxyestradiol | 66.8 | 1 052.2 |
| 2-Hydroxyestrone | 38.0 | 795.0 |
| 4-Hydroxyestrone | 12.8 | 159.0 |

A. LANGNER; H.-H. BORCHERT; S. PFEIFER

**Table 16.** Kinetic parameters of the arylamine N-methyltransferase from the rabbit liver for some substrates (according to LYON and JAKOBY, 1981)

| Substrate | Km (mM) | Vmax (nmol · mg⁻¹ · min⁻¹) |
|---|---|---|
| Tryptamine | 0.1 | 50 |
| N-Methyltryptamine | 0.09 | 6.9 |
| L-Tryptophanmethylester | 0.2 | 43 |
| Serotonin | 0.5 | 3.2 |
| Aniline | 1.8 | 44 |
| Imidazole | 0.6 | 23 |
| Histamine | 1.6 | 3.0 |

**Table 17.** Kinetic parameters of the rat liver thiol S-methyltransferase for some substrates (according to WEISIGER and JAKOBY, 1981)

| Substrate | Km (mM) | Vmax (nmol · mg⁻¹ · min⁻¹) |
|---|---|---|
| p-Chlorothiophenol | 0.00054 | 6.2 |
| Phenyl sulfide | 0.0011 | 6.1 |
| 4-Nitrothiophenol | 0.0028 | 7.5 |
| Diethylthiocarbamyl sulfide | 0.012 | 3.5 |
| 2-Thioacetanilide | 0.043 | 7.6 |
| Hydrogen sulfide | 0.064 | 7.8 |
| 2-Benzimidazole thiol | 0.11 | 2.4 |
| Thioglycolic acid | 0.19 | 3.7 |
| L-Cysteine methyl ester | 0.21 | 1.4 |
| Methane thiol | 0.24 | 0.9 |
| N-Acetyl-L-cysteine | 0.40 | 1.0 |
| 6-Propyl-2-thiouracil | 1.0 | 6.4 |
| 1-Methylimidazole-2-thiol | 1.4 | 2.0 |
| 2,3-Dimercaptopropanol | 1.6 | 5.4 |
| 3-Mercaptopropionic acid methyl ester | 4.7 | 3.2 |
| 2-Mercaptoethanol | 8.1 | 5.4 |

secondary and tertiary amines (CALDWELL, 1982b). Kinetic parameters for some substrates by arylamine N-MT from rabbit liver are given in Table 16.

Studies on stereospecificity of N-methylation have been carried out by CUNDY et al. (1985). They reported that R-(+)-nicotine is a substrate of N-MT from guinea pig lung, whereas S-(−)-nicotine is a competetive inhibitor of the N-methylation of the R-(+)-isomer.

The thiol S-MT exhibits a broad specificity for thiol substrates (WEISIGER and JAKOBY, 1980). As shown in Table 17 the thiol S-MT methylates a large variety of lipophilic compounds. The naturally occuring hydrophilic thiols, glutathione and cysteine, act neither as substrates nor as inhibitors (WEISIGER and JAKOBY, 1979).

According to HIEMKE and GHRAF (1983), there is a decrease in maximal velocities of methylation and an increase in the apparent Km values of the reaction with increasing polarity of thiol substrates.

The thiopurine S-methylase catalyzes the S-methylation of thiopurines, thiopurine ribonucleotides and thiopyrimidines. For the methylation of 6-mercaptopurin, for example, Km values of 2.4, 1.8, 1.5 and 0.81 mM have been found in rat blood, intestine, and spleen and human lymphocytes respectively (WALKER et al., 1981).

### 3.2.4.3. Mechanism of the methyltransfer

MTs catalyze the transfer of a methyl group from SAM to the corresponding substrate resulting in the formation of the methylated product and S-adenosyl-homocysteine (SAH).

The kinetic mechanism of COMT has previously been investigated using partially purified enzyme preparations and different mechanisms have been advanced. A random-order mechanism in which a quaternary complex of

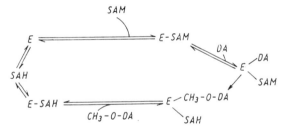

**Fig. 5.** Compulsory-order mechanism of COMT from human brain (according to JEFFERY and ROTH, 1985). DA = dopamine; E = enzyme.

substrate, SAM, magnesium and the enzyme is formed was described by COWARD et al. (1973). A double-displacement (substituted enzyme) mechanism was proposed by BORCHARDT (1973) from the results of studies with dead-end inhibitors. WOODARD et al. (1980) found that the methyltransfer from SAM to the catechol proceeded with an inversion of the configuration of the methyl group.

JEFFERY and ROTH (1985) proposed a compulsory-order mechanism in which a quarternary complex is formed with SAM as preferentially bound substrate (Fig. 5). The same mechanism is hypothesized by TUNNICLIFF et al.

A. LANGNER; H.-H. BORCHERT; S. PFEIFER

(1983) for COMT from rat brain. For the arylamine N-MT a rapid equilibrium random Bi Bi mechanism was proposed (LYON and JAKOBY, 1981). Kinetic studies investigating the mechanism of thiol S-MTs have been carried out by WEISIGER and JAKOBY (1980). They postulated that the enzyme-catalyzed reaction is biphasic.


### 3.2.5. Regulation of methyltransferase activity

Little data are available about regulating mechanisms of the activity of MTs. Steroids appear to have a regulatory role. In pregnancy a two-fold increase in COMT activity in the uterus was observed. Hypophysectomy reduces the activity in the rat liver (BORCHARDT, 1980). Administration of testosterone leads to a decrease of the COMT activity in the brain (PARVEZ et al., 1984). Furthermore, the authors reported an increase of the COMT activity in the brain after adrenalectomy and castration.

The activity of HMT is reduced by castration in male rats and testosterone administration reverses this effect. The administration of corticosteroids increases the activity of HMT in guinea pig stomach (BORCHARDT, 1980).

The thiopurine S-MT was reported to be regulated by testosterone (WOODSON et al., 1981).

The physiological and pathological state may influence the enzyme activity. In brain of adult rats stress produces a significant change in HMT activity in the cerebral cortex (BORCHARDT, 1980). BARTH et al. (1977) have shown that patients with duodenal ulcers have significantly lowered gastric mucosa activity of HMT in comparison to healthy subjects. After hydronephrosis a significant increase of enzyme activity in rabbit kidney has been observed (BARTH, 1975).

The level of human erythrocyte COMT activity is regulated by inheritance. The human genetics of erythrocyte COMT have been studied by WEINSHILBOUM (1974). It was observed that the frequency distribution of erythrocyte COMT activity in a randomly selected population is bimodal. The trait for low erythrocyte enzyme activity is inherited in an autosomal recessive fashion with a gene frequency of approximately 0.5. A pair of alleles at a single locus, $COMT^L$ for low activity and $COMT^H$ for high activity, is responsible for most of the variance in erythrocyte COMT (WEINSHILBOUM and RAYMOND, 1977). Individuals homozygous for $COMT^L$ have low enzyme activity, those homozygous for $COMT^H$ have high activity, and heterozygous subjects have intermediate levels of activity. The erythrocyte enzyme from individuals with low activity is thermolabile in comparison to individuals with high activity, suggesting inherited differences in the structure of the protein molecule.

SLADEK-CHELGREN and WEINSHILBOUM (1981) have observed that the COMT activity of human erythrocytes and lymphocytes is very similar and concluded that the genetic polymorphism which regulates the COMT activity in erythrocytes may also regulate the level of human lymphocyte COMT activity. A common genetic polymorphism has been discovered that is responsible for wide variations in human erythrocyte thiol S-MT activity. KEITH et al. (1983 b) investigated the heredity of human erythrocyte membrane thiol S-MT and found the frequency distribution of enzyme activities was unimodal. These results suggested that inheritance is the primary factor regulating the 5-fold variation in erythrocyte thiol S-MT activity.

WALKER et al. (1981) investigated the polymorphism of the thiopurine S-MT from rat erythrocytes. A 4-fold variation in the enzyme activity has been observed resulting from monogenetic heredity. The frequency distribution of this enzyme activity was trimodal. 90% of the individuals of the population investigated were homozygous for an allele for high enzyme activity, and about 10% were heterozygous for the allele for high enzyme activity and for the allele for low enzyme activity.

### 3.2.6. Inhibition and induction of methyltransferases

The COMT requires $Mg^{2+}$ for optimum activity. However, $Mg^{2+}$ concentrations greater than 2 mM cause inhibition (BORCHARDT, 1981).

The enzyme is very sensitive to inhibition by SAH (WEISIGER and JAKOBY, 1980).

The methylation is inhibited by various phenolic and polyphenolic compounds, as well as structurally related compounds, e.g., tropolene, 8-hydroxy-quinoline, 3-hydroxy-4-pyrone. The mechanism of inhibition is probably one of a competitive manner (BORCHARDT, 1980; BORCHARDT, 1981; MULDER, 1982).

The inhibition of COMT in vivo may be a means for control of the endogenous catecholamine metabolism or for preventing the inactivation of catechols administered exogenously as drugs. A COMT inhibitor, for example, may be used for increasing the central action of L-dopa. This concept has been used in patients with PARKINSON disease (RECHES et al., 1981). BORCHARDT et al. (1982a) and BORCHARDT and BHATIA (1982) investigated some compounds which are potent inhibitors in vitro of COMT from rat liver. These compounds include 5-substituted 3-hydroxy-4-methoxy-benzoic acids and 5-substituted 3-hydroxy-4-methoxy-benzaldehydes. These substances which were strong inhibitors in vitro failed to influence the enzyme activity in vivo. 2,3-Dihydroxy-pyridine was observed by RAXWORTHY et al. (1983) to inhibit the COMT activity from pig liver.

WOODSON et al. (1983) investigated some benzoic acid derivatives as inhibitors of the human kidney thiopurine S-MT. The $IC_{50}$ values ranged from 20 μM for 3,4-dimethoxy-5-hydroxy-benzoic acid to 2.1 mM for acetylsalicylic acid. The mechanism of inhibition was estimated to be noncompetitive or mixed. Structure-activity relationship analysis demonstrated that the benzoic acid structure was important for inhibitory activity, and that inhibition was enhanced by the addition of methoxy and/or phenolic hydroxyl groups to the ring. Similar results have been obtained by KEITH et al. (1983a) who found methoxy derivatives of benzoic acid to be strong inhibitors of the erythrocytes thiopurine S-MT.

There are no data available about specific inducers of MTs.

### 3.2.7. Concluding remarks

Methyltransferases (MTs) provide a major route of catabolism for many endogenous compounds and some xenobiotics. However, the methylation does not lead to more soluble or less toxic metabolites in each case.

While O- and N-MTs methylate xenobiotics similarly to endogenous substrates for these enzymes, S-MTs catalyze the methyl transfer to various drugs with SH-groups. These findings suggest the occurrence of a broad spectrum of thiol MTs.

Toxicological importance seems apparent in the N-MTs from the brain of man, monkey, mouse, rabbit and rat and from rabbit liver which are able to N-methylate 4-phenyl-1,2,3,6-tetrahydropyridine and other exogenous pyridino compounds to N-methylpyridinium ions. Some of such ions are neurotoxins (ANSHER et al., 1986). The methylation to quaternary amines is suggested as a means by which lipophilic compounds having gained entrance to the cell are converted to charged species that efflux much less readily. In this respect N-methylpyridinium compounds possess a long biological half-life which is caused by amine N-MT as well as their possible toxicity.

MTs seem to be important in cancerogenity. As mentioned above, SILVER-MAN and LOWE (1982) have reported on methylation of polycyclic aromatic hydrocarbons and the relationship to cancerogenity caused by these metabolites. The cancerogenity is dependent on the position of the methyl group. Unsubstituted compounds are relatively inactive. The DNA methylation is considered as one possible gene control mechanism of transcription in eukaryotic cells. PFOHL-LESZKOWICZ et al. (1986) have found that the methylation of DNA was decreased if DNA formed an adduct with mutagenic compounds. This could be a possible mechanism of mutagenesis and cancerogenesis.

The thiopurine S-MT also has toxicological relevance. The studies of WALKER et al. (1981) show that the S-methylation of 6-mercaptopurine forms a product acting as a strong inhibitor of purine biosynthesis.

Another viewpoint of the investigation of MTs is the problem of influencing the enzyme activity. This may lead to the possibility of controlling the duration of action of endogenous compounds and drugs with respect to the therapeutic aim.

---

## 3.3. Acetyltransferases

### 3.3.1. Introduction

Although it occurs in many tissues, acetylation is a minor pathway of metabolism. Substrates of acetyltransferases (AT) are aromatic primary amines, aliphatic amines, amino acids, hydrazines and hydrazides. Interest in the formation and disposition of the acetylated compounds is due to the recognition response to corresponding agents by serving to reduce the levels of metabolites with undesirable properties (e.g., arylhydroxylamines) or to provide substrates (e.g., arylhydroxamic acids) for metabolic pathways which generate products that can adversely affect biological systems (KING and GLOWINSKY, 1983).

ATs are important for detoxication and metabolism of both endogenous compounds and xenobiotics. While the acetyl transfer to xenobiotics mostly occurs at the N-atom, for endogenous compounds O-, C- or S-acetylation has been observed (BARMAN, 1969).

The enzymatic N-acetylation of serotonin occurs in many mammalian tissues and is, in general, considered as a common mechanism for the inactivation of serotonin. In the pineal gland the N-acetylation of this amine is important because this enzymatic step initiates the generation of melatonine (KLEIN and KAMBOODIRI, 1981; MOELLER et al., 1987). Some biogenic amines, such as tryptamine, noradrenaline, histamine, phenylethanolamine and tyramine, are inactivated by N-acetylation (ANDRES et al., 1983a).

O-AT are involved in the biosynthesis of the mediator PAF-acether (ALBERT and SNYDER, 1983; BAN et al., 1986). Endogenous thiols, e.g., co-enzyme A, and alcohols, e.g., choline, may also be metabolized by ATs (CALDWELL, 1982b). Endogenous compounds are also activated by ATs; for example, the formation of the neurotransmitter acetylcholine is catalyzed by the choline AT (HARRIS, 1987).

In the metabolism of xenobiotics, the acetylation does not always lead to the production of agents with increased water solubility and less toxicity (CALDWELL, 1982b). Acetylation is a primary route of biotransformation of sulfonamides and isoniazid. The acetyl derivatives of some sulfonamides are not only less lipid soluble, but also less water soluble than their parent compounds. Therefore, injury to the urinary tract may result from the precipitation of the conjugated sulfonamides within renal passage ways as the kidney concentrates the urine and it becomes insoluble in more acidic solution. Acetylation may also lead to toxification of xenobiotics. Various N-acetyl conjugates are known to exhibit pharmacological activity, e.g., the N-acetyl metabolites of sulfanilamide and procainamide (CALDWELL, 1978). Some drugs, e.g., isoniazid, may be transformed to toxic intermediates by N-acetylation (see PFEIFER and BORCHERT, 1975 – 1983).

A. LANGNER; H.-H. BORCHERT; S. PFEIFER

### 3.3.2. Localization and multiplicity of acetyltransferases in mammals

ATs are detected as cytosolic enzymes in many tissues of various species with predominance in cells of the liver and intestinal mucosa. In the rabbit, for example, about 80% of the acetylating activity can be determined in the liver and gut (WEBER and KING, 1981). The distribution in the gastrointestinal tract is heterogenous. High AT activity is found in the small intestine, whereas the stomach and colon exhibit low activities. Low activities have been found also in brain, kidney, lung, spleen, testes and thymus of rabbits (HEARSE and WEBER, 1973). In liver cytosol of hamster, guinea pig, mouse, rat and man, high activities for N-acetylation of arylamines could be demonstrated (KING and GLOWINSKY, 1983). MANDELBAUM-SHAVIT and BLOUDHEIM (1981) have studied the acetylation of p-aminobenzoic acid by human blood. They found that the N-AT activity in blood is 100-fold higher than in erythrocytes.

In Table 18 the distribution of the arylhydroxamic acid N,O-acetyltransferase (AHAT) of some species is shown. This enzyme seems to occur also in the microsomal fraction and was classified according to the results of gel filtration of solubilized preparations (Table 19). Types I, II and III could be distinguished by differences in size, by their sensitivity to the inhibitor diethyl-p-nitrophenylphosphate and by differences in their relative abilities to hydrolyse arylhydroxamic acids and amides.

**Table 18.** Relative cytosolic arylhydroxamic acid-N,O-acetyltransferase activities (according to KING and ALLABEN, 1980)*

| Tissue | Relative activity at | | | | | | | |
|---|---|---|---|---|---|---|---|---|
| | rat | hamster | rabbit | guinea pig | monkey | human | mouse | dog and goat |
| Liver | 111 | 278 | 371 | 9 | 56 | 12 | 5 | 2 |
| Kidney | 29 | 11 | 4 | 12 | | | 2 | 2 |
| Small intestine | 36 | 118 | 43 | 12 | 20 | 17 | 2 | 2 |
| Colon | 38 | 31 | 6 | 10 | 2 | 5 | 2 | 2 |
| Stomach | 24 | 36 | 2 | 14 | 2 | | 2 | 2 |
| Lung | 13 | 18 | 3 | 3 | 2 | 2 | 2 | 2 |
| Mammary gland | 10 | | | | | | | |
| Spleen | 7 | 4 | 2 | 3 | 2 | | 2 | 2 |
| Brain | 3 | 6 | 2 | 4 | | | | |
| Uterus | | | | 10 | | | | |
| Bladder | | | | 20 | | | | |

* The assay involved incubation of N-hydroxy-2-acetylaminofluorene and tRNA with cytosol equivalent.

**Table 19.** Types of arylhydroxamic acid-N,O-acetyltransferase from cytosol and microsomes (according to KING and GLOWINSKI, 1983)

| Enzyme source | Type | | |
|---|---|---|---|
| Rat | | | |
| Liver cytosol | | II | III |
| Liver microsomes | | II | |
| Mammary gland cytosol | | II | III |
| Guinea pig | | | |
| Liver cytosol | I | II | III |
| Liver microsomes | I | II | |

Cytosolic AT is inhibited by diethyl-p-nitrophenylphosphate. It is possible that type II is a cytosolic form which may occur also in the microsomal fraction. Therefore, it can be concluded that type II exists in cytosolic and microsomal form too.

Studies on the AHAT activity in the cytosol of rat small intestine have shown that two forms of this enzyme exist, which could be differentiated by molecular weight, relative ability to use arylhydroxamic acid as substrate and immunochemical properties. Multiple forms of cytosolic AHAT which were separated by gel filtration could also be found in the liver and small intestine of the hamster (KING and ALLABEN, 1978) but not in other tissues. In Table 20 ATs important for the metabolism of xenobiotics are shown.

AHATs are involved in forming reactive electrophilic products from many arylhydroxamic acids. The enzymatic reaction catalyzed by the AHAT includes two successive steps, N-deacylation and O-acylation (WEBER and KING, 1981):

$$\underset{\overset{|}{OH}}{Aryl-N-Ac} \xrightarrow{AHAT} \underset{\overset{|}{O-Ac}}{Aryl-N-H}$$

Arylhydroxamic acid        N-Acetoxyarylamine

From studies on N-AT and AHAT from rabbit liver GLOWINSKY et al. (1980) suggested that N-acetylation and the intramolecular N,O-acetyl transfer by AHAT were properties of the same enzyme. These suggestions have also been made for the enzymes from rat liver cytosol by SAITO et al. (1986). The N-AT and the AHAT from mouse liver, however, have not been found to be identical proteins (KING and GLOWINSKY, 1983).

**Table 20.** Acetyltransferases important for the metabolism of xenobiotics

| Enzyme | Source | References |
|---|---|---|
| Arylamine-N-AT | liver of mouse, pigeon | KING and GLOWINSKY, 1983; ANDRES et al., 1983b |
| Arylhydroxamic acid-N,O-AT | liver of rat | WEBER and KING, 1981 |
| N-Hydroxylamin-O-AT | liver of hamster | SAITO et al., 1986 |
| Cysteine-S-conjugate-N-AT | kidney of rat | DUFFEL and JAKOBY, 1985 |
| Antibiotics-modifying AT | bacteria | HOOD, 1982 |

**Table 21.** Physico-chemical properties of some acetyltransferases

| Enzyme | Source | Molecular weight | Isoelectric point | Optimum pH value | References |
|---|---|---|---|---|---|
| AHAT | liver of rat | 37,000 38,500 | 4.9; 4.9−5.0; 4.5 | 7.0 | ALLABEN and KING, 1984; KING and ALLABEN, 1978, 1980 |
| N-AT (identical with AHAT) | liver of rabbit | 33,000 | | 5−9 | WEBER and KING, 1981; ALLABEN and KING, 1984; SAITO et al., 1986 |
| Arylamine N-AT | liver of pigeon | 31,000 33,000 | 4.8 | | ANDRES et al., 1983a, 1983b |
| N-Hydroxy-arylamine-O-AT (identical with AHAT and N-AT) | liver of hamster | 33,000 | | | SAITO et al., 1986 |

A N-hydroxylamine O-AT activity in hamster liver cytosol was described by SAITO et al. (1986). However, it was suggested that this enzyme was identical with AHAT and arylamine N-AT also found in hamster liver cytosol. The enzyme catalyzing the N-acetylation of cysteine-S-conjugates is found in microsomes and has been isolated and purified from microsomes of rat kidney (DUFFEL and JAKOBY, 1985; OKAJAMA et al., 1984).

### 3.3.3. Physico-chemical properties

In Table 21 some physico-chemical properties of important ATs are listed.
ANDRES et al. (1983b) analysed the primary structure of the arylamine AT
isolated from pigeon liver. This enzyme consists of 52 mol% of hydrophilic
and 38 mol% of hydrophobic amino acids. The ATs from hamster, rabbit and
pigeon liver possess sulfhydryl groups essential for catalyzing the acetyl
transfer (ANDRES et al., 1983b; SAITO et al., 1986; KING and ALLABEN, 1980;
GLOWINSKY et al., 1983) because the AT activity is inhibited by reagents that
react with sulfhydryl groups, such as p-chloromercuri-benzoic acid, N-ethylmaleimide or iodoacetamide. The SH-groups cause a high instability of most
of the ATs. The N-AT and the AHAT activities from rabbit liver are inhibited
by some ions, such as $Cu^{2+}$, $Zn^{2+}$, $Mn^{2+}$ or $Ni^{2+}$ (WEBER and KING, 1981).
JENDRYCZKO et al. (1985) described the action of some metal ions in the N-AT
activity of the hamster liver. The inhibition caused by $Zn^{2+}$ was reversed by
dimercaptosuccinate. The inhibition caused by $Cu^{2+}$ was not influenced by
this compound but was reversed by EDTA. These results suggest the important role of the sulfhydryl groups.


### 3.3.4. Biochemical characteristics

#### 3.3.4.1. Cosubstrates

The physiological cosubstrate of the acetylation is acetyl CoA. The corresponding activating system in the soluble cell fraction activates the acetic
acid by ATP and coenzyme A:

$$R\text{-}COOH + ATP \rightarrow R\text{-}CO\text{-}AMP + PP$$

$$R\text{-}CO\text{-}AMP + CoA\text{-}SH \rightarrow R\text{-}CO\text{-}S\text{-}CoA$$

The acetyl group of the acetyl CoA is transferred by the AT to the substrate.
The concentration of acetyl CoA is rate limiting in this reaction. Low concentrations decrease the rate of the acetyl transfer (OLSEN and MØRLAND,
1983). These authors have also found that acute administration of ethanol
enhanced the elimination of both sulfadimidine and procainamide in humans.
They suggested that ethanol could enhance the speed of the acetylation since
ethanol is metabolized to acetate and may increase the formation and concentration of acetyl CoA. Furthermore, other substances being precursors of
acetyl CoA, such as citrate and pyruvate, enhanced the acetylation. Acetate
treatment increased both Km and Vmax of both sulfanilamide and procainamide acetylation (OLSEN, 1982). It was observed that in rat liver homogenates

Table 22. Km values of the arylamine-N-acetyltransferase from pigeon liver for various acetyl donors (according to ANDRES et al., 1983a)

| Acetyl donor | Acceptor amine | Km value (nm) |
| --- | --- | --- |
| Acetyl Co A | p-nitroaniline | 0.007 |
| Acetylpantetheine | p-nitroaniline | 0.23 |
| N,S-Diacetylcysteamine | p-nitroaniline | 13 |
| S-Acetylmercaptoethanol | p-nitroaniline | 100 |
| p-Nitrophenylacetat | aniline | 0.3 |
| p-Nitroacetanilide | aniline | 1.0 |

acetyl CoA increased the rate of sulfanilamide acetylation in a dose-dependent manner.

Acetyl CoA is the most potent acetyl donor. Besides this compound, acetyl-thiocholine and N-diacetylcystamine are acetyl donors (WEBER and KING, 1981). However, the rate of N-acetylation of isoniazid with each of these donors is lower than with acetyl CoA. ANDRES et al. (1983b) described substituted aromatic amines serving as acetyl donors. In Table 22 Km values of the arylamine N-AT from pigeon liver for various acetyl donors are shown. From these data it will be seen that with decreasing substrate analogy the Km values increase.

3.3.4.2. Substrate specificity

The AHAT of the rabbit liver cytosol catalyzes the acetyl CoA dependent N-acetylation of xenobiotics with $-NH_2$ or $-NH-NH_2$ groups bound directly or bound over a short aliphatic carbon sidechain to the aromatic system. Substrates with such structure are 2-aminofluorene, sulfonamides, isoniazid, hydralazine, phenelzine and procainamide. Aliphatic amines, such as histamine, tyramine and 5-hydroxytryptamine, are also N-acetylated. p-Nitroaniline, phenylalanine, cyclohexylamine and glucosamine do not serve as substrates for the AT from the rabbit liver cytosol (WEBER and KING, 1981). Secondary and tertiary amines are not acetylated. HEIN and WEBER (1982) determined the Km values of the N-AT of rabbit liver (fast acetylator) for some arylamine and hydrazine compounds (Table 23). From these results it can be concluded that there is a high affinity of the N-AT to the arylamines whereas the affinity to hydrazines is lower.

ANDRES et al. (1983a) have studied the relationship between the structure of the substrates and the affinity of arylamine N-AT of the pigeon liver. In Table 24 the Km values of this enzyme for some xenobiotics are shown.

**Table 23.** Km values of the N-acetyltransferase from rabbit liver (rapid acetylators) for some arylamines and hydrazines (according to HEIN and WEBER, 1982)

| Compound | Km (mM) | Compound | Km (mM) |
|---|---|---|---|
| Benzidine | 0.0059 | α-Naphthylamine | 0.081 |
| 2-Aminofluorene | 0.0075 | Hydralazine | 0.36 |
| p-Aminosalicylic acid | 0.038 | Isoniazid | 0.89 |
| β-Naphthylamine | 0.040 | Monoacetylhydrazine | 1.3 |
| Sulfamethazine | 0.042 | Phenelzin | 4.2 |
| p-Aminosalicylic acid | 0.055 | | |

**Table 24.** Km values of the arylamine N-acetyltransferase from pigeon liver for various acceptor amines (according to ANDRES et al., 1983a)

| Acceptor amine | Km (mM) | Acceptor amine | Km (mM) |
|---|---|---|---|
| Aniline | 0.005 | Pentylamine | 0.7 |
| Toluidine | 0.005 | Decylamine | 0.5 |
| Phenylendiamine | 0.003 | Hydrazine | 0.57 |
| Benzidine | 0.001 | Hydroxylamine | 0.42 |
| p-Aminobenzoic acid | 0.23 | Semicarbazide | 3.8 |
| Anthranilic acid | 25 | 2-Phenylethylamine | 0.7 |
| 4-Amino-4'-nitroazobenzene | 0.001 | Benzylamine | 0.6 |
| p-Nitroaniline | 0.15 | Serotonin | 0.17 |
| o-Nitroaniline | —* | Tryptamine | 0.17 |
| Methylamine | 4.5 | Noradrenaline | 0.27 |
| Ethylamine | 60 | Histamine | 0.41 |
| Propylamine | 100 | Tyramine | 2.6 |
| Butylamine | 14 | | |

* No reaction.

Benzidine has a very low Km value suggesting that the biphenyl structure causes a strong binding between substrate and enzyme. From the comparison of the Km values for aniline, benzylamine and phenylethylamine, ANDRES et al. (1983a) concluded that the optimum activity of the enzyme occurs if the $NH_2$ group is bound to a carbon of the aromatic ring.

The cysteine-S-conjugate N-AT isolated from rat kidney microsomes catalyzes the acetylation of lipophilic thioethers of L-cysteine (DUFFEL and JAKOBY, 1985). Neither L-cysteine nor polar thioether conjugates may serve as

**Table 25.** Kinetic parameters of the cysteine-S-conjugate-N-acetyltransferase from rat kidney (according to DUFFEL and JAKOBY, 1985)

| Acetyl acceptor | Km (μM) | Vmax (nmol · min$^{-1}$ · mg$^{-1}$) |
|---|---|---|
| S-Benzyl-L-cysteine | 140 | 440 |
| S-Butyl-L-cysteine | 63 | 320 |
| S-Propyl-L-cysteine | 670 | 330 |
| S-Ethyl-L-cysteine | 7,600 | 360 |
| O-Benzyl-L-serine | 2,100 | 210 |

acetyl acceptors. Kinetic parameters of the cysteine-S-conjugate N-AT for some cysteine conjugates are given in Table 25.

These results clearly demonstrate that the enzyme activity is increased with the lipophilic nature of the substrates. Substitution of sulfur by oxygen as in O-benzyl-L-serine leads to a strong increase of the Km value. However, the Vmax value is hardly influenced.

### 3.3.4.3. Mechanism of the acetyltransfer

A nonsequential (ping-pong) reaction mechanism was proposed by ANDRES et al. (1983a) for the arylamine N-AT of the pigeon liver. In this mechanism the first product is formed and released prior to the reaction of the enzyme with the second substrate (FROMEN, 1979). The first step is the reaction of the enzyme with the acetyl donor forming an acetyl-enzyme intermediate, and the second step is the transfer of the acetyl group from the acetyl-enzyme intermediate to the acceptor substrate as follows:

Acetyl donor + HS-enzyme → acetyl-S-enzyme + deacetylated donor

Acetyl-S-enzyme + acceptor substrate → HS-enzyme + acetylated acceptor

The essential role of the sulfhydryl group of the enzyme is apparent in this mechanism. The acetyl-enzyme intermediate could be isolated by gel filtration; however, after denaturation with urea and iodoacetamide, the isolation of the acetylated enzyme was not possible in the active form. The ping-pong kinetics of the AT described for pigeon was also observed for N-AT from human, monkey, rat and mouse (ANDRES et al., 1983a). Many authors have described an inhibition of ATs by the products.

ANDRES et al. (1983a) have observed that the first product, co-enzyme A, inhibited the activity of the arylamine N-acetyltransferase in a noncompetitive manner. Benzyl-N-acetyl-L-cysteine a product of the cysteine-S-conjugate N-AT reaction was described to be a strong competitive inhibitor of the enzyme (OKAJIMA et al., 1984). A suicidal inactivation of hamster hepatic AHAT was observed by HANNA et al. (1982). They used N-arylhydroxamic acids as suicide substrates in a promising technique for probing the mechanism of AHAT mediated reactions.

### 3.3.5. Genetic regulation of the acetyltransferase activity

The activity of ATs vary greatly between individuals. Genetic studies in man and in rabbits identified two distinct acetylation phenotypes, the so-called "fast or rapid" and "slow" acetylators (cf. MEISEL et al., 1986). The genetic basis for the polymorphic acetylation capacity resides with the existence of two major alleles at a single autosomal gene locus governing the production of the hepatic N-AT. Rapid acetylators are either homozygous or heterozygous for a codominant rapid acetylator gene, while slow acetylators are homozygous for the slow acetylator gene (HEIN et al., 1982).

In rabbits, besides the liver N-AT polymorphism, an extrahepatic polymorphism for p-aminobenzoic acid acetylation was described (McQUENN and WEBER, 1980) being in relation to the acetylation phenotype. It was observed that rapid isoniazid acetylators have low N-AT activity for p-aminobenzoic acid in blood while the slow acetylator phenotype exhibits a high p-aminobenzoic acid N-AT activity in the blood.

The genetically dependent acetylation polymorphism in rabbits seems to be responsible for acyl transfer forming reactive products from arylhydroxamic acids. WEBER et al. (1983) assayed partially purified liver preparations from 17 inbred and partially inbred rabbit strains for polymorphic N-AT and AHAT activities. The strains identified as rapid acetylators had high enzyme activity, and the strains identified as slow acetylators had low enzyme activity. These findings provide additional evidence for the hypothesis that polymorphic N-AT and AHAT are properties of the same enzyme in this species.

Genetic variations are also known for hamster, mouse and rat. HEIN et al. (1982, 1985) reported on the N-acetylation polymorphism in the inbred hamster.

Not all drugs that undergo biological N-acetylation express genetic variability in each species related to the acetylator status. Thus, substrates such as isoniazid and procainamide that exhibit large differences in N-acetylation capacity between rapid and slow acetylators, are classified as polymorphic, whereas substrates like p-aminobenzoic acid which exhibit smaller differences between rapid and slow acetylators are classified as monomorphic (WEBER and HEIN, 1985). In Table 26 some polymorphic and monomorphic substrates in various species are listed. The biochemical basis for these observations is

**Table 26.** Polymorphic (PM) and monomorphic (MM) substrates for N-acetyltransferases of various species

| Substrate | Human | Rabbit | Mouse | Hamster | References |
|---|---|---|---|---|---|
| Isoniazid | PM | PM | PM | MM | CALDWELL, 1982b; HEIN et al., 1982, 1985 |
| Hydralazine | PM | | | | CALDWELL, 1982b |
| Sulfamethazine | PM | PM | | MM | CALDWELL, 1982b; HEIN et al., 1982 |
| Dapsone | PM | | | | CALDWELL, 1982b |
| Procainamide | PM | PM | PM | MM | CALDWELL, 1982b; HEIN et al., 1982, 1985 |
| p-Aminobenzoic acid | MM | | MM | PM | WEBER et al., 1978; HEIN et al., 1982, 1985 |
| p-Aminosalicyclic acid | MM | | | PM | CALDWELL, 1982b; HEIN et al., 1982 |
| Sulfanilamide | MM | | | | CALDWELL, 1982b |
| 2-Aminofluorene | | PM | | MM | HEIN et al., 1982 |

different depending on species. In both the rabbit and the mouse a single cytosolic N-AT is responsible for catalyzing the N-acetylation of both monomorphic and polymorphic substrates (CALDWELL, 1982b). In the inbred hamster liver cytosol, in contrast, HEIN and WEBER (1982) reported two separable forms of liver N-AT activity. The results of this investigation characterize the substrate-dependent acetylation of both monomorphic and polymorphic substrates by the two enzymes in each acetylator gene type (HEIN et al., 1985). More recently, studies were performed to clarify the molecular basis of the genetically polymorphic N-AT activities in humans and rabbit. Differences with regard to chromatographic behaviour, pH optimum, thermal stability, and substrate affinity, were not detected in respect of enzymes of slow and rapid acetylator phenotypes. However, from immunochemical studies, there is good evidence that liver N-ATs from rapid and slow acetylating rabbits can be regarded as isozymes (ANDRES et al., 1983b; McQUENN and WEBER, 1980).

WEBER and KING (1981) have also suggested that there are molecular differences between N-AT of slow and rapid acetylators and that the phenotypic differences do not result from different levels of the same enzyme.

The acetylation polymorphism is of great therapeutic significance with respect to arylamine and hydrazine drug pharmacokinetics and toxicity. For example, slow acetylators are more susceptible to the toxic side effects of isoniazid on the central nervous system, and to drug-induced lupus erythematosus from isoniazid, hydralazine and procainamide (HEIN and WEBER, 1982). Isoniazid is an inhibitor of the phenytoin metabolism. In slow acetylators, a cumulation of phenytoin after isoniazid treatment was observed (PFEIFER et al., 1988).

Furthermore, a higher predominance of arylamine-induced bladder cancer was found among slow acetylators (HEIN and WEBER, 1982).

### 3.3.6. Toxicological aspects of acetyltransferase reactions

Aromatic amines are believed to induce tumors as a consequence of their N-oxidation catalyzed by microsomal monooxygenases followed by a second metabolic transformation to derivatives reacting with nuleic acids (KING and ALLABEN, 1980). Among the following steps of activation forming reactive products the N,O-acetyl transfer catalyzed by AHAT is considered as one of the most important enzymatic reactions (SAITO et al., 1986). The mechanism of this activation can be explained as follows (see MANGOLD and HANNA, 1982):

A. LANGNER; H.-H. BORCHERT; S. PFEIFER

First of all the aromatic amine (29) is N-acetylated by the N-AT followed by the cytochrome P-450 dependent N-hydroxylation. In the next step the acetyl group is transferred by the AHAT from the arylhydroxamic acid (31) to the oxygen of the arylhydroxylamine intermediate (32) forming N-acetoxy-arylamine (33). This compound is hydrolysed and a resonance-stabilized aryl-nitrenium ion (34) is formed which can react with biological macromolecules (c) such as DNA, RNA or proteins and causes, furthermore, the irreversible inactivation of the enzyme AHAT (d). AHAT also catalyzes the transfer of the acetyl groups from an arylhydroxamine acid to aromatic amines (a) form-ing stable amides (35). One can see that the AHAT uses the aromatic hydro-xamic acid as an acetyl donor. The arylhydroxylamine formed in this donor reaction is O-acetylated by the same enzyme (b) to the reactive N-acetoxy-arylamine (33). Metabolic activation of arylhydroxamic acids by N,O-acetyl-transfer may be responsible for some aromatic amine-induced tumors. It is suggested that tumors of the mammary glands can readily be induced on administration of carcinogenic aromatic amines, such as N-OH-2-acetylamino-fluorene to female rats (KING et al., 1979; SHIRAI et al., 1981). In the metabolic activation of some amines, besides the acetylation reaction sulfate conjugation is also involved (KING et al., 1983). Arylhydroxylamines, which are inter-mediates in the activation mechanism, may be mutagenic compounds.

Acetylation reactions seem to be involved in toxification of some drugs. Rapid acetylators were reported to be more sensitive to isoniazid induced hepatitis. But isoniazid, itself, is not hepatotoxic. It firstly must be meta-bolically activated to its hepatotoxic compound(s). The acetylation seems to be the rate-limiting step in forming the metabolite acetylhydrazine capable of inducing hepatitis (NEBERT, 1980). The risk of dihydralazine hepatitis was reported to be 5-fold higher in low acetylators in comparison to rapid acetyla-tors (BAUMGARTEN et al., 1988).

### 3.3.7. The importance of acetyltransferases in development of resistance of bacteria to antibiotics

Bacterial enzymes which inactivate antibiotics have been recognized since the beginning of the antibiotic era. The major antibiotics which act as targets for these enzymes are aminoglycosides, chloramphenicol and $\beta$-lactams (HOOD, 1982). In the inactivation of aminoglycosides and chloramphenicol N-ATs and O-ATs are involved. Aminoglycoside-modifying enzymes are widely distributed among bacteria. For example, an in vitro evaluation of 32 gentamicin-resistant, Gram-negative bacteria performed by KETTNER et al. (1984) revealed that 24 strains produced gentamicin-modifying enzymes. It is known that amino-glycosides are inactivated by three enzymatic transferase reactions: by phos-phorylation or adenylylation of hydroxylgroups or by acetylation of their amine groups (COOMBE and GEORGE, 1982).

In kanamycine (**36**) the enzymes aminoglycoside-AT (AAT), aminoglycoside-adenyltransferase (AAD) and aminoglycoside-phosphotransferase (APH) degrade the molecule at the following positions:

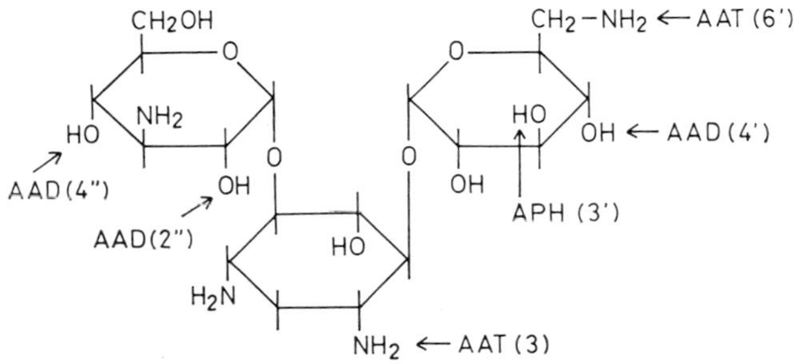

<u>36</u>

N-ATs modify aminoglycosides in position 6' and 3'. Accordingly 2-ATs exist in terms of the stereo- and regiospecificity. Furthermore various iso-enzymes are known. Recently five isoenzymes of aminoglycoside 3-AT have been detected. The production of a chloramphenicol AT has been demonstrated in many strains of chloramphenicol resistent bacteria. The biochemical mechanism of bacterial resistance to chloramphenicol (**37**) is commonly that of inactivation by O-acetylation of the antibiotic, a reaction catalyzed by the chloramphenicol AT.

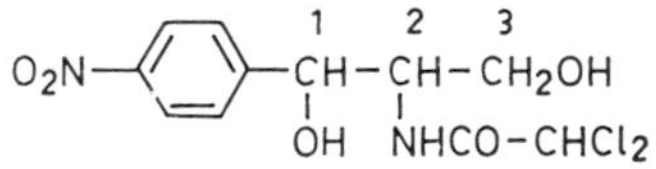

<u>37</u>

Acetylation at position 3 is followed by chemical rearrangement resulting in transfer of the acetyl group to position 1, thus allowing further enzymatic acetylation at position 3 to give the diacetate. Both the mono- and the diacetate are inactive (HOOD, 1982). The forward reaction proceeds to the formation of a ternary complex by a rapid-equilibrium mechanism wherein the addition of substrates may be random by a preference for acetyl CoA as the leading substrate can be detected (KLEANTHOUS and SHAW, 1984). The enzyme is cytoplasmic and tetrameric consisting of four identical catalytic subunits with the molecular weight of about 25,000 in all bacterial species. The plasmid associated genes for chloramphenicol AT common to the intestinal bacteria specify three main enzyme types I, II and III (KLEANTHOUS ˙ and SHAW, 1984).

A. LANGNER; H.-H. BORCHERT; S. PFEIFER

### 3.3.8. Concluding remarks

Acetyltransferases (ATs) are an important group of metabolizing enzymes. However, the biotransformation of drugs by acetylation is an insignificant way of drug metabolism because the termination of action and the excretion of the drugs is hardly influenced. For a few drugs, such as sulfonamides, isoniazid and tiaramide, the acetylation may be a major route of biotransformation (IWASAKI et al., 1983; OLSEN and MØRLAND, 1981; MCLEAN et al., 1983; PFEIFER et al., 1988), but in general the physico-chemical properties and the biological activity are not changed drastically.

The involvement of AT reactions in toxicological processes is of great importance. In a multistep process for the metabolic activation of carcinogenic arylamines to form DNA-binding species, the enzymatic acetyl transfer follows the initial N-hydroxylation by a microsomal mixed function oxidase as one of the important enzymatic reactions leading to the formation of highly reactive products capable of binding covalently to tissue macromolecules. These reactive intermediates may be involved in cancerogenesis and mutagenesis.

Another mechanism being of toxicological significance is the polymorphism leading to adverse reactions to drugs which are mainly degraded by acetylation in many subjects. On the other hand the genetic variance may lead to a failing of the pharmacological effect.

ATs are not yet characterized completely. There are difficulties in isolating and purifying the enzyme proteins because of the high instability of the enzymes caused mainly by oxidation of sulfhydryl groups. Therefore, little is known about specific inhibitors and inducers of the enzyme activity.

The great toxicological importance of ATs requires a better knowledge of these enzymes and the possibilities of influencing their activities.

# 4.   References

ABENHEIM, L., Y. ROMAIN, and O. KUCHEL (1981), J. Physiol. Pharmacol. **59**, 300—306.
ALBERT, D. H. and F. SNYDER (1983), J. Biol. Chem. **258**, 97—102.
ALLABEN, W. T. and C. M. KING (1984), J. Biol. Chem. **259**, 12128—12134.
ANDERSON, R. J. and B. L. JACKSON (1984), Clin. Chim. Acta **138**, 185—196.
ANDRES, H. H., H. J. KOLB, R. J. SCHREIBER, and L. WEISS (1983a), Biochim. Biophys. Acta **746**, 193—201.
ANDRES, H. H., H. J. KOLB, and L. WEISS (1983b), Biochim. Biophys. Acta **746**, 182—192.
ANSHER, S. S., J. L. CADET, W. B. JAKOBY, and J. R. BAKER (1986), Biochem. Pharmacol. **19**, 3359—3363.
AXELROD, J. and H. WEISSBACH (1960), Science **131**, 1312—1315.
AXELROD, J. (1982), J. Pharmacol. Exp. Ther. **138**, 28—33.

BAKKE, J. E. (1986), in: Advances in Xenobiotic Conjugation Chemistry, (G. D. PAUL-SON, J. CALDWELL, D. H. HUTSON, J. J. MENN, eds.), ACS, Symposium Series, Washington, p. 301.

BAN, C., M. M. BILLAH, C. T. TRUONG, and J. M. JOHNSTON (1986), Arch. Biochem. Biophys. **246**, 9—18.

BARAŃCZYK-KUŹMA, A., R. T. BORCHARDT, C. SCHASTEEN, and C. L. PINNICK (1981), Psychopharmacol. Bull. **17**, 50—51.

BARAŃCZYK-KUŹMA, A., R. T. BORCHARDT, and C. L. PINNICK (1985), Acta Biochim. Pol. **32**, 35—45.

BARAŃCZYK-KUŹMA, A. and T. SZYMCZYK (1987), Biochem. Pharmacol. **36**, 3141—3146.

BARMAN, T. E. (1969), Enzyme Handbook, Bd. 1, Springer-Verlag, Berlin, Heidelberg, New York, p. 499.

BARTH, H., K. NABER, K. E. BARTHEL, I. NIEMEYER, and W. LORENZ (1975), Agents Act. **5**, 442—443.

BARTH, H., H. TROIDL, W. LORENZ, H. ROHDE, and R. GLASS (1977), Agents Act. **7**, 75—79.

BAUMANN, E. (1876), Pflügers Arch. ges. Physiol. Menschen und Tiere **12**, 63, 69.

BAUMANN, E. (1876), Pflügers Arch. ges. Physiol. Menschen und Tiere **13**, 285—308.

BAUMANN, E. (1876), Ber. dtsch. chem. Ges. **9**, 54.

BAUMGARTEN, R., W. SIEGMUND, J.-D. FENGLER, G. FRANKE, R. REICHARDT, and R. KRÜGER (1988), Z. klin. Med. **43**, 87—89.

BONHAM CARTER, S. M., G. REIN, V. GLOVER, and M. SANDLER (1983), Br. J. Clin. Pharmacol. **15**, 323—330.

BORCHARDT, R. T. (1973), J. Med. Chem. **16**, 387—391.

BORCHARDT, R. T. (1980), in: Enzymatic Basis of Detoxication, Vol. II, (W. B. JAKOBY, ed.), Academic Press, New York, pp. 43—61.

BORCHARDT, R. T. (1981), in: Methods in Enzymology, Vol. 77, (W. B. Jakoby, ed.), Academic Press, 267—272.

BORCHARDT, R. T. and C. S. SCHASTEEN (1982), Biochim. Biophys. Acta **708**, 272—279.

BORCHARDT, R. T. and P. BHATIA (1982), J. Med. Chem. **25**, 263—277.

BORCHARDT, R. T., J. H. HUBER, and M. HOUSTON (1982a), J. Med. Chem. **25**, 258 to 263.

BORCHARDT, R. T., C. S. SCHASTEEN, and S.-E. WU (1982b), Biochim. Biophys. Acta **708**, 280—293.

BOUTHILLIER, M., A. CHAPDELAINE, G. BLEAU, and K. D. ROBERTS (1981), Steroids **38**, 523—535.

BOUTHILLIER, M., G. BLEAU, A. CHAPDELAINE, and K. D. ROBERTS (1984a), Biol. Reprod. **31**, 936—941.

BOUTHILLIER, M., G. BLEAU, A. CHAPDELAINE, and K. D. ROBERTS (1984b), Can. J. Biochem. Cell Biol. **63**, 71—76.

BOUTHILLIER, M., G. BLEAU, A. CHAPDELAINE, and K. D. ROBERTS (1985), J. steroid Biochem. **22**, 733—738.

BROOKS, S. C., B. A. PACK, J. ROZHIN, and C. CHRISTENSEN (1982), in: Sulfate Metabolism and Sulfate Conjugation, (G. J. MULDER, J. CALDWELL, G. M. J. VAN KEMPEN, J. VONK, eds.), Taylor and Francis, London, 173—180.

BUTLER, P. R., R. J. ANDERSON, and D. L. VENTON (1983), J. Neurochem. **41**, 630—639.

CALDWELL, J. (1978), in: Conjugation Reactions in Drug Biotransformation, (A. AITO, ed.), Elsevier/North Holland Biomedical Press, Amsterdam, 477—485.

CALDWELL, J. (1980), in: Concepts in Drug Metabolism, Part A, (P. JENNER, B. TESTA, eds.), Marcel Dekker, New York, 211.

CALDWELL, J. (1982a), in: Metabolic Basis of Detoxication-Metabolism of Functional Groups, (W. B. JAKOBY, ed.), Academic Press, New York, 271—290.

CALDWELL, J. (1982b), in: Metabolic Basis of Detoxication — Metabolism of Functional Groups, (W. B. JAKOBY, ed.), Academic Press, New York, 291—306.

CALDWELL, J., J. C. HOBART, and J. A. COTGREAVE (1986), in: Development of Drugs and Modern Medicines, (J. W. CORROD, G. G. GIBSON, M. MITCHARD, eds.), Verlag Chemie, Weinheim, 267.

CAMPBELL, N. R. C., J. H. DUNNETTE, G. MWALUKO, J. VAN LOON, and R. M. WEINSHILBOUM (1984), Clin. Pharmacol. Ther. **35**, 55—63.

CAMPBELL, N. R. C., R. S. SUNDARAM, P. G. WERNESS, J. VAN LOON, and R. M. WEINSHILBOUM (1985), Clin. Pharmacol. Ther. **37**, 308—315.

CHEN, L. J. (1982), Biochim. Biophys. Acta **717**, 316—321.

CHEN, L. J. and J. H. SEGEL (1985), Arch. Biochem. Biophys. **241**, 371—379.

COHN, R. (1893), Hoppe Seyler's Z. physiol. Chem. **17**, 274.

CONNETT, R. J. and N. KIRSHNER (1970), J. Biol. Chem. **245**, 329—334.

COOMBE, R. G. and A. M. GEORGE (1982), Biochemistry **21**, 871—875.

COWARD, J. K., E. P. SLISZ, and F. Y. H. WU (1973), Biochemistry 2291—2297.

COWARD, J. K. (1981), in: Studies in Organic Chemistry, Vol. 10, (B. S. GREEN, Y. ASHANI, D. CHIPMAN, eds.), Elsevier Scientific Publishing Company, Amsterdam, 189—199.

CUNDY, K. C., P. A. CROOKS, and C. S. GODIN (1985), Biochem. Biophys. Res. Commun. **128**, 312—316.

DAMANI, L. A., M. S. SHAKER, and C. S. GODIN (1986), J. Pharm. Pharmacol. **38**, 547—550.

DAWSON, J. R., K. NORBECK, and P. MOLDEUS (1983), Biochem. Pharmacol. **32**, 1789 to 1791.

DUFFEL, M. W. and W. B. JAKOBY (1981), J. Biol. Chem. **256**, 11123—11127.

DUFFEL, M. W. and W. B. JAKOBY (1985), in: Methods in Enzymology, Vol. 113, (A. MEISTER, ed.), Academic Press, New York, 516—520.

EADSFORTH, C. V. and D. H. HUTSON (1984), in: Foreign Compound Metabolism, (J. CALDWELL, G. D. PAULSON, eds.), Taylor and Francis, London, 171.

FEDERSPEIL, M. J. and S. S. SINGER (1981), Comp. Biochem. Physiol. **69B**, 511 to 516.

FONG, K.-L. and B.-H. HWANG (1983), Biochem. Pharmacol. **32**, 2781—2786.

FOYE, W. W. and V. KULAPADITHAROM (1985), J. Pharm. Sci. **74**, 355—358.

FROMEN, H. J. (1979), in: Methods in Enzymology, Vol. **63**, (D. L. PURICH, ed.), Academic Press, New York, 42—53.

GLOWINSKI, I. B., and W. W. WEBER (1980), J. Biol. Chem. **255**, 7883—7890.

GLOWINSKI, I. B., W. W. WEBER, J. M. FYSH, J. B. VAUGHT, and C. M. KING (1980), J. Biol. Chem. **255**, 7883—7890.

GORDONSMITH, R. H., M. J. RAXWORTHY, and P. A. GULLIVER (1982), Biochem. Pharmacol. **31**, 433—437.

GREEN, J. M. and S. S. SINGER (1983), Can. J. Biochem. Cell Biol. **61**, 15—21.

GROSSMAN, M. H., C. R. CREVELING, R. RYBCZYNSKI, M. BRAVERMAN, C. ISERSKY, and X. C. BREAKEFIELD (1985), J. Neurochem. **44**, 421—432.

GUCHHAIT, R. B. (1979), in: Transmethylation, (E. USDIN, R. T. BORCHARDT, C. R. CREVELING, eds.), Elsevier/North Holland Biomedical Press, Amsterdam, 9—18.

HANNA, P. E., R. B. BANKS, and V. C. MARHEVKA (1982), Molec. Pharmacol. **21**, 159—165.

HARRIS, J. B. (1987), J. Neurochem. **48**, 702—708.

HEARSE, D. J. and W. W. WEBER (1973), Biochem. J. **132**, 519—526.

HEIN, D. W. and W. W. WEBER (1982), Drug Metab. Dispos. **10**, 225—229.

HEIN, D. W., J. G. OMICHINSKI, J. A. BREWER, and W. W. WEBER (1982), J. Pharmacol. Exp. Ther. **220**, 8—15.

HEIN, D. W., W. G. KIRLIN, R. J. FERGUSON, and W. W. WEBER (1985), J. Pharmacol. Exp. Ther. **234**, 358—364.
HEYDORN, W. E., G. J. CREED, C. R. CREVELING, and D. M. JACOBOWITZ (1986), Neurochem. Int. **8**, 581—586.
HIEMKE, C. and R. GHRAF (1983), J. Neurochem. **40**, 592—594.
HIS, W. (1887), Naunyn-Schmiedeberg's Arch. exp. Pathol. Pharmakol. **22**, 253.
HOBKIRK, R. (1985), Can. J. Biochem. Cell Biol. **63**, 1127—1144.
HOBKIRK, R. and C. A. CARDY (1985), Can. J. Biochem. Cell Biol. **63**, 785—791.
HOOD, J. D. (1982), Spec. Publ. Soc. Gen. Microbiol. **6**, 131—145.
HORTON, J. K., M. J. MEREDITH, and J. R. BEND (1987), J. Pharmacol. Exp. Ther. **240**, 376—380.
HOSWITZ, J. P., V. K. IYER, H. B. VASDAM, J. COROMBOS, and S. C. BROOKS (1986), J. Med. Chem. **29**, 692—698.
ISHIZAKA, T. and K. ISHIZAKA (1984), Prog. Allergy **34**, 188—235.
IWASAKI, K., T. SHIRAGA, K. NODA, K. TADA, and H. NOGUCHI (1983), Xenobiotica **13**, 273—278.
IWASAKI, K., T. SHIRAGA, K. NODA, K. TADA, and H. NOGUCHI (1986), Xenobiotica **16**, 651—659.
JAKOBY, W. B. (1978), Adv. Enzymol. **46**, 383—414.
JAKOBY, W. B., R. D. SEKURA, E. S. LYON, C. J. MARCUS, and J. WANG (1980), in: Enzymatic Basis of Detoxication, Vol. 2, (W. B. JAKOBY, ed.), Academic Press, New York, 199—227.
JAKOBY, W. B., E. S. LYON, G. MARCUS, and G. L. WAND (1981), in: Methods in Enzymology, Vol. 77, (W. B. JAKOBY, ed.), Academic Press, New York, 206—213.
JEFFERY, D. R. and J. A. ROTH (1985), J. Neurochem. **44**, 881—885.
KEITH, R. A., R. T. ABRAHAM, P. PAZMINO, and R. M. WEINSHILBOUM (1983a), Clin. Chim. Acta **131**, 257—272.
KEITH, R. A., J. VAN LOON, L. F. WUSSOW, and R. M. WEINSHILBOUM (1983b), Clin. Pharmacol. Ther. **34**, 521—528.
KELLER, W. and F. WÖHLER (1842), Liebigs Ann. Chem. **43**, 108.
KETTNER, M., J. NAVAROVA, G. LEBEK, and V. KRCMERY (1984), Zbl. Bakt. Hyg. A **257**, 372—382.
KIM, D.-H., L. KONISHI, and K. KOBASHI (1986), Biochim. Biophys. Acta **872**, 33 to 41.
KING, C. M. and W. T. ALLABEN (1978), in: Conjugation Reactions in Drug Biotransformation, (A. AITO, ed.), Elsevier/North Holland Biomedical Press, Amsterdam, 431—441.
KING, C. M., N. R. TRAUB, Z. M. LORTZ, and M. R. THISSEN (1979), Cancer Res. **39**, 3369—3372.
KING, C. M. and W. T. ALLABEN (1980), in: Enzymatic Basis of Detoxication, Vol. 2, (W. B. JAKOBY, ed.), Academic Press, New York, 187—197.
KING, C. M., and I. B. GLOWINSKI (1983), Environm. Health Perspect. **49**, 43—50.
KING, C. M., C. Y. WANG, M.-S. LEE, J. B. VAUGHT, M. HIROSE, and K. C. MORTON (1983), in: Extrahepatic Drug Metabolism and Chemical Carcinogenesis, (J. RUDSTRÖM, J. MONTELIUS, M. BENGTSSON, eds.), Elsevier Science Publishers B.V., Amsterdam, 557—566.
KIRKPATRICK, R. B., and R. A. BELSAAS (1985), J. Lipid. Res. **26**, 1431—1437.
KLEANTHOUS, C. and W. V. SHAW (1984), Biochem. J. **223**, 211—220.
KLEIN, D. C. and M. A. A. KAMBOODIRI (1981), in: Function and Regulation of Monoamine Enzymes: Basic and Clinical Aspects, (E. USDIN, N. WEINER, M. B. H. YOUDIM, eds.), Macmillan, London, 711—722.
KOSTER, H., E. SCHOLTENS, and G. J. MULDER (1979), Med. Biol. **57**, 340—344.

KOSTER, H., I. HALSEMA, E. SCHOLTENS, M. KNIPPERS, and G. J. MULDER (1981), Biochem. Pharmacol. 30, 2569—2575.

KRIJGSHELD, K. R., E. SCHOLTENS, and G. J. MULDER (1981), Biochem. Pharmacol. 30, 1973—1979.

KUCHEL, O., N. T. BUU, P. HAMET, P. LAROCHELLE, M. BOURGUE, and J. GENEST (1984), J. Lab. Clin. Med. 104, 238—244.

LEWIS, W. G., K. WITT, and S. S. SINGER (1981), Proc. Soc. Exp. Biol. Med. 166, 70—76.

LOVENBERG, W. (1982), in: Biochemistry of S-Adenosylmethionine and related Compounds, (E. USDIN, R. T. BORCHARDT, C. R. CREVELING, eds.), Macmillan, London, 427—436.

LYMAN, S. D. and A. POLAND (1983), Biochem. Pharmacol. 32, 3345—3350.

LYON, E. S. and W. B. JAKOBY (1981), in: Methods in Enzymology, Vol. 77, (W. B. JAKOBY, ed.), Academic Press, New York, 263—266.

MANDELBAUM-SHAVIT, F. and S. H. BLOUDHEIM (1981), Biochem. Pharmacol. 30, 65—69.

MANGOLD, B. L. K. and P. E. HANNA (1982), J. Med. Chem. 25, 630—638.

MARIO, A., E. SALGADO, M. TRUEBA, and J. H. MACARULLA (1986), Comp. Biochem. Physiol. 85B, 795—803.

MARZULLO, G. and A. J. FRIEDHOFF (1979), in: Transmethylation, (E. USDIN, R. T. BORCHARDT, C. R. CREVELING, eds.), Elsevier/North Holland Biomedical Press, Amsterdam, 277—286.

MARZULLO, G. and J. FRIEDHOFF (1981), in: Function and Regulation of Monoamine Enzymes: Basic and Clinical Aspects, (E. USDIN, N. WEINER, M. B. H. YOUDIM, eds.), Macmillan, London, 665—673.

MCLEAN, S., H. GALLOWAY, S. BUTLER, and D. WHITTLE (1983), Xenobiotica 33, 81—85.

MCQUENN, C. and W. W. WEBER (1980), Biochem. Genet. 18, 889—904.

MEERMAN, J. H. N., A. B. D. VAN DOORN, and G. J. MULDER (1980), Cancer Res. 40, 3772—3779.

MEERMAN, J. H. N., F. A. BELAND, and G. J. MULDER (1981), Carcinogenesis 2, 413 to 416.

MEISEL, M., T. SCHNEIDER, W. SIEGMUND, S. NIKSCHICK, R.-J. KLEBINGAT, and A. SCHERBER (1986), Biol. Res. Pregn. 7, 74—76.

MILLER, J. A. and E. C. MILLER (1976), IARC Sci. Publ. 12, 153—176.

MIZUMA, T., M. HAYASHI, and S. ANAZU (1983), Pharmacobio-Dyn. 6, 851—858.

MOELLER, M., S. REUSS, J. OLCESE, J. STEHLE, and H. VOLLRATH (1987), Experientia 43, 186—188.

MOUELHI, M. E. and F. C. KAUFFMAN (1986), Hepatology 6, 450—456.

MULDER, G. J. (1979), Trends Biochem. Sci. 4, 86—90.

MULDER, G. J. (1982), in: Metabolic Basis of Detoxication — Metabolism of Functional Groups, (W. B. JAKOBY, J. R. BEND, J. CALDWELL, eds.), Academic Press, New York, 247—269.

MULDER, G. J. (1986), Fed. Proc. 45, 2229—2234.

MUNK, I. (1876), Pflügers Arch. ges. Physiol. Menschen und Tiere 12, 146.

NEBERT, D. W. (1980), in: Enzymatic Basis of Detoxication, Vol. 1, (W. B. JAKOBY, ed.), Academic Press, New York, 25—68.

NISSINEN, E. (1984), Biochem. Pharmacol. 33, 3105—3108.

OKAJIMA, K., M. INOUE, and K. ITCH (1984), Bur. J. Biochem. 142, 281—286.

OLSON, H., and J. MØRLAND (1981), Acta pharmacol. toxicol. 49, 102—109.

OLSON, H., J. MØRLAND, and M. A. ROTHSCHILD (1981), Acta pharmacol. toxicol. 49, 438—446.

OLSON, H. (1982), Acta pharmacol. toxicol. **50**, 67—74.
OLSON, H. and J. MØRLAND (1983), Pharmacol. Biochem. Behav. **18**, Suppl. 1, 295—300.
OTTERNESS, D. M., R. A. KEITH, A. L. KERREMAUS, and R. M. WEINSHILBOUM (1986), Drug Metab. Dispos. **14**, 680—688.
PARVEZ, H., M. BASTART-MALSOT, and S. PARVEZ (1984), Neuroendocrinol. Lett. **6**, 187—192.
PEARSON, R. K., T. P. MAUS, R. J. ANDERSON, L. C. WOODSON, C. REITER, and R. M. WEINSHILBOUM (1981), in: Function and Regulation of Monoamine Enzymes: Basic and Clinical Aspects, (E. USDIN, N. WEINER, B. H. YOUDIM, eds.), Macmillan, London, 735—743.
PENNINGS, E. J. M., J. L. VAN BRUSSEL, J. ZANEN, and G. M. J. VAN KEMPEN (1981), in: Phenolsulfotransferases in Mental Health Research, (M. SANDLER, E. USDIN, eds.), Macmillan, London, 29—43.
PENNINGS, E. J. M. and G. M. J. VAN KEMPEN (1982), in: Sulfate Metabolism and Sulfate Conjugation, (G. J. MULDER, J. CALDWELL, G. M. J. VAN KEMPEN, R. J. VONK, eds.), Taylor and Francis, London, 37—44.
PFEIFER, S. and H.-H. BORCHERT (1975, 1977, 1979, 1981, 1983), Biotransformation von Arzneimitteln, Vol. 1—5, VEB Verlag Volk und Gesundheit, Berlin; Verlage Chemie, Weinheim—New York—Deerfield Beach, Florida—Basel, 1977—1983.
PFEIFER, S., P. PFLEGEL, and H.-H. BORCHERT (1988), Grundlagen der Biopharmazie, VEB Verlag Volk und Gesundheit, Berlin.
PFOHL-LESZKOWICZ, A., E. HEBERT, G. SAINT-RUF, M. LENG, and G. DIRHEIMER (1986), Cancer Lett. **32**, 65—71.
PICOTTI, G. B., A. M. CESURA, R. KETTLER, and M. D. PRADE (1981a), Psychopharmacoll. Bull. **17**, 45—46.
PICOTTI, G. B., A. M. CESURA, M. D. GALVA, P. MANTEGAZZA, R. KETTLER, and M. D. PRADA (1981b), in: Phenolsulfotransferase in Mental Health Research, (M. SANDLER, E. USDIN, eds.), Macmillan, London, 44—54.
QUICK, A. J. (1927), J. Biol. Chem. **77**, 581.
RAMLI, J. B. and J. F. WHELDRAKE (1981), Comp. Biochem. Physiol. **69C**, 379—381.
RAXWORTHY, M. J. and P. A. GULLIVER (1982), J. steroid Biochem. **17**, 17—21.
RAXWORTHY, M. J., I. R. YOUDE, P. A. GULLIVER (1983), Biochem. Pharmacol. **32**, 1361—1364.
RAXWORTHY, M. J. and P. A. GULLIVER (1986), Biochim. Biophys. Acta **870**, 417—425.
RAXWORTHY, M. J., I. R. YOUDE, and P. A. GULLIVER (1986), Xenobiotica **16**, 47—52.
RECHES, A., D. JIANG, and S. FAHN (1981), in: Function and Regulation of Monoamine Enzymes: Basic and Clinical Aspects, (E. USDIN, N. WEINER, M. B. H. YOUDIM, eds.), Macmillan, London, 683—689.
REIN, G., V. GLOVER, and M. SANDLER (1981), in: Phenolsulfotransferase in Mental Health Research, (M. SANDLER, E. USDIN, eds.), Macmillan, London, 98—126.
REIN, G., V. GLOVER, and M. SANDLER (1982), Biochem. Pharmacol. **31**, 1893—1897.
REIN, G., V. GLOVER, and M. SANDLER (1984), J. Neurochem. **42**, 80—85.
REITER, C. and R. WEINSHILBOUM (1982), Clin. Pharmacol. Ther. **32**, 612—621.
REITER, C., G. MWALUKO, J. DUNNETTE, J. VAN LOON, and R. M. WEINSHILBOUM (1983), Naunyn-Schmiedeberg's Arch. Pharmacol. **324**, 140—147.
REVELY, A. M., S. M. B. CARTER, M. A. REVELEY, and M. SANDLER (1982), J. Psychiatr. Res. **17**, 303—307.
ROE, D. A. (1982), in: Sulfate Metabolism and Sulfate Conjugation, (G. J. MULDER, J. CALDWELL, G. M. J. VAN KEMPEN, R. VONK, eds.), Taylor and Francis, London, 163.
SAITO, K., A. SHINOHARA, T. KAMATAKI, and R. KATO (1986), J. Biochem. **99**, 1689 to 1697.

SANDLER, M. (1981), in: Phenolsulfotransferase in Mental Health Research, (M. SANDLER, E. USDIN, eds.), Macmillan, London, 1—7.
SANDLER, M., V. GLOVER, S. M. BONHAM CARTER, J. LITTLEWOOD, and G. REIN (1981), in: Phenolsulfotransferase in Mental Health Research, (M. SANDLER, E. USDIN, eds.), Macmillan, London, 186—206.
SCHMIEDEBERG, O. and H. MEYER (1879), Hoppe Seyler's Z. physiol. Chem. **3**, 422.
SEKURA, R. D. and W. B. JAKOBY (1979), J. Biol. Chem. **254**, 5658—5663.
SEKURA, R. D. and W. B. JAKOBY (1981), Arch. Biochem. Biophys. **211**, 352—359.
SEKURA, R. D., R. SATO, H. J. CAHNMANN, J. ROBBINS, and W. B. JAKOBY (1981a), Endocrinology **108**, 454—456.
SEKURA, R. D., M. W. DUFFEL, and W. B. JAKOBY (1981b), in: Methods in Enzymology, Vol. 77, (W. B. JAKOBY, ed.), Academic Press, New York, 197—206.
SELLINGER, O. Z., R. A. SCHATZ, and W. G. OHLSON (1978), J. Neurochem. **30**, 437—445.
SHERWIN, C. P. (1922), Physiol. Rev. **2**, 264.
SHIBATA, K. (1986), Agric. Biol. Chem. **50**, 1489—1493.
SHIRAI, T. and C. M. KING (1982), Carcinogenesis **3**, 1385—1391.
SILVERMAN, B. D. (1982), Cancer Biochem. Biophys. **6**, 89—94.
SINGER, S. S., A. MOSHTAGHIE, A. LEE, and T. KUTZER (1980), Biochem. Pharmacol. **29**, 3181—3188.
SINGER, S. S. and A. MOSHTAGHIE (1981), Biochim. Biophys. Acta **666**, 212—215.
SINGER, S. S. (1982), in: Biochemical Actions of Hormones, Vol. **9**, (G. LITWACK, ed.), Academic Press, London, New York, 271—303.
SINGER, S. S. and B. BRILL (1982), Biochim. Biophys. Acta **712**, 590—596.
SINGER, S. S., M. J. FEDERSPIEL, J. GREEN, W. LEWIS, and V. MARTIN (1982), Biochim. Biophys. Acta **700**, 110—117.
SINGER, S. S. (1984), Biochem. Soc. Transact. **12**, 35—39.
SINGER, S. S., A. Z. ANSEL, N. VAN BRUNT, J. TORRES, and E. G. GALASKA (1984), Biochem. Pharmacol. **33**, 3485—3490.
SINGER, S. S., E. G. GALASKA, T. A. FEESER, R. L. BENAK, A. Z. ANSEL, and A. MOLONY (1985), Can. J. Biochem. Cell Biol. **63**, 23—32.
SLADEK-CHELGREN, S. and R. M. WEINSHILBOUM (1981), Biochem. Genet. **19**, 1037—1053.
THITHAPAUDHA, A. and V. H. COHN (1978), Biochem. Pharmacol. **27**, 263—271.
TONDA, K. and M. HIRATA (1983), Chem.-Biol. Interact. **47**, 277—287.
TOTH, L. A., M. C. SCOTT, and M. A. ELCHISAK (1986), Life Sci. **39**, 519—526.
TSENG, L., L. Y. LEE, and J. MAZELLA (1985), J. steroid Biochem. **22**, 611—615.
TUNNICLIFF, G. and T. T. NGO (1983), Int. J. Biochem. **15**, 733—738.
URE, A. (1841), London Med. Gaz. **27**(I), 73.
VAN KEMPEN, G. M. J. and E. J. M. PENNINGS (1981), Psychopharmacol. Bull. **17**, 46—47.
VAN KEMPEN, G. M. J. and E. J. M. PENNINGS (1982), in: Sulfate Metabolism and Sulfate Conjugation, (G. J. MULDER, J. CALDWELL, G. M. J. VAN KEMPEN, R. J. VONK, eds.), Taylor and Francis, London, 77—83.
VAN LOON, J. A. and R. M. WEINSHILBOUM (1982), Biochem. Genet. **20**, 637—658.
VAN LOON, J. A. and R. M. WEINSHILBOUM (1984), Biochem. Genet. **22**, 997—1014.
WALKER, R. C., L. C. WOODSON, and R. M. WEINSHILBOUM (1981), Biochem. Pharmacol. **30**, 115—121.
WATANABE, H. K. and M. MATSUI (1984), J. Pharmacobio-Dyn. **7**, 641—647.
WEBER, W. W., D. W. HEIN, M. HIRATA, and E. PATTERSON (1978), in: Conjugation Reactions in Drug Biotransformation, (A. AITO, ed.), Elsevier/North Holland Biomedical Press, Amsterdam, 145—153.
WEBER, W. W., and C. M. KING (1981), in: Methods in Enzymology, Vol. 77, (W. B. JAKOBY, ed.), Academic Press, New York, 272—280.

WEBER, W. W., D. W. HEIN, I. B. GLOWINSKI, C. M. KING, and R. R. FOX (1983), IARC Sci. Publ. **39**, 405—412.

WEBER, W. W. and D. W. HEIN (1985), Pharmacol. Rev. **37**, 25—79.

WEINSHILBOUM, R. M. (1974), Nature **252**, 490—501.

WEINSHILBOUM, R. M. and F. A. RAYMONT (1977), Am. J. Hum. Genet. **29**, 125.

WEINSHILBOUM, R. M. and R. J. ANDERSON (1981), in: Phenolsulfotransferase in Mental Health Research, (M. SANDLER, E. USDIN, eds.), Macmillan, London, 8—28.

WEINSHILBOUM, R. M. (1986), Fed. Proc. **45**, 2223—2228.

WEISIGER, R. A. and W. B. JAKOBY (1979), Arch. Biochem. Biophys. **196**, 631—637.

WEISIGER, R. A. and W. B. JAKOBY (1980), in: Enzymatic Basis of Detoxication, Vol. **II**, (W. B. JAKOBY, ed.), Academic Press, New York, 131—140.

WEISIGER, R. A., L. M. PINKUS, and W. B. JAKOBY (1980), Biochem. Pharmacol. **29**, 2885—2887.

WEISIGER, R. A. and W. B. JAKOBY (1981), in: Methods in Enzymology, Vol. **77**, (W. B. JAKOBY, ed.), Academic Press, New York, 257—262.

WESTLEY, J. (1980), in: Enzymatic Basis of Detoxication, Vol. **II**, (W. B. JAKOBY, ed.), Academic Press, New York, 245.

WHITTEMORE, R. M., L. B. PEARCE, and J. A. ROTH (1985), Biochemistry **24**, 2477 to 2482.

WHITTEMORE, R. M., L. B. PEARCE, and J. A. ROTH (1986), Arch. Biochem. Biophys. **249**, 464—471.

WOLLENBERG, P. and W. RUMMEL (1985), Naunyn-Schmiedeberg's Arch. Pharmacol. **329**, 195—200.

WONG, K. P. (1982), Biochem. Pharmacol. **31**, 59—62.

WONG, K. P. and T. YEO (1982), Biochem. Pharmacol. **31**, 4001—4003.

WONG, K. P., T. YEO, and W. F. TSOI (1984), Biog. Amines **1**, 319—328.

WOODARD, R. W., M. FSAI, H. G. FLORS, P. A. CROOKS, and J. K. COWARD (1980), J. Biol. Chem. **255**, 9124—9127.

WOODSON, L. C., T. P. MAUS, C. REITER, and R. M. WEINSHILBOUM (1981), J. Pharmacol. Ther. **218**, 734—738.

WOODSON, L. C. and R. M. WEINSHILBOUM (1983), Biochem. Pharmacol. **32**, 819—826.

WOODSON, L. C., M. M. AMES, C. D. SELASSIE, C. HANSCH, and R. M. WEINSHILBOUM (1983), Molec. Pharmacol. **24**, 471—478.

YOUDE, I. R., M. J. RAXWORTHY, P. A. GULLIVER, D. DIJKSTRA, and A. S. HORN (1984), J. Biochem. Pharmacol. **36**, 309—313.

YOUNG, W. F., H. OKAZAKI, E. R. LAWS, and R. M. WEINSHILBOUM (1984), J. Neurochem. **43**, 706—715.

YOUNG, W. F., E. R. LAWS, F. W. SHARBROUGH, and R. M. WEINSHILBOUM (1985), J. Neurochem. **44**, 1131—1137.

YU, P. H., B. ROZDILSKY, and A. A. BOULTON (1985), J. Neurochem. **45**, 836—843.

# Chapter 7

# Structure and Regulation
# of UDP Glucuronosyltransferases

P. I. MACKENZIE

# 1.   Introduction

UDP Glucuronosyltransferases are a family of integral proteins of the endoplasmic reticulum which conjugate thousands of chemicals and endogenous metabolites to the acidic sugar, UDP glucuronic acid. The classical approach to the study of these proteins, employing enzyme kinetics and protein purification, has demonstrated the diversity of this family in terms of substrate specificities, tissue distribution, perinatal development and induction by foreign chemicals. However, much of these data have not been definitive. This may be attributed to the complexity of the enzymatic profile of the endoplasmic reticulum and the intrinsic problem of obtaining a pure enzyme whose catalytic activities are unaffected by either its preparation or its removal from a membrane environment. The recent advent of recombinant DNA technology has provided a means of circumventing these problems. This review will summarize the structure and regulation of UDP glucuronosyltransferases and will emphasize the contribution of studies involving cDNA clones in establishing the complete amino acid sequences of several forms of transferase. These data provide a framework for analysis of their structural features, substrate preferences, and regulation. As the only definitive data currently available is for rat cDNAs, this review will focus on studies with this species.

Unfortunately, space limitations do not allow the citation of many original articles, but several of these can be found in the excellent reviews of other authors (DUTTON, 1980; KASPER and HENTON, 1980; ARMSTRONG, 1987; BOCK et al., 1987; BURCHELL et al., 1987a, b, c; SIEST et al., 1987).

# 2.   Function of UDP glucuronosyltransferases

Living organisms are constantly exposed to fat-soluble chemicals from their environment, such as food preservatives, combustion products, drugs, pesticides, carcinogens, cosmetics, dyes and solvents. These will often accumulate in the body to toxic levels unless enzymatically modified. One major modification process in vertebrates is glucuronidation: the transfer of glucuronic acid from UDP glucuronic acid to an oxygen, nitrogen, sulphur or carbon atom of an acceptor molecule. This reaction is catalyzed by UDP glucuronosyltransferase (EC. 2.4.1.17), a family of membrane-bound enzymes in the endoplasmic reticulum of the liver and other organs. When UDP glucuronic acid binds to the active site of this enzyme, the $\alpha$-bond between its C-1 atom and UDP is weakened, thereby permitting the attack of a nucleophilic aglycone from the opposite plane of the molecule to form a $\beta$-linkage between glucuronic acid and the acceptor (a $\beta$-D-glucuronide) and UDP (ARMSTRONG, 1987). The increased water-solubility of the aglycone, which now contains a charged

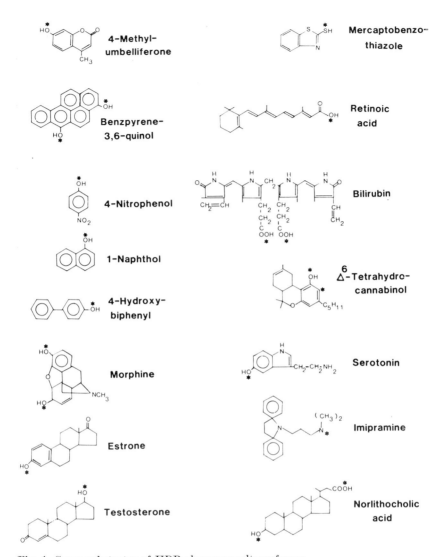

**Fig. 1.** Some substrates of UDP glucuronosyltransferase.
Sites of attachment of UDP glucuronic acid are marked with an asterix. Data are compiled from the following sources: DUTTON (1980); FENSELAU et al. (1980); MILLER and DELUCA (1986); LEHMAN et al. (1983); LEVY et al. (1978) and SHATTUCK et al. (1986).

sugar acid, alters its biological reactivity and aids in its elimination from the body. As well as the many thousands of foreign chemicals processed in this manner, many endogenous metabolites are also glucuronidated. These include bilirubin, steroids, biogenic amines, fat-soluble vitamins and bile acids. An

P. I. MACKENZIE

example of the diversity of chemical structures and functional groups glucuronidated is illustrated in Figure 1.

Conjugation with glucuronic acid in general reduces the biological activity
of a compound, a property concordant with its major role as a detoxification
system. This is important in pharmacology and toxicology as the potency
of a drug, carcinogen or endogenous effector is mediated by a balance between
its uptake and activation by, for example, the monooxygenase complex, and
its removal facilitated by glucuronidation or other conjugating systems. Redox
cycles between polycyclic aromatic hydrocarbon quinols and quinones, which
generate toxic, reactive oxygen species, may also be prevented by removal
of the phenol as a glucuronide (LILIENBLUM et al., 1985). Some glucuronides
however, are more biologically reactive than their parent aglycone. This is
exemplified by the electrophilic modification of protein sulphydryl groups by
acyl-linked glucuronides of certain hypolipidemic drugs (STOGNIEW and
FENSELAU, 1982) and the potent cholestatic effects of steroid D-ring glucuronides (MEYERS et al., 1981). The N-O-glucuronide of N-hydroxy-2-acetyl-
aminofluorene and its deacylated product, the O-glucuronide of N-hydroxy-2-
aminofluorene readily form adducts with guanine residues of nucleic acids
(COMMONER et al., 1974). This "toxification" role of glucuronidation is also
evident in the induction of bladder cancer by aromatic amines. These compounds are rapidly inactivated in the liver by glucuronidation and transported to the kidney. As their N-glucuronides, in contrast to most other
glucuronides, are acid-labile, a low urinary pH will lead to the release of the
parent aglycone and subsequent initiation of bladder carcinogenesis (KASPER
and HENTON, 1980).

UDP glucuronosyltransferase also slowly catalyses the hydrolysis of glucuronides to aglycone and glucuronic acid at acid pH (PETERS et al., 1986)
and the hydrolysis of UDP glucuronic acid to UDP and glucuronic acid
(HOCHMAN and ZAKIM, 1984). The biological significance of these minor reactions is unknown but may become significant in diseased states where the
intracellular pH may be low.

## 3.   Forms of UDP glucuronosyltransferases

A serious problem encountered with some drug therapies is the competition
between drugs and endogenous substrates for the active sites of UDP glucuronosyltransferases. For example, the inhibition of bilirubin transferase by the
antibiotic, novobiocin, leads to jaundice (BURCHELL et al., 1987a). Steroid
hormone and fat-soluble vitamin imbalances may also result from competition
with drugs that are glucuronidated by steroid transferases. Thus, the substrate preferences of each form of transferase must be delineated using both
endogenous and relevant exogenous aglycones as a first step in predicting

**Table 1.** Forms of rat liver UDP glucuronosyltransferase

| Form | MW (Kd) | Inducer | Substrate | Glyco-protein | cDNA clone | Other names |
|------|---------|---------|-----------|---------------|------------|-------------|
| $GT_1A$ | 55/56 | 3-methylcholanthrene | 4-nitrophenol 1-naphthol | yes | 4NPGT | $GT_1$ |
| $GT_1B$ | 53/54 | 3-methylcholanthrene | 1-naphthol | — | — | phenol GT |
| $GT_1C$ | 61 | 3-methylcholanthrene | — | — | — | |
| $GT_2A$ | 53 | phenobarbital | testosterone 4-OH biphenyl | yes | UDPGTr-2 | $GT_2$ |
| $GT_2B$ | 56 | phenobarbital | morphine | — | — | |
| $GT_3A$ | 50 | — | testosterone | no | UDPGTr-3 RLUG38 | 17-hydroxy steroid GT |
| $GT_3B$ | 50 | — | testosterone | no | UDPGTr-5 | |
| $GT_3C$ | 53 | — | androsterone bile acids | yes | UDPGTr-4 RLUG23 | 3-hydroxy-androgen GT |
| $GT_4$ | — | — | acetaminophen | — | — | |
| $GT_5$ | — | arochlor | estrone | — | — | |
| $GT_6$ | 53 | clofibrate | bilirubin | — | — | |
| $GT_7$ | — | pregnenolone-16-α carbonitrile | digitoxigenin monodigitoxide | — | — | |

potential adverse side effects and as a basis for the design of more potent drugs with fewer side effects.

The total number of forms of transferase glucuronidating the vast array of structurally dissimilar endogenous metabolites and foreign compounds is unknown. Studies on the catalytic activities, purification, immunochemistry and cloning of transferases published to date provide evidence for a minimum of 12 forms (Table 1.). Substrate specificities have classically been determined for purified enzymes. Data from these studies are often difficult to evaluate as the enzyme's catalytic properties (e.g., affinity towards its substrate) are often affected by the detergent used during purification and the physical and chemical properties of the lipids used in its reconstitution (HOCHMAN and ZAKIM, 1983; THOMASSIN et al., 1986). An alternative approach involves the synthesis of pure enzyme from a transferase cDNA transfected

P. I. MACKENZIE

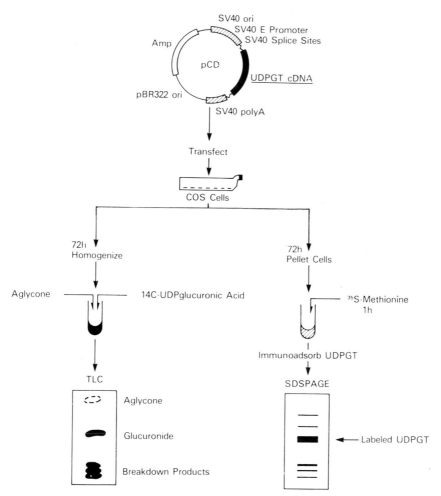

**Fig. 2.** Expression of UDP glucuronosyltransferase cDNAs in cell culture. The transferase cDNA expression vector is transfected into COS cells where it is amplified 100- to 1000-fold. The cDNA is transcribed from the SV40 early region promotor and the transcribed RNA is polyadenylated using SV40 polyadenylation signals. The RNA is translated in the cell and nascent enzyme labelled with [35]S-methionine and analyzed on SDS-PAGE gels or assayed by TLC using a battery of substrates and [14]C-UDP glucuronic acid.

into transferase-deficient cells. One scheme, which has been used successfully, is illustrated in Figure 2. (MACKENZIE 1986a, b, 1987). The nascent enzyme is integrated into the endoplasmic reticulum where it can be radiolabelled and assayed with a battery of substrates in the absence of detergents and synthetic phospholipids. Ideally, each enzyme form should be defined by a unique amino acid sequence. In several cases, amino acid sequence data are unavailable so that correlations between cloned and purified enzymes may change as more data are published. The transferase forms described below have been named after a major substrate as recommended (BOCK et al., 1983). Their names may, therefore, change as their substrate preferences are characterized more thoroughly. Eventually a nomenclature based on primary amino acid sequences, as proposed for cytochromes P-450 (NEBERT et al., 1987) would be more definitive.

## 3.1. 4-Nitrophenol UDP glucuronosyltransferase

The amino acid sequence of a form of transferase which glucuronidates 4-nitrophenol (4NPGT), has been deduced from its corresponding cDNA (IYANAGI et al., 1986). The unprocessed form of this enzyme consists of 529 residues whereas the cleaved form (see section 4) has 505 residues and a calculated molecular weight of 57K daltons. This does not agree well with the molecular weight of the purified enzyme of 55K daltons (GREEN et al., 1987b). The latter was determined by SDS-PAGE analysis and included endoglycosidase H-sensitive carbohydrate moieties. This discrepancy between the calculated and experimentally determined molecular weight values of different forms of transferase is also characteristic of cytochromes P-450 and other integral proteins of the endoplasmic reticulum (GONZALEZ et al., 1985).

4NPGT is found in the liver, kidney, lung, small intestinal mucosa, spleen and skin. Its mRNA is more abundant in the kidney and spleen than in the liver and jejunum. In contrast to other forms of transferase, 4NPGT is preferentially present in the centrilobular region of the liver and is markedly increased in early focal lesions formed after administration of a variety of carcinogens (ROY CHOWDHURY et al., 1987; KNAPP et al., 1987; BOCK et al., 1987). This form appears to be the major 3-methylcholanthrene-inducible form in rat liver. The levels of 4NPGT mRNA are elevated 10- to 15-fold after this treatment in the liver but only 3-fold in the kidney. However, after induction, the mRNA levels for this form were similar in both tissues (IYANAGI et al., 1986; ROY CHOWDHURY et al., 1987). Catalytic and immunochemical evidence indicate that other inducers of this form were B-naphthoflavone, ethoxyquin, trans-stilbene oxide, aroclor 1254, 2-acetylaminofluorene, diethylnitrosamine, the antihistamic drug, methapyrilene and aflatoxin B1 (LILIEN-BLUM et al., 1982).

In addition to 4-nitrophenol, this enzyme glucuronidates simple phenols such as 1-naphthol and 4-methylumbelliferone and phenols of polycyclic hydrocarbons such as benzopyrene and dibenzanthracene, where it displays regiospecificity. The better polycyclic aromatic hydrocarbon substrates were

---

short ($< 1.3$ nm) and broad ($> 1.1$ nm) when viewed from the hydroxyl group along the axis of the C-O bond to the plane of the molecule (LILIENBLUM et al., 1987). Benzpyrene 3,6 quinol was a more specific substrate as reflected in the respective 10- and 40-fold increase in the formation of its mono- and di-glucuronides after 3-methylcholanthrene treatment (LILIENBLUM et al., 1985). In contrast, the glucuronidation of 4-nitrophenol was only increased 5-fold, probably because other forms of transferase metabolize this substrate. Other less specific substrates of this enzyme include 1- and 2-naphthylamine (LILIEN-BLUM and BOCK, 1984; GREEN and TEPHLY, 1987a). An endogenous substrate for 4NPGT has not yet been discovered. As determined by Northern analysis with an oligomer DNA probe (MACKENZIE, unpublished data), the levels of 4NPGT mRNA and/or the mRNAs of other closely related forms in rats were elevated before birth, a finding consistent with the late fetal increases in the rates of glucuronidation of 2-aminophenol, phenol, 4-nitrophenol and 1-naph-thol (see DUTTON, 1980).

### 3.2. Phenol UDP glucuronosyltransferase

A second 3-methylcholanthrene-inducible form of transferase (phenol GT) with a molecular weight of 53K daltons has been detected in rat liver using anti-bodies raised against a purified kidney enzyme (COUGHTRIE et al., 1987b) and the 55K dalton 4NPGT from liver (KOSTER et al., 1986; COFFMAN et al., 1987). Whether this 53K dalton protein has a unique amino acid sequence or is related to the other form through posttranslational modifications is unknown. Phenol GT has a tissue distribution and substrate specificity similar to the 55K dalton 4NPGT. However, an accurate assessment of its substrate pre-ference will depend on its purification to homogeneity or the expression of its cDNA in tissue culture. It is absent from Gunn rat liver and kidney and is not inducible by 3-methylcholanthrene in these animals (COUGHTRIE et al., 1987a). It is conceivable that the cloned 3-methylcholanthrene form of trans-ferase corresponds to this phenol transferase rather than to the 55K dalton protein described above as the first 20 amino acids deduced from the cDNA vary at two positions to that determined from the purified 55K dalton 4NPGT (IYANAGI et al., 1986; GREEN and TEPHLY, 1987b).

### 3.3. 4-Hydroxybiphenyl UDP glucuronosyltransferase

A form of transferase designated UDPGTr-2 has been cloned, sequenced and expressed in tissue culture cells. Although 4-hydroxybiphenyl was a major exogenous substrate, this form was also active in the glucuronidation of chloramphenicol, 4-methylumbelliferone, phenolphthalein, and the endogenous substrates, testosterone, dihydrotestosterone, estradiol and to a small extent, etiocholanolone (MACKENZIE, 1986a, 1987). The enzyme, which is N-glycosylated, contains 529 residues with a calculated molecular weight of about 60K daltons. Nascent mature enzyme expressed from UDPGTr-2 cDNA transfected into cultured mammalian cells had a molecular weight of 53K daltons when analysed by SDS -PAGE (MACKENZIE, 1986a).

The levels of UDPGTr-2 mRNA were elevated more than 5-fold by phenobarbital, whereas 3-methylcholanthrene and other inducers of drug-metabolizing enzymes such as dexamethasone, pregnenolone-16α-carbonitrile, imidazole, rifampicin, methylpyrazole and pyrazole were without significant effect (MACKENZIE and HAQUE, 1987). In rats, UDPGTr-2 mRNA levels rise after birth (MACKENZIE unpublished data), consistent with the neonatal rise in the rates of glucuronidation of testosterone and chloramphenicol reviewed previously (DUTTON, 1980). This enzyme is probably a member of the "GT2" class (BOCK et al., 1980).

### 3.4. Morphine UDP glucuronosyltransferase

A labile, 56K dalton form of transferase which glucuronidates morphine and apomorphine has been isolated from female Wistar rat livers (PUIG and TEPHLY, 1986). It was competitively inhibited by codeine and did not use 4-hydroxybiphenyl, 4-nitrophenol, testosterone, androsterone, estrone, bilirubin, 4-aminobiphenyl or 1-naphthylamine as substrates. The cDNA to this form has not been synthesized, but based on enzyme assays (DUTTON, 1980), its mRNA probably appears after birth and is elevated by phenobarbital. This form is perhaps another member of the "GT2" group of enzymes.

### 3.5. 3-Hydroxyandrogen UDP glucuronosyltransferase

The glucuronidation of bile acids and the 3-hydroxy group of androgenic steroids appears to be mostly effected by a relatively abundant form of transferase which has been purified to apparent homogeneity (KIRKPATRICK et al., 1984). The purified enzyme is a glycoprotein with a molecular weight of 52K daltons on SDS polyacrylamide gels. The cDNA of this enzyme (designated UDPGTr-4 or RLUG23) has been cloned and sequenced (MACKENZIE, 1986b; JACKSON and BURCHELL, 1986) demonstrating that the unprocessed enzyme consists of 530 residues with a calculated molecular weight of about 60K daltons. However its molecular weight expressed from cDNA in cell culture is 53K, which is in agreement with that determined for the enzyme purified from rat liver microsomes. The purified enzyme and that expressed from cDNA are both active in the glucuronidation of etiocholanolone, androsterone, and the conventional bile acids, lithocholate and ursodeoxycholate. The 3-hydroxyl groups of 5β short-chain (C20−C23) bile acids are also glucuronidated by this form which, in contrast, is not active toward their side-chain carboxyl groups (RADOMINSKA-PYREK et al., 1986). The enzyme carries out the N-glucuronidation of aniline, 4-aminobiphenyl, 1-naphthylamine and 2-naphthylamine (GREEN and TEPHLY, 1987b). As these aromatic amines often induce tumours via their N-hydroxylated derivatives, N-glucuronidation, a competing pathway, may influence their carcinogenic potential.

The finding that the glucuronidation of androsterone is low at birth and surges to adult levels in 20-day-old rats (MATSUI and WATANABE, 1982) was confirmed at the mRNA level using UDPGTr-4 cDNA probes (MACKENZIE, unpublished data).

## 3.6. 17-Hydroxysteroid UDP glucuronosyltransferase

A 50K dalton form of transferase which lacks N-linked glycosyl groups and glucuronidates the 17β-hydroxy moiety of the steroids, testosterone and estradiol, as well as 4-nitrophenol and 1-naphthol, has been purified. Activity towards 4-methylumbelliferone, estrone, morphine and etiocholanolone was not detected in the purified enzyme preparation (FALANY and TEPHLY, 1983). The N-terminal 19 amino acids of this purified protein corresponds exactly to that deduced from a transferase cDNA, UDPGTr-3 (MACKENZIE, 1987) and the same cDNA called RLUG38 cloned independently (HARDING et al., 1987). The polypeptide encoded by these cDNAs is 530 residues in length with a calculated molecular weight of about 60K daltons. Expression of UDPGTr-3 cDNA in tissue culture produced a 50K dalton transferase which was not N-glycosylated and which also glucuronidated testosterone and estradiol. However, significant activity towards 1-naphthol and 4-nitrophenol was not detected (MACKENZIE, 1987). The discrepancy between the substrate profiles of the purified enzyme and that synthesized from cDNA may result from contamination of the purified enzyme by small amounts of another form with high activity towards 1-naphthol and 4-nitrophenol or a detergent-induced alteration in the conformation of the enzyme during its solubilization and purification that renders it inactive towards these substrates. Alternatively, the membrane environment of the green monkey kidney cell transfected with the cDNA may affect the substrate preference of the nascent enzyme so that it no longer recognizes these exogenous substrates. This can easily be tested by transfection of the cDNA into another recipient cell, such as yeast. A further possibility is that the cDNA encoded transferase is another form of 17β-hydroxysteroid UDP glucuronosyltransferase, with a similar amino acid sequence to the enzyme purified from microsomes.

### 3.7. Testosterone UDP glucuronosyltransferase

The cDNA to a form of transferase (UDPGTr-5), active in the glucuronidation of testosterone, has been cloned (MACKENZIE, unpublished results). The encoded protein consists of 530 residues, does not contain potential N-linked glycosylation sites and is 93% similar in sequence to UDPGTr-3, a 17$\beta$-hydroxy steroid transferase. More experimental data are required to determine whether this form is an allelic variant of UDPGTr-3 or a unique enzyme with a different substrate preference.

### 3.8. Estrone UDP glucuronosyltransferase

Evidence for a separate form of transferase which glucuronidates estrone and the 3-hydroxy group of estradiol has been provided by catalytic, induction and chromatofocussing studies (FALANY et al., 1983). Estrone transferase has been purified from rabbit liver (TUKEY et al., 1982). However, the purification of this form to homogeneity and the cloning of its cDNA have not been reported for the rat.

### 3.9. Digitoxigenin-monodigitoxoside UDP glucuronosyltransferase

A form specific for the glucuronidation of the cardiac glycoside, digitoxigenin monodigitoxide, has been isolated from male Wistar rats (VON MEYERINCK et al., 1985). Although the enzyme preparation contained three protein bands when subjected to SDS-PAGE, it was not active towards substrates such as 4-nitrophenol, 4-methylumbelliferone, androsterone, bile acids, testosterone, morphine, 4-hydroxybiphenyl, estrone and bilirubin, which are typical substrates of other forms of transferase. This form is induced 13−20-fold by pregnenolone-16$\alpha$-carbonitrile, troleandomycin and dexamethasone (WATKINS et al., 1982; SCHEUTZ et al., 1986), and 4-fold by spironolactone (SCHMOLDT and PROMIES, 1982). It is first detected 20 days after birth in rats and is apparently absent in the livers of Gunn rats (WATKINS and KLAASSEN, 1982).

### 3.10. Bilirubin UDP glucuronosyltransferase

A 53K dalton protein which catalyzes the glucuronidation of bilirubin has been purified from rat liver (ROY CHOWDHURY et al., 1986a; BURCHELL and BLANCKAERT, 1984). The purified enzyme catalyzed the conjugation of the C-8 isomer of bilirubin and could also convert the monoglucuronide to either the diglucuronide in the presence of UDP glucuronic acid at pH 7.8 or to bilirubin in the presence of UDP at pH 5.5 (ROY CHOWDHURY et al., 1986a). Glucosides and xylosides of bilirubin were also formed by this enzyme when the appropriate UDP sugars were present, although the rates of conjugation were lower with these sugar nucleotides than with UDP glucuronic acid. These UDP-sugars and arylcarboxylic acids such as triphenylacetic acid (FOURNEL et al., 1986), were competitive inhibitors of bilirubin glucuronidation. However, other carboxylic acid-containing aglycones such as clofibric acid, were not substrates of this enzyme. Although apparently inactive towards phenols and steroids, bilirubin transferase, in common with other forms, glucuronidated the carcinogenic metabolite, 4-hydroxydimethylaminoazobenzene (ROY CHOWDHURY et al., 1986a). The purified protein had N-linked high mannose residues and retained its catalytic activity after deglycosidation with endoglycosidase F (BURCHELL et al., 1987a). The glucuronidation of bilirubin is specifically induced by the hypolipidemic drug, clofibrate, and chlorophenoxyacetic acid. Other arylcarboxylic acids such as phenylpropionic acid, and phenobarbital were less specific and increased the glucuronidation of a broad range of substrates (FOURNEL et al., 1985).

Gunn rats cannot glucuronidate bilirubin and therefore exhibit severe unconjugated hyperbilirubinaemia throughout life (BILLING, 1987). Immunochemical evidence suggested that bilirubin transferase was present in these animals leading to the conclusion that a defect in the enzyme rendered it catalytically inactive towards bilirubin (ROY CHOWDHURY et al., 1985). However, the enzyme could accept bilirubin, in which the C8 position was already esterified, as substrate (BURCHELL and ODELL, 1981). The alternative view, that the enzyme is absent in Gunn rats was supported recently by studies with a human cDNA probe which demonstrated that the enzyme protein and mRNA were absent in Gunn rat liver and extrahepatic tissues (BURCHELL et al., 1987a), perhaps due to a deletion in the Gunn rat genome. Conclusive evidence of the presence or absence of bilirubin UDP glucuronosyltransferase in Gunn rats must await the cloning of its cDNA from normal rats and preparation of a form-specific DNA probe.

## 3.11. Other forms of UDP glucuronosyltransferase

A recent report provides evidence for the existence of an enzyme which glucuronidates acetaminophen as this drug is not glucuronidated by other purified forms of transferase. The enzyme can be separated from estrone UDP glucuronosyltransferase by chromatofocussing (COFFMAN et al., 1987).

Immunochemical studies which provided evidence for the existence of two forms of 3-methylcholanthrene-inducible transferase, also suggested the presence of a third form with a molecular weight of 61K daltons (BOCK and LILIENBLUM, 1984). This form was detected in a Reuber hepatoma cell line and in small amounts in microsomal preparations from rat liver and kidney. It may represent an aberrantly glycosylated form of transferase already described.

# 4. Sequence and structure of UDP glucuronosyltransferases

The cloning of UDP glucuronosyltransferase cDNAs has established for the first time, the primary amino acid sequences of the different forms and has provided insights into their structural interrelationships. A comparison of the cDNA-deduced amino acid sequences of five forms of transferase (Table 2) demonstrates that UDP glucuronosyltransferases are encoded by a super family of genes comprising at least two families (i.e. the sequence of a member of one family is less that 50% similar to that of a member of the second family). One family is represented by the 3-methylcholanthrene-inducible form glucuronidating 4-nitrophenol, whereas the second family is represented by the other four forms which glucuronidate steroids. Furthermore, the latter family can be further subdivided into two subfamilies (less than 75% similarity between members of different subfamilies); one containing pheno-barbital-inducible UDPGTr-2 and the other containing the constitutive forms

**Table 2.** Percent similarity of the deduced amino acid sequence of five UDP glucurono-syltransferase cDNAs*

|            | UDPGTr-2 | UDPGTr-3 | UDPGTr-4 | UDPGTr-5 | 4NPGT |
|------------|----------|----------|----------|----------|-------|
| UDPGTr-2   | 100      |          |          |          |       |
| UDPGTr-3   | 68       | 100      |          |          |       |
| UDPGTr-4   | 68       | 82       | 100      |          |       |
| UDPGTr-5   | 61       | 93       | 84       | 100      |       |
| 4NPGT      | 43       | 45       | 43       | 42       | 100   |

\* Compiled from IYANAGI et al., 1986 and MACKENZIE, 1986a, b, 1987.

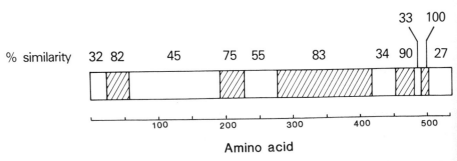

**Fig. 3.** Comparison of the coding regions between UDPGTr-2 and UDPGTr-4 cDNAs Regions of high similarity are hatched.

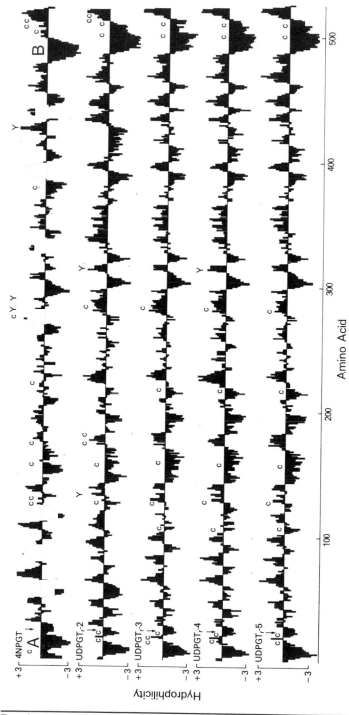

**Fig. 4.** Hydrophilicity profiles of UDP glucuronosyltransferases.
The hydrophilicity profiles were determined for a sliding window of seven amino acids (KYTE and DOOLITTLE, 1982). Positive numbers indicate hydrophilic regions whereas hydrophobic areas are denoted by negative values. The N-terminal leader sequence (A), potential cleavage site (↓), C-terminal hydrophobic segment (B), cysteine residues (C) and potential N-linked glycosylation sites (Y) are indicated.

UDPGTr-3, -4 and -5. Presumably other forms of transferases which have not been cloned as yet, would belong to these and other families and sub-families, revealing a superfamily to rival the cytochrome P-450 superfamily in complexity.

Despite the wide divergence in amino acid sequence between forms, certain regions within the sequences are more similar than other regions. For example, a comparison of the sequences of UDPGTr-2 and UDPGTr-4 demonstrates that, although their overall similarity is 68%, a region in the amino-terminal halves of the proteins encompassing 150 residues is only 45% similar whereas a more conserved (83% similar) region of similar size is found in the carboxy-halves of the polypeptides (Fig. 3). All forms cloned to date have a potential amino-terminal leader sequence and a carboxy-terminal hydrophobic segment bounded by hydrophilic regions. These regions can be seen in a hydrophilicity plot of the various forms (Fig. 4). As also shown in this figure, cysteine residues are predominantly located in the amino-terminal half of the protein and in the carboxy-terminal 20−30 residues. Potential asparagine-linked glyco-sylation sites are found in some forms and are absent in others. These regions are discussed below.

## 4.1. Leader peptide

All forms of transferase have an amino terminal sequence characteristic of a signal peptide (Fig. 5). This sequence consists of a hydrophobic stretch of 13 amino acids preceded by a short region which is either uncharged or con-tains a positive charged residue and ends with a segment containing at least one helix-breaking residue (proline, glycine). Transcription of transferase

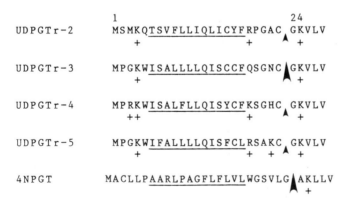

**Fig. 5.** N-Terminal leader sequences of UDP glucuronosyltransferases.
The hydrophobic segment is underlined. The established (▲) or potential (▲) sites of cleavage to the mature form are indicated.

cDNA in vitro and translation of its complementary RNA in the presence and absence of dog pancreatic microsomes demonstrated that the newly synthesized protein was cleaved and inserted into the microsomal membrane. Only the cleaved enzyme was integrally associated with the membrane as unprocessed enzyme remained in the supernatant after removal of the membranes by centrifugation (MACKENZIE, 1986a, 1987). Cleavage of the nascent enzyme to a mature form was also demonstrated with mRNA isolated from rat liver (MACKENZIE and OWENS, 1984). The sites of cleavage have still not been delineated for all transferases (Fig. 5), even after a comparison with the amino-terminal sequences of four purified forms (GREEN and TEPHLY, 1987b; IYANAGI et al., 1986). The sequence of one form starts with glycine whereas another starts with aspartate. The amino-terminal residues of the other two forms were not reported. It is also possible that cleavage occurs at a common site to yield an N-terminal lysine residue and the glycine or aspartate residue is added posttranslationally.

## 4.2. Glycosylation

A comparison of the molecular weights of transferases synthesized in vitro with those present in microsomal membranes suggested that some forms of transferase were glycosylated (MACKENZIE and OWENS, 1984). This was directly demonstrated by the positive staining of purified forms with periodic acid/Schiff reagent (ROY CHOWDHURY et al., 1986b). In addition, treatment of purified forms, or forms isolated by immunoadsorption with endo-B-N-acetylglucosaminidase H or F, demonstrated that the glycosyl moieties were of the simple high-mannose type and were apparently unmodified by processing enzymes of the Golgi (MACKENZIE et al., 1984b; GREEN et al., 1987; BURCHELL et al., 1987a). O-Linked sugars may also be present (ROY CHOWD-HURY et al., 1986). Glycosylation of UDP glucuronosyltransferase forms could also be demonstrated when RNA transcribed from specific transferase cDNA was translated in vitro in the presence of dog pancreatic microsomes or translated in situ, from cDNA transfected into cultured mammalian cells (MACKEN-ZIE, 1986a, unpublished). Potential N-linked glycosylation sites, characterized by the sequence Asn-X-Thr/Ser where X represents any amino acid, are present in the primary amino acid sequences of some forms of transferase but are absent in other forms (see Fig. 4). All potential glycosylation sites may not be utilized, as UDPGTr-2 with two potential glycosylation sites and UDPGTr-4 with only one site, have the same molecular weights both before and after processing (MACKENZIE, unpublished results).

The function of carbohydrate which is covalently bound to the enzyme is unknown. Potential functions include a role in catalysis or a role in stabilizing the protein so that it is more resistant to proteolysis. Glycosylation is not a prerequisite for catalytic activity as some unglycosylated forms (e.g. UDPGTr-3) are active. In addition, the possibility that a hydrophilic glycosyl moiety may determine the aglycone substrate preference, is not supported by preliminary experiments with some forms of deglycosylated enzyme (MACKENZIE, unpublished data). Carbohydrate may, however, modify the UDP glucuronic acid binding site and affect rates of catalysis or the choice of sugar nucleotide substrate. The expression of cDNAs to individual transferase forms in cultured mammalian cells and the ability to manipulate their state of glycosylation by tunicamycin should provide further data on these possible roles of glycosylation.

## 4.3. Catalytic domains

The amino acid sequences of various forms of UDP glucuronosyltransferase, as deduced from sequencing their cDNAs, have provided a basis for determining the nature of their catalytic and regulatory domains. It would appear reasonable to presume that different enzyme forms would contain highly conserved regions involved in binding the common cosubstrate, UDP glucuronic acid and effectors such as UDP N-acetylglucosamine or any putative regulatory factor mediating the stimulation of transferase by this latter compound. By the same reasoning, one may presume that the peptide domain involved in binding the aglycone, which varies from form to form, would be less conserved. These considerations and a computer search for global and local similarities between transferases and other proteins in the protein data bank, have pointed to the following potential functional domains.

### 4.3.1. UDP-Glucuronic acid binding site

A region which is highly conserved between all transferase sequences currently available encompasses residues 337−485 in the carboxy-terminal halves of the proteins. Whereas the amino acid sequences of 4-nitrophenol UDP glucuronosyltransferase and UDPGTr-3, for example, are only 43% similar overall, this conserved segment is 63% similar between these proteins and includes a stretch of 35 residues (350 to 385) which are even more conserved (93% similarity). Interestingly, amino acids 364 to 437 are encoded by a single exon in the UDPGTr-4 gene (HAQUE and MACKENZIE, unpublished data). When this conserved segment was compared to protein sequences in the Genbank protein data bank, local homologies to several proteins which bind sugars, nucleotides or phosphate groups, were found (Fig. 6A). This comparison suggests that the conserved sequence in the various forms of transferase delineated above may be involved in the binding of the common cosubstrate, UDP glucuronic acid. Cysteine residues are absent from this region, which is a feature charac-

## A

```
                                    365                                              412
UDPGTr-3                    GHPKTKAFVTHSGANGVYEAIYHGIPMVGIPMFGEQHDNIAHMVAKGAA
UDPGTr-4                    ************************+***+****+ *************
UDPGTr-5                    *****************+*********+*******************
UDPGTr-2                    ******************+*****+******+*+*+***********+****
4NPGT                      ****  +****+*******  ***  ***  +*****+  **+  *  +*++
Arabinose binding pro.     +  +  +*+**+  +  +  *  *  +**    +     ++  +  + +*+  + +++
Pyruvate kinase            +    *+*+**  +  +*+*     +*+  +  *+  + +  + +*  +  ++
ATPase (yeast)             +  +  +*++++*+++  + *+    +  +  +  +  +  +++++++  +  ++
Strept-kinase              **++*+*++*****                                *++  +  ++
Mos-kinase                 ++    +  ++ **  +*+
Aspartokinase              +  ++  **+  ++***+**+***++*            +*++*+***  *** +**
Inorganic pyrophosphatase                            **+***++*+**  *  *+++++**  ++*  **
Enolase (yeast)                                       **+****  *    *  ++*++***+
Enolase (rat)
```

## B

```
UDPGTr-2                         MVGIPMFG
UDPGTr-3                         MIGIPLFG
UDPGTr-4                         IVGIPLFA
UDPGTr-5                         MIGIPMFG
4NPGT                            MVMMPLFG

Mercuric reductase
NADP-binding domain              V-GSSVVA
FAD-binding domain               VIGSGGAA
```

Fig. 6. Possible UDP glucuronic acid-binding site. UDP glucuronosyltransferase residues 365 to 412 are compared to regions of other proteins which bind sugars, nucleotides or phosphate groups (Part A). The sequence of UDPGTr-3 is shown. Conserved residues (*) or residues with similar physical properties (+) of proteins which aligned with transferase sequences using the FASTP program (LIPMAN and PEARSON, 1985) are shown. Non-conserved residues are indicated by gaps. The residues indicated by a solid line in A are shown in B, together with the residues of mercuric reductase known to be involved in the binding of the pyrophosphate groups of FAD and NADP (MOLLER and AMONS, 1985).

teristic of the nucleotide-binding domain of elongation factor, EF-Tu (MC-CORMICK et al., 1985).

A stretch of residues, typical of those which bind pyrophosphate groups, is also found within this conserved region. The pyrophosphate-binding sequences of several nucleotide-binding proteins have been identified (MOLLER and AMONS, 1985). This sequence is characterized by two hydrophobic amino acids followed by two glycine residues which are separated by one amino acid. The second glycine residue is less well conserved and may provide space between more rigid secondary structures such as $\beta$-sheets and $\alpha$-helices, for the approaching phosphate groups of nucleotides. Hence the substitution of this glycine with another small residue such as serine in the NADPH-binding site of mercuric reductase, apparently still provides the necessary entrance space. In the UDP glucuronosyl transferase sequence proposed to correspond to pyrophosphate-binding sequences, the second glycine is replaced by a proline residue which would also induce a flexible loop for entrance of the pyrophosphate group (Fig. 6 B).

The region encompassed by amino acids 337 and 485 is hydrophilic in nature and contains several conserved arginine and histidine residues, features which would be compatible for binding UDP glucuronic acid. Kinetic evidence involving the use of 2,3-butanedione, strongly suggests that an arginine residue is involved in electrostatic interaction with the glucuronyl-carboxyl group of this cosubstrate (ZAKIM et al., 1983). The amide group of a histidine residue may also be important in the binding of the uridyl moiety of UDP glucuronic acid as diethylpyrocarbonate, which destroys histidine residues, renders UDP glucuronosyltransferase inactive (ARION et al., 1984).

### 4.3.2. Aglycone-binding site

If the UDP glucuronic acid binding site is localized to the carboxy-terminal half of the protein, the amino-terminal half may contain the aglycone-binding site. This region is less conserved between forms, which would be expected as different forms have different substrate specificities. It also contains several cysteine residues. Sulphydryl groups are necessary for catalytic activity as sulphydryl blocking agents (e.g. p-chloromercuribenzoate) were able to inhibit 4-nitrophenol and o-aminophenol glucuronidation (ISSELBACHER et al., 1962). Experimental evidence directly implicating the amino- and carboxy-terminal domains of transferase in the binding of aglycone and UDP glucuronic acid respectively is lacking. However, the above comparisons provide a basis for targeting residues involved in catalysis. Experimental confirmation of their importance may then be obtained by site-directed mutagenesis of transferase cDNA and the expression of the mutated protein in transfected cells.

## 4.4. Potential transmembrane region

All UDP glucuronosyltransferases whose amino acid sequences are known have a region rich in hydrophobic amino acids near their carboxy-terminus (Fig. 7). This region contains 17 hydrophobic amino acids and is bounded by hydrophilic residues. It is sufficiently large to traverse a lipid bilayer, is capable of forming an α-helix and is the only such region common to all transferases. This domain may be important in maintaining the enzyme in the endoplasmic reticulum. The signals or mechanisms which preferentially retain transferase in the endoplasmic reticulum and prevent its transport via the Golgi to other

```
          491                                         529
UDPGTr-2  HSLD  VIGFLLLCVVGVVFIIT  KFCLFCCRKTANMGKKKKE
          -                        +    ++      ++++-

UDPGTr-3  HSLD  VIGFLLTCSAVIAVLTV  KCFLFIYRLFVKKEKKMKNE
          -                        +     +   ++-++ + -

UDPGTr-4  HSLD  VIGFLLTCFAVIAALTV  KCLLFMYRFFVKKEKKMKNE
          -                        +     +   ++-++ + -

UDPGTr-5  HSLD  VIGFLLACLAVIAALAV  KCFLFIYRFFAKKQKKMKNK
          -                        +     +   ++ ++ + +

4NPGT     HSLD  VIGFLLAIVLTVVFIVY  KSCAYGCRKCFGGKGRVKKSHKSKTH
          -                        +     ++    +   ++   + +
```

**Fig. 7.** Potential transmembrane regions of UDP glucuronosyltransferase. The carboxy-terminal sequences are shown. Hydrophobic segments are underlined and charged amino acids denoted.

cell organelles are unknown. Studies on another resident protein of the endoplasmic reticulum, the adenovirus E19 protein, strongly suggest that a short sequence in the carboxy-terminus is the signal for retention in this organelle (PAABO et al., 1987). This region contains numerous basic residues (in common with transferase) and when shortened from 15 to 7 residues, results in transport of the E19 protein to the cell surface. The 20 residue region at the carboxy-terminus of UDPGTr-2, -3, -4 and -5 and the corresponding 26 residues of 4-nitrophenol UDP glucuronosyltransferase may also be involved in some signalling or regulatory function analogous to the phosphorylation-dephosphorylation of amino acids in the cytoplasmic domains of surface-membrane receptors. However, only the carboxy-terminus of the 4-nitrophenol transferase contains serine residues which may be potential sites for phosphorylation by protein kinases.

# 5. Synthesis and topology of UDP glucuronosyltransferase

UDP glucuronosyltransferase is synthesized on membrane-bound polysomes (MACKENZIE, unpublished data) and is cotranslationally inserted into the microsomal membrane during synthesis (MACKENZIE, 1986a, 1987). During this process a signal peptide is removed from the amino terminus and some forms of transferase are glycosylated. The cleaved peptide, as demonstrated by in vitro translation experiments, is integrated into the membrane. Uncleaved enzyme, in contrast, does not become membrane associated. These data indicate that transferase is made by a mechanism similar to that used by other proteins of the endoplasmic reticulum, Golgi and plasma membrane (WALTER and LINGAPPA, 1986). The synthesis of these proteins is initiated on free polysomes. As synthesis proceeds, an amino-terminal leader sequence becomes accessible to a signal recognition particle in the cytoplasm, which mediates attachment of the translation complex to the endoplasmic reticulum via a docking protein. The growing polypeptide chain is extruded into the lumen of the endoplasmic reticulum where cleavage of the leader peptide and the attachment of complex carbohydrate chains to asparagine residues may occur. This transfer process continues until a "halt transfer" signal, characterized by a hydrophobic stretch of amino acids surrounded by hydrophilic sequences, is synthesized (SABATINI et al., 1982). The polypeptide chain becomes anchored to the membrane by this segment leaving the remainder of the chain exposed to the cytoplasm. Cleavage of the leader sequence by a luminally-located signal peptidase places the amino terminus of the nascent peptide in the lumen of the endoplasmic reticulum. The amount of protein exposed to the lumen depends upon the position of the halt transfer signal along the polypeptide chain.

As the putative halt transfer sequence of UDP glucuronosyltransferase appears to be located at the carboxy terminus, more than 95% of the polypeptide chain, including the catalytic site, would be located on the luminal side of the endoplasmic reticulum leaving only $19-26$ amino acids facing the cytoplasm. The amino-terminal half of the molecule, which is highly variable between the different forms of transferase, may be embedded in the inner leaflet of the lipid bilayer to allow access of hydrophobic substrates to the active site and to permit modulation of catalytic activity by the membrane microenvironment. This domain contains several cysteine residues which may form disulfide bonds and stabilize any tertiary structure in the hydrophobic environment. Disulfide bonds are formed on the luminal surface of the endoplasmic reticulum and for this reason, are generally confined to secretory proteins or domains of integral proteins exposed to the lumen of intracellular membranes or to the outside of the cell. Disulphide bridges appear to stabilize the structure of membrane-bound phospholipase A2 (MARAGANORE, 1987). Any glycine residue on the amino-terminus of mature forms of transferase may be conjugated with myristic acid which would anchor this region to the lipid bilayer in a manner similar to other myristylated proteins (WOLD, 1986). The carboxy-terminal half of the polypeptide is more conserved between forms and, as it is postulated to contain the UDP sugar binding site, may be more exposed to the lumen. This model of the topology of transferase is illustrated in Figure 8.

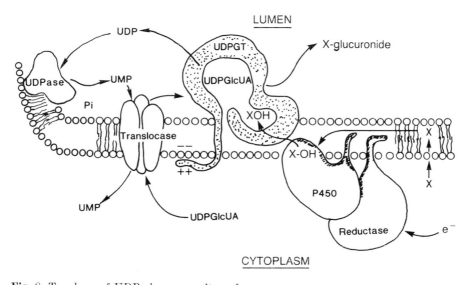

**Fig. 8.** Topology of UDP glucuronosyltransferase.
A model, based on previous proposals (HALLINAN, 1978), for the orientation of transferases in the endoplasmic reticulum and their relation to other drug-metabolizing enzymes and putative sugar transporters, is illustrated. A lipophilic compound (X) reaches the active site of cytochrome P-450 from the membrane or cytoplasm, and is hydroxylated. The hydroxylated compound (X-OH) is subsequently glucuronidated and released into the lumen. Liberated UDP is hydrolyzed by nucleoside diphosphatase to prevent end product inhibition of UDP glucuronosyltransferase. UDP glucuronic acid, which is synthesized in the cytoplasm, is transported to the lumen and the active site of transferase by a putative translocase. This protein, as an antiporter, may also transport UMP from the lumen to the cytoplasm where it can be rephosphorylated.

Any disruption in the normal process of synthesis may result in the aberrant positioning of transferase in other cellular membranes or compartments. For example, a defect in the signal peptide or translocation apparatus may result in transferase polypeptides remaining in the cytoplasm. A mutated halt transfer signal or defects in the membrane-anchoring domain or the domain specifying endoplasmic reticulum-residency, in contrast, would result in the translocation of transferase to the plasma membrane or its secretion from the cell. The re-

---

ported presence of transferase on the surface of some cultured cells, as demonstrated by the glucuronidation of aglycone covalently attached to Sephadex beads (ROTH and LEVINE, 1983), may reflect such an abnormality.

A model depicting transferase compartmented or localized on the lumen of the endoplasmic reticulum has been proposed before to explain the activation of the enzyme by agents such as detergents which, by disrupting the lipid barrier, allow greater access of the charged common cosubstrate, UDP glucuronic acid, to the catalytic site (BERRY et al., 1975; HALLINEN, 1978). Several independent studies suggest that transferase is luminally-located. During in vitro translation in the presence of dog pancreatic membranes, the antigenic determinants on a processed form of transferase become inaccessible to externally added polyclonal antibody. Access of antibody to the enzyme was demonstrated only when detergent was present to disrupt the microsomal membrane (MACKENZIE and OWENS, 1984). In addition, transferase in its native environment is not accessible to the protein modifying agents, N-ethylmaleimide and diazobenzene sulphonate (HAEGER et al., 1980). However, more definitive evidence on the orientation of the active sites of transferase in microsomes will be required in order to validate this model.

Models depicting the active site exposed to the cytoplasm have been proposed based on kinetic data (ZAKIM and VESSEY, 1982). The premise of these is that alterations in kinetic parameters caused by membrane perturbants are the result of changes in the enzyme's interaction with its membrane environment rather than changes in the access of substrate to an active site behind a membrane barrier. These models, however, would be counter to the structural features of transferases and to the experimental evidence outlined above.

The localization of UDP glucuronosyltransferase on the luminal side of the endoplasmic reticulum poses several problems concerning the access of substrates and protection and removal of products from the cell.

### 5.1.  Access of UDP glucuronic acid and removal of UDP

The location of UDP glucuronosyltransferase on the luminal side of the endoplasmic reticulum would necessitate the presence of a transporter to carry the membrane-impermeable UDP glucuronic acid from its site of synthesis in the cytoplasm to the catalytic site on the enzyme. Several possibilities can be considered as to the nature of this sugar transporter.

Microsomes from rat lung can synthesize the lipid-linked disaccharide, GlcUA-GlcNAc-P-P-dolichol which is though to serve as a donor of disaccharide residues in the biosynthesis of the carbohydrate side chains of proteoglycans (TURCO and HEATH, 1977). It is feasible that a similar lipid-linked product may also donate the glucuronosyl moiety to aglycones in intact microsomes. The activating effect of UDP N-acetylglucosamine on transferase in native

P. I. MACKENZIE

microsomes may reflect its participation in the formation of this intermediate, rather than any direct effect on the enzyme or a sugar transporter.

The presence in the Golgi of nucleotide sugar transporters involved in delivering sugars to the luminal enzymes which glycosylate proteins, has been well established. By use of similar techniques, carrier-mediated translocation of sugar nucleotides into the lumen of the endoplasmic reticulum has also been detected (LENNARZ, 1987; HIRSCHBERG and SNIDER, 1987). Uptake of UDP glucuronic acid was saturable, time and temperature dependent and exhibited specificity, as the related compounds, UDP, UMP, uridine, D-glucuronic acid and UDP galactose, were without effect on uptake (HAUSER et al., 1985). These experimental data should be viewed with caution however, as an unidentified lipophilic acceptor of UDP glucuronic acid in microsomal membranes may contribute to the apparent "uptake" of this sugar nucleotide (MACKENZIE, 1986a; PUHAKAINEN and HÄNNINEN, 1974). The physiological activator UDP N-acetylglucosamine is thought to enhance catalytic activity by binding to this sugar transporter and increasing access of UDP glucuronic acid rather than binding directly to the enzyme (BURCHELL et al., 1983). Another possible mechanism of UDP glucuronic acid movement through the membrane may involve "mobile pores" formed by intramembraneous reverse micelles or the formation of other non-bilayer structures such as the inverted type II hexagonal phase, a hexagonal assembly of tubes in which the lipid head groups surround a narrow aqueous channel in the membrane (KRUIJFF, 1987).

UDP is a potent end-product inhibitor of glucuronidation and should be removed for efficient catalysis. Possible mechanisms of removal may involve coupled exchange with UDP glucuronic acid via an antiport protein as has been shown for the uptake of UDP glucose and UDP N-acetylglucosamine (HIRSCHBERG and SNIDER, 1987) or hydrolysis to UMP by nucleoside diphosphatidase on the luminal side of the endoplasmic reticulum. UMP, unlike UDP, can penetrate the membrane and is rephosphorylated by kinases located in the cytoplasm (KUHN and WHITE, 1977).

## 5.2. Protection of glucuronides from hydrolysis

$\beta$-Glucuronidase is synthesized and core-glycosylated in the rough endoplasmic reticulum and sequestered in the lumen of this organelle before it is transported to the lysosomes (PAIGEN, 1979). While present in the endoplasmic reticulum, $\beta$-glucuronidase which is complexed to the protein, ergasyn, could hydrolyse the glucuronides formed by transferase and set up a nonproductive cycle of conjugation and hydrolysis within the lumen of the endoplasmic reticulum. It is clear that $\beta$-glucuronidase complexed to ergasyn is active when assayed in vitro, in gels. However, whether the tetrameric enzyme complexed to ergasyn in its luminal location in the endoplasmic reticulum in vivo is also active remains to be established. Perhaps the luminal microenvironment (for example, the conditions of pH and ionic composition and the presence of the endogenous inhibitor, D-glucaro-1,4-lactone) maintains the enzyme in an inactive conformation thus preventing hydrolysis of newly synthesized glucuronides. Alternatively, the enzyme complex is inactive when bound to membrane during its passage to the lysosomes, where it is released from the membrane and activated in the acidic environment maintained by this organelle. A similar phenomenon has been observed for calcium-dependent proteases which, when bound to membrane, may be inactivated by complexing with a specific inhibitor protein. Removal of this complex from the membrane results in its dissociation and the subsequent activation of the protease (MELLGREN, 1987). If the catalytic site of UDP glucuronosyltransferase is demonstrated unequivocally to be on the luminal side of the endoplasmic reticulum, the regulation of $\beta$-glucuronidase in the same compartment will need to be investigated.

## 5.3. Secretion of glucuronides from the cell

The mode of egress of glucuronides from their site of synthesis in the hepatocyte to the bile or blood is unknown. One possibility involves a vesicular transport mechanism (COLEMAN, 1987). In this scenario, vesicles containing glucuronides bud off from the endoplasmic reticulum and are transported to the bile canaliculus where fusion and release of vesicle contents into the bile occurs. The transport of glucuronides in vesicles, rather than associated with cytosolic proteins, is supported by experiments demonstrating disruption of bilirubin glucuronide transport into the bile by colchicine, a microtubular depolymerizing drug which interferes with vesicle movement (CRAWFORD et al., 1986). All glucuronides may be transported in this way, but the smallest may leak back into the plasma across the tight junctions which act as molecular sieves, increasingly restricting molecules of increasing molecular weight. Hence glucuronides in the urine may consist of these small molecules or those derived from extrahepatic sources, whereas larger conjugates are excreted in the bile. If vesicles are also directed to the sinusoidal plasma membrane, glucuronides would be released into the plasma. A transport mechanism, similar to that for bile acids, would be needed to transport the larger conjugates released into the plasma back into the bile.

As outlined above, this vesicular mechanism would be consistent with the proposed topology of UDP glucuronosyltransferase in the endoplasmic reticulum, where glucuronides are formed in the lumen of this organelle. If glucuronides are formed on the cytoplasmic side of the endoplasmic reticulum membrane, then their removal from the cell would probably involve intracellular movement associated with cytoplasmic proteins such as ligandin and their transport from the cell surface via anion transporters in the plasma and bile cannalicular membranes.

# 6.  Mechanism of regulation

Studies on the glucuronidation of various substrates have demonstrated that UDP glucuronosyltransferases exhibit tissue specificity and are differentially regulated by chemicals and during perinatal development (DUTTON, 1980). One component of this regulation involves physiochemical changes in the enzymes' lipid microenvironment and the supply of cofactors, inhibitors and other allosteric effectors. These have been reviewed (DUTTON, 1980). Another component involves changes in the specific content of various forms of transferase. These changes have been measured with catalytic assays and antibodies. The interpretation of catalytic and immunochemical data, however, is often difficult due to the ability of several forms to glucuronidate common substrates and the lack of mono-specific antibodies. With the development of DNA probes and the refinement of molecular biological techniques, more definitive data can be obtained as has been detailed in Part 3. In addition, the regulation of the mRNA and genes of this superfamily of enzymes can now be studied.

## 6.1.  mRNA regulation

Each UDP glucuronosyltransferase form, for which a DNA probe is available, is encoded by a mRNA of approximately 2.3K bases (MACKENZIE et al., 1984; IYANAGI, 1986). Larger transcripts encoding transferase forms have also been detected (MACKENZIE et al., 1984; BURCHELL et al., 1987a). These appear to result from cleavage and polyadenylation of a primary transcript at a secondary site which is further downstream (764 bp in the case of UDPGTr-4 and 1530 bp in the case of UDPGTr-5) to the major site of cleavage and polyadenylation (MACKENZIE, 1986b, unpublished data). The function of these longer transcripts, which in two cases encode proteins identical to those encoded by more abundant shorter transcripts, is unknown. Preliminary data suggest that these longer transcripts are more readily degraded than their more abundant 2.3K base counterparts (MACKENZIE, unpublished data). Some inducers of drug-metabolizing enzymes such as pregnenolone-16α-carbonitrile, increase the steady state levels of these larger transcripts while having little effect on the 2.3K base mRNA (MACKENZIE and HAQUE, 1987). It is thus conceivable that these inducers cause a temporary elevation of mRNA and protein by stabilizing a readily degradable transcript. This mechanism would be appropriate for the handling of acute changes in aglycone levels that would presumably occur after food ingestion or drug therapy. Transcripts larger than 2.3K bases, may also encode less abundant forms of transferase. Based on an analysis of RNA prepared from the livers of Gunn rats and their normal counterparts, it was suggested that bilirubin UDP glucuronosyltransferase was encoded by a 2.7K base mRNA (BURCHELL et al., 1987a). However, the absence of this size mRNA class in Gunn rat may merely reflect its degradation during preparation or alterations in the processing of its precursor RNA to the 2.7K base transcript.


## 6.2. Gene regulation

As well as mRNA stabilization, changes in the transcription rates of UDP glucuronosyltransferase genes in response to chemical or hormonal stimuli may be a major factor in their regulation. Although changes in transcription rates have been demonstrated to be of major importance in the control of cytochrome P-450 levels (GONZALEZ et al., 1984), studies showing this for transferases have not been reported. Nevertheless, preliminary data indicate the importance of genomic regulation in the control of glucuronidation. For example, Wistar rats exhibit genetic differences in the rates of hepatic glucuronidation of androsterone (MATSUI and WATANABE, 1982). Rats with low activity (LA) lack androsterone UDP glucuronosyltransferase protein and mRNA. This inherited deficiency was thought to be a result of a deletion in the androsterone transferase gene, as restriction fragment lengths observed on Southern blotting of genomic DNA from LA Wistar rats were different from their normal (HA) counterparts (CORSER et al., 1987). In order to investigate genomic regulation, it is therefore necessary to isolate and sequence UDP glucuronosyltransferase genes and to determine the genomic elements associated with those genes which play a role in their regulation. The gene for androsterone UDP glucuronosyltransferase is currently being sequenced and consists of six exons extending over approximately 15K base pairs of DNA (HAQUE and MACKENZIE, unpublished data). These data should provide a basis for analysing the defect in the LA Wistar rat gene in more detail.

Isolation of the bilirubin, digitoxigenin monodigitoxide and phenol (53K dalton) UDP glucuronosyltransferase genes from normal rats should enable elucidation of the molecular processes underlying their absence in Gunn rats. It is conceivable that a single genomic deletion or rearrangement may have a deleterious effect on these three enzymes, especially if their genes are localized on the same chromosome. Gene localisation studies have already demonstrated that the androsterone transferase gene and other members of that subfamily reside on chromosome 5 in the mouse (KRASNEWICH et al., 1987). However, the chromosomal localization of the genes encoding bilirubin, digitoxigenin monodigitoxide and phenol transferases are unknown. UDP glucuronosyl-

---

P. I. MACKENZIE

transferase genes may be under coordinate regulation with P-450 genes (NEBERT and GONZALEZ, 1987). Studies with 3-methylcholanthrene indicate that the levels of $P_1$-450 and 4-nitrophenol transferase ($GT_1$) appear to be regulated by the Ah locus (OWENS, 1977), whereas studies with clofibrate indicate a common mechanism in the regulation of bilirubin transferase and the P-450 catalyzing lauric acid 4-hydroxylation (FOURNEL et al., 1985). There is also evidence that cytochrome P-450p and digitoxigenin mono-digitoxide transferase are both regulated by a nonclassical glucocorticoid receptor (SCHUETZ et al., 1986). It will be interesting to determine the chromosomal location of these and other transferase genes and to determine whether they are linked to cytochrome P-450 subfamilies, which have already been mapped to different chromosomal locations (NEBERT and GONZALEZ, 1987).

Ultimately, analysis of the molecular mechanisms of UDP glucuronosyltransferase regulation will require identification and characterization of interactions between proteins and specific genomic DNA segments or chromatin domains which mediate the regulatory processes. The isolation and sequencing of transferase genes will provide a basis for these future studies.

## 7. Horizons

I hope it is apparent from this review completed in December, 1987, that DNA recombinant technology has opened up new and exciting avenues of research on this multigene family. It has provided a way of characterizing UDP glucuronosyltransferases in terms of their primary amino acid sequences and substrate specificities and has avoided the problems inherent in purifying and characterizing catalyticallyactive membrane-bound enzymes.

A major focus of future research will involve the isolation and cloning of cDNAs to new forms of transferase. Future experiments will also most likely utilize the in situ expression of mutagenized and chimeric transferase cDNA clones to provide clues as to the amino acids and peptide domains essential for catalysis. Thus, the UDP glucuronic acid binding site will be identified and analysed to determine, for example, those parameters in bilirubin transferase permitting conjugation of bilirubin to glucose and xylose as well as glucuronic acid. The aglycone binding site will be localized and the residues specifying aglycone preference identified. Modification of these residues by mutagenesis may also perhaps lead to the production of an enzyme which is catalytically more efficient or has changed sugar-nucleotide binding specificities. As well as providing basic information on the biology of this important family of drug-metabolizing enzymes, this knowledge will be of immense aid in the design of therapeutic drugs and in the prevention of adverse reactions resulting from competition between drugs and endogenous compounds for common active sites. Those sequences in the protein conferring residency in the endoplasmic reticulum will probably be identified and perhaps modified to deliver the enzyme to novel intracellular sites. The isolation and sequencing of transferase genes and their flanking regions will also be a major area of research in the future. This will enable a study of the genetic elements and cellular factors underlying the various modes of regulation of transferase forms and will perhaps identify the genetic changes responsible for inherited defects in glucuronidation. These data will be important in potential gene replacement therapies involving Criggler-Najjar patients, who lack bilirubin transferase, and in determining the susceptibility of different organs to toxic agents as a function of age and previous exposure to drugs and dietary components.

Finally, characterization of the structure and regulation of transferase genes in other vertebrates may provide clues as to the evolutionary origin of various UDP glucuronosyltransferases.

## Acknowledgements

The author is grateful for the support of Dr. Dan NEBERT, Dr. Frank GONZALEZ, Professor Don BIRKETT and colleagues and the financial assistance of the National Institutes of Health, U.S.A. and the National Health and Medical Research Council of Australia. The author is a NH & MRC Research Fellow.

## 8. References

ARION, W. J., B. BURCHELL, and A. BURCHELL (1984), Biochem. J. **220**, 835—842.

ARMSTRONG, R. N. (1987), CRC Critical reviews in Biochemistry **22**, 39—88.

BERRY, G., A. STELLON, and T. HALLINAN (1975), Biochim. Biophys. Acta **403**, 335 to 344.

BILLING, B. H. (1987), in: Diseases of the Liver, 6th edition (L. SCHIFF and E. R. SCHIFF, eds.). J. B. Lippincott Co. Philadelphia, pp. 103—127.

BOCK, K. W., U. G. V. CLAUSBRUCH, R. KAUFMANN, W. LILIENBLUM, F. OESCH, H. PFEIL, and K. L. PLATT (1980), Biochem. Pharmacol. **29**, 495—500.

BOCK, K. W., B. BURCHELL, G. J. DUTTON, O. HANNINEN, G. J. MULDER, I. S. OWENS, G. SIEST, and T. R. TEPHLY (1983), Biochem. Pharmacol. **32**, 953—955.

BOCK, K. W., and W. LILIENBLUM (1984), in: Advances in glucuronide conjugation. (S. MATERN, K. W. BOCK, and W. GEROK, eds.), MTP Press Ltd. Lancaster, pp. 51—57.

BOCK, K. W., W. LILIENBLUM, G. FISCHER, G. SCHIRMER, and B. S. BOCK-HENNIG (1987), Arch. Toxicol. **60**, 22—29.

BURCHELL, B. (1981), Rev. Biochem. Toxicol. **3**, 1—32.

BURCHELL, B. and G. B. ODELL (1981), FEBS Letts. **135**, 304—308.

BURCHELL, B., P. J. WEATHERILL, and C. BERRY (1983), Biochim. Biophys. Acta **735**, 309—313.

BURCHELL, B. and N. BLANCKAERT (1984), Biochem. J. **223**, 461—465.

BURCHELL, B., M. W. H. COUGHTRIE, M. R. JACKSON, S. R. P. SHEPHERD, D. HARDING, and R. HUME (1987a), Molec. Aspects Med. **9**, 429—455.

BURCHELL, B., M. R. JACKSON, R. B. CORSER, M. W. H. COUGHTRIE, D. HARDING, S. H. SHEPHERD, and S. M. WILSON (1987b), Biochem. Soc. Trans. 15, 581—584.
BURCHELL, B., M. R. JACKSON, M. W. H. COUGHTRIE, D. HARDING, S. WILSON, and J. R. BEND (1987c), in Drug Metabolism from Molecules to Man. (in Press).
COFFMAN, B. L., M. D. GREEN, Y. M. IRSHAID, and T. R. TEPHLY (1987), Fed. Proc. 46, Abstr. No. 3188.
COFFMAN, B. L., M. D. GREEN, and T. R. TEPHLY (1987), The Pharmacologist 29, 207 ASPET Abstr.
COLEMAN, R. (1987), Biochem. J. 244, 249—261.
COMMONER, B., A. J. VITHAYATHIL, and J. I. HENRY (1974), Nature 249, 850—852.
CORSER, R. B., M. W. H. COUGHTRIE, M. R. JACKSON, and B. BURCHELL (1987), FEBS Letts. 213, 448—452.
COUGHTRIE, M. W. H., B. BURCHELL, I. M. SHEPHERD, and J. R. BEND (1987a), Mol. Pharm. 31, 585—591.
COUGHTRIE, M. W. H., B. BURCHELL, and J. R. BEND (1987b), Biochem. Pharmacol. 36, 245—251.
CRAWFORD, J. M., S. V. WESTMORELAND, and J. L. GOLLAN (1986), Hepatology 6, AASLD Abstr. No. 297.
DUTTON, G. J. (1980), in: Glucuronidation of Drugs and Other Compounds. CRC Press, Inc. Boca Raton, Florida, 1—180.
FALANY, C. N. and T. R. TEPHLY (1983), Arch. Biochem. Biophys. 227, 248—258.
FALANY, C. N., J. ROY CHOWDHURY, N. ROY CHOWDHURY, and T. R. TEPHLY (1983), Drug Metab. Disp. 11, 426—432.
FENSELAU, G. and L. P. JOHNSON (1980), Drug Metab. Disp. 8, 274—283.
FOURNEL, S., J. MAGDALOU, J. THOMASSIN, J. VILLOUTREIX, G. SIEST, J. CALDWELL, and J. ANDRE (1985), Biochim. Biophys. Acta 842, 202—213.
FOURNEL, S., B. GREGOIRE, J. MAGDALOU, M. CARRE, C. LAFAURIE, G. SIEST, and P. CAUBERE (1986), Biochim. Biophys. Acta 883, 190—196.
GONZALEZ, F. J., R. H. TUKEY, and D. W. NEBERT (1984), Mol. Pharm. 26, 117-121.
GONZALEZ, F. J., D. W. NEBERT, J. P. HARDWICK, and C. B. KASPER (1985), J. Biol. Chem. 260, 7435—7441.
GREEN, M. D., B. L. COFFMAN, J. F. PUIG, and T. R. TEPHLY (1987), The Pharmacologist 29, 208 ASPET Abstr.
GREEN, M. D. and T. R. TEPHLY (1987a), Cancer Res. 47, 2028—2031.
GREEN, M. D. and T. R. TEPHLY (1987b), Fed. Proc. 46, Abstr. No. 3189.
HAEGER, B., R. DE BRITO, and T. HALLINAN (1980), Biochem. J. 192, 971—974.
HALLINAN, T. (1978), in: Conjugation reaction in drug biotransformation (A. AITIO, ed.), Elsevier/North-Holland, Amsterdam, pp. 257—267.
HARDING, D., S. M. WILSON, M. R. JACKSON, B. BURCHELL, M. D. GREEN, and T. R. TEPHLY (1987), Nuc. Acids Res. 15, 3936.
HAUSER, S. C., J. C. ZIURYS, and J. L. GOLLAN (1985), Hepatology 5, AASLO Abstr. No. 359.
HIRSCHBERG, C. B. and M. D. SNIDER (1987), Ann. Rev. Biochem. 56, 63—87.
HOCHMAN, Y. and D. ZAKIM (1983), J. Biol. Chem. 258, 4143—4146.
HOCHMAN, Y. and D. ZAKIM (1984), J. Biol. Chem. 259, 5521—5525.
ISSELBACHER, K. J., M. F. CHRABAS, and R. C. QUINN (1962), J. Biol. Chem. 237, 3033—3036.
IYANAGI, T., M. HANIU, K. SOGAWA, Y. FUJII-KURIYAMA, S. WATANABE, J. E. SHIVELY, and K. F. ANAN (1986), J. Biol. Chem. 261, 15607—15614.
JACKSON, M. R. and B. BURCHELL (1986), Nuc. Acids Res. 14, 779—785.
KASPER, C. B. and D. HENTON (1980), in: Enzymatic Basis of Detoxification, Vol. II, Academic Press. pp. 3—34.

KUHN, N. J. and A. WHITE (1977), Biochem. J. **168**, 423—433.

KIRKPATRICK, R. B., C. N. FALANY, and T. R. TEPHLY (1984), J. Biol. Chem. **249**, 6176—6180.

KNAPP, S. A., M. D. GREEN, T. R. TEPHLY, and J. BARON (1987), Fed. Proc. Abstr. No. 3192. (in Press).

KOSTER, A. Sj., G. SCHIRMER, and K. W. BOCK (1986), Biochem. Pharmacol. **35**, 3971—3975.

KRASNEWICH, D., C. A. KOZAK, D. W. NEBERT, and P. I. MACKENZIE (1987), Somatic Cell Mol. Genetics **13**, 179—182.

KRUIJFF, B. DE (1987), Nature **329**, 587—588.

KYTE, J. and R. F. DOOLITTLE (1982), J. Mol. Biol. **157**, 105—132.

LEHMAN, J. P., C. FENSELAU, and J. R. DEPAULO (1983), Drug Metab. Disp. **11**, 221 to 225.

LENNARZ, W. J. (1987), Biochemistry **26**, 7205—7210.

LEVY, S., B. YAGEN, and R. MECHOULAM (1978), Science **200**, 1391—1392.

LILIENBLUM, W., A. K. WALLI, and K. W. BOCK (1982), Biochem. Pharmacol. **31**, 907—913.

LILIENBLUM, W. and K. W. BOCK (1984), Biochem. Pharm. **33**, 2041—2046.

LILIENBLUM, W., B. S. BOCK-HENNIG, and K. W. BOCK (1985), Mol. Pharm. **27**, 451 to 458.

LILIENBLUM, W., K. L. PLATT, G. SCHIRMER, F. OESCH, and K. W. BOCK (1987), Mol. Pharm. **32**, 173—177.

LIPMAN, D. J. and W. R. PEARSON (1985), Science **227**, 1435—1441.

MACKENZIE, P. I. and I. S. OWENS (1984), Biochem. Biophys. Res. Commun. **122**, 1441—1449.

MACKENZIE, P. I., F. J. GONZALEZ, and I. S. OWENS (1984a), J. Biol. Chem. **259**, 12153—12160.

MACKENZIE, P. I., F. J. GONZALEZ, and I. S. OWENS (1984b), Arch. Biochem. Biophys. **230**, 676—680.

MACKENZIE, P. I. (1986a), J. Biol. Chem. **261**, 14112—14117.

MACKENZIE, P. I. (1986b), J. Biol. Chem. **261**, 14112—14117.

MACKENZIE, P. I. (1987), J. Biol. Chem. **262**, 9744—9749.

MACKENZIE, P. I. and S. J. HAQUE (1987), in Lever Cells and Drugs (ed. A. GUILLOUZO) INSERM/John Libbey.

MARAGANORE, J. M. (1987), TIBS **12**, 176—177.

MATSUI, M. and H. K. WATANABE (1982), Biochem. J. **202**, 171—174.

McCORMICK, F., B. F. C. CLARK, T. F. M. LA COUR, M. KJELDGAARD, L. NORSKOV-LAURITSEN, and J. NYBROG (1985), Science **230**, 78—82.

MELLGREN, R. L. (1987), FASEB J. **1**, 110—115.

MEYERS, M., W. SLIKKER, and M. VORE (1981), J. Pharmacol. Exp. Ther. **218**, 63—73.

MILLER, D. A. and H. F. DELUCA (1986), Arch. Biochem. Biophys. **244**, 179—186.

MOLLER, W. and R. AMONS (1985), FEBS Letts. **186**, 1—7.

NEBERT, D. W. and F. J. GONZALEZ (1987), Ann. Rev. Biochem. **56**, 945—993.

NEBERT, D. W., M. ADESNIK, M. J. COON, R. W. ESTABROOK, and F. J. GONZALEZ (1987), DNA **6**, 1—11.

OWENS, I. S. (1977), J. Biol. Chem. **252**, 2827—2833.

PAABO, S., B. M. BHAT, W. S. M. WOLD, and P. A. PETERSON (1987), Cell **50**, 311 to 317.

PAIGEN, K. (1979), Ann. Rev. Genet. **13**, 417—466.

PETERS, W. H. P., P. L. M. JANSEN, H. T. M. CUYPERS, R. A. DE ABREU, and H. NAUTA (1986), Biochim. Biophys. Acta **873**, 252—259.

PUHAKAINEN, E. and O. HÄNNINEN (1974), FEBS Letts. **39**, 144—148.

Puig, J. F. and T. R. Tephly (1986), Mol. Pharm. 30, 558—565.
Radominska-Pyrek, A., D. Green, R. Lester, and T. Tephly (1986), Fed. Proc. 45, Abstr. No. 4527.
Roth, S. and J. B. F. Levine (1983), Exp. Cell Res. 143, 217—225.
Rothman, J. E. (1987), Cell 50, 521—522.
Roy Chowdhury, J., N. Roy Chowdhury, P. M. Novikoff, and I. M. Arias (1985), in: Advances in glucuronide conjugation (S. Matern, K. W. Bock, and W. Gerok, eds.), MTP Press, Lancaster, Boston pp. 33—40.
Roy Chowdhury, N., R. Stockert, S. Srinivasan, M. Lederstein, and J. Roy Chowdhury (1986), pp. 1719.
Roy Chowdhury, N., I. M. Arias, M. Lederstein, and J. Roy Chowdhury (1986a), Hepatology 6, 123—128.
Roy Chowdhury, J., N. Roy Chowdhury, C. N. Falany, T. R. Tephly, and I. M. Arias (1986b), Biochem. J. 233, 827—837.
Roy Chowdhury, N., M. Saber, P. Mackenzie, and J. Roy Chowdhury (1987), Gastroenterology 92, AGA/AASLD Abstr. No. 1768.
Sabatini, D. D., G. Kreibich, T. Morimoto, and M. Adesnik (1982), J. Cell Biol. 92, 1—22.
Schmoldt, A. and J. Promies (1982), Biochem. Pharmacol. 31, 2285—2289.
Schuetz, E. G., G. A. Hazelton, J. Hall, P. B. Watkins, C. D. Klaassen, and P. S. Guzelian (1986), J. Biol. Chem. 261, 8270—8275.
Shattuck, K. E., A. Radominska-Pyrek, P. Zimniak, E. W. Adcock, R. Lester, and J. St. Pyrek (1986), Hepatology 6, 869—873.
Siest, G., B. Antoine, S. Fournel, J. Magdalou, and J. Thomassin (1987), Biochem. Pharmacol. 36, 983—989.
Stogniew, M. and C. Fenselau (1982), Drug Metab. Disp. 10, 609—613.
Thomassin, J., S. Dragacci, B. Faye, J. Magdalou, and G. Siest (1986), Comp. Biochem. Physiol. 83C, 127—131.
Tukey, R. H., R. Robinson, B. Holm, C. N. Falany, and T. R. Tephly (1982), Drug Metab. Disp. 10, 97—101.
Turco, S. J. and E. C. Heath (1977), J. Biol. Chem. 252, 2918—2928.
Von Meyerinck, L., B. L. Coffman, M. D. Green, R. B. Kirkpatrick, A. Schmoldt, and T. R. Tephly (1985), Drug Metab. Disp. 13, 700—704.
Walter, P. and V. R. Lingappa (1986), Ann. Rev. Cell Biol. 2, 499—516.
Watkins, J. B., Z. Gregus, T. N. Thompson, and C. D. Klaassen (1982), Toxicol. Appl. Pharmacol. 64, 439—446.
Watkins, J. B. and C. D. Klaassen (1982), Drug Metab. Disp. 10, 590—594.
Wold, F. (1986), TIBS 11, 58—59.
Zakim, D. and D. A. Vessey (1982), in: Membranes and Transport (A. N. Martonosi, ed.), Plenum Publishing Corp., New York 1, 269—273.
Zakim, D., Y. Hochman, and W. C. Kenney (1983), J. Biol. Chem. 258, 6430—6434.

# Chapter 8

# Glutathione Transferases

B. KETTERER and J. B. TAYLOR

# 1. Introduction

The glutathione transferases are multifunctional enzymes associated with the detoxication of drugs and carcinogens, the reduction of organic hydroperoxides and nitrates, the binding and intracellular transport of lipophiles and the biosynthesis of the local hormone leukotriene C. In what follows we give a broad account of reactions with GSH which are catalysed by GSH transferases, the enzymes which carry out this catalysis, their distribution in the living world, their structure and, finally, what is known at present of their genetic regulation.

# 2. Glutathione

Glutathione (GSH), the tripeptide $\gamma$-glutamyl cysteinyl glycine, is one of the most abundant small organic molecules in the cell, its concentration lying between 1 and 10 mM according to the tissue. One of its principal biological functions depends on nucleophilic thiol group of its cysteinyl residue (CHASSEAUD, 1979), some of the reactions of which are shown below.

## 2.1. Attack on electrophilic carbon

a) nucleophilic displacement at saturated carbons

$$RCH_2Cl + GSH \rightarrow RCH_2SG \tag{1}$$

b) nucleophilic displacement at aromatic carbons

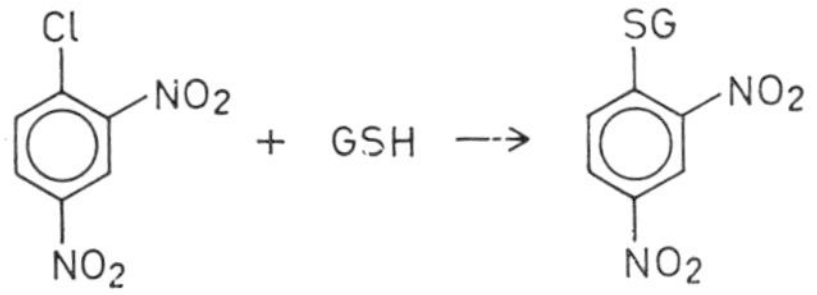

$$\tag{2}$$

c) nucleophilic attack on strained oxirane rings

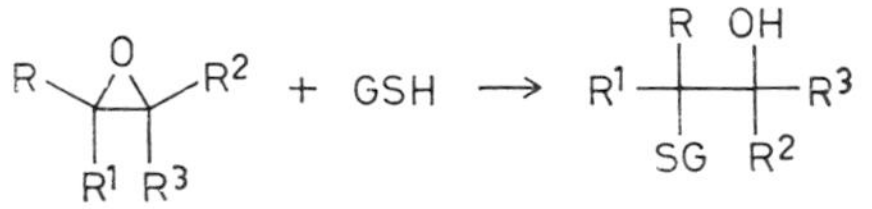

$$\tag{3}$$

d) MICHAEL addition to $\alpha, \beta$ unsaturated compounds

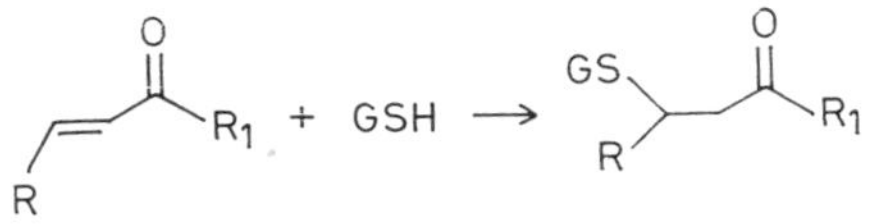

$$\tag{4}$$

## 2.2. Attack on electrophilic sulphur

$$GSH + R'SSR' \rightleftharpoons GSSR' + R'SH \tag{5}$$

## 2.3. Attack on electrophilic nitrogen

$$GSH + RNO \rightarrow R\overset{\overset{O}{\uparrow}}{N}SG \rightarrow R\overset{\overset{O}{\uparrow}}{N}SG \tag{6}$$

$$GSH + RONO_2 \rightarrow ROH + [GSNO_2]$$
$$[GSNO_2] + GSH \rightarrow GSSG + HNO_2 \tag{7}$$

## 2.4. Attack on electrophilic oxygen

$$ROOH + GSH \rightarrow ROH + [GSOH]$$
$$[GSOH] + GSH \rightarrow GSSG + H_2O \tag{8}$$

## 2.5. GSH dependent isomerisation

$$\tag{9}$$

$\Delta^5$ androstene-3,17-dione $\qquad$ $\Delta^4$ androstene-3,17-dione

In all these reactions it is probable that GS⁻ rather than GSH is the reactive species. This has been demonstrated to be the case in the reactions 2 to 5 (CHASSEAUD, 1979).

# 3. Glutathione transferases

Since many electrophiles may be cytotoxic and/or genotoxic and therefore mutagenic and carcinogenic, catalysis of their reaction with GSH in vivo is important for their detoxication. The enzymes which do this are GSH transferases (E.C. 2.5.1.18). Whereas GSH transferase is the only term used to describe catalysis of reactions 1—6, the enzymic catalysis of reaction 7 has also been called nitrate reductase, a reaction important in the metabolism of certain vasodilatory drugs (see below) and that of reaction 8 has also been called Se-independent GSH peroxidase (Se-dependent GSH peroxidase being a different protein, catalysing the reduction of $H_2O_2$ in addition to organic peroxides). In both reactions 7 and 8 the initial GSH conjugates are thought to be [GSNO₂] and [GSOH] respectively which then react with more GSH to give GSSG.

## 4. Soluble glutathione transferases in the rat

### 4.1. Multiple isoenzymes, nomenclature and genetic relationships

Most of the exploratory work on GSH transferases has been done in the rat and, in terms of structure, function and tissue distribution, they are the best understood and therefore provide the standards by which others are often compared. All soluble GSH transferases in the rat are dimers. At least ten subunits have been characterized, although more are known to exist. These ten well defined subunits, together with some of their distinguishing characteristics, are shown in Table 1. Two nomenclatures are shown. The one used in this chapter is numerical and based on the chronological order of their characterization; the other is based on their mobility on sodium dodecyl sulphate polyacrylamide gel electrophoresis (JAKOBY et al., 1984).

**Table 1.** GSH Transferase subunits in the rat[1]

| Nomenclature | | Some physical characteristics | | |
|---|---|---|---|---|
| chronological | mobility on SDS page | app mol wt $\times 10^{-3}$ | isoelectric point[2] | retention time on reverse phase h.p.l.c. (min)[3] |
| 1 | Ya | 25 | 10 | 50 |
| 2 | Yc | 28 | 9.8 | 33 |
| 3 | Yb₁ | 26.5 | 8.5 | 24 |
| 4 | Yb₂ | 26.5 | 6.9 | 27 |
| 5 | NN | 26.5 | 7.3 | 33 |
| 6 | Yn | 26 | 5.8 | 37 |
| 7 | Yf or Yp | 24 | 7.0 | 31 |
| 8 | Yk | 24.5 | 6.0 | 54 |
| 9 | Yn | 24 | 5.8 | 28 |
| 10 | NN | 25.5 | 9.6 | 33 |

[1] Some of these figures are unpublished results from E. LALOR, B. COLES, D. J. MEYER, K. H. TAN, and B. KETTERER;
[2] isoelectric points refer to dimers of the subunit in question;
[3] see OSTLUND FARRANTS et al., 1987;
NN — not named.

Complete primary structures, deduced from cDNA sequences, are known for subunits 1 (PICKETT et al., 1984; LAI et al., 1984; TAYLOR et al., 1984), 2 (TELAKOWSKI-HOPKINS et al., 1985), 3 (DING et al., 1985), 4 (ROTHKOPF et al., 1986) and 7 (SUGUOKA et al., 1985; TAYLOR et al., 1987). Based on these sequences the above subunits fall into three categories comprising 1 and 2; 3 and 4; and 7. There is 69% identity in full length sequence between 1 and 2, 77% identity between 3 and 4 and very little identity between the three categories; this implies that these categories are multigene families. They have been named alpha (1 and 2), mu (3 and 4) or pi (7) respectively (MAN-NERVIK et al., 1985a, b). On the basis of present evidence, involving complete or partial amino acid sequences, enzymic properties and immuno-logical cross reactivity, the genetic relationship between all ten subunits is believed to be as follows: alpha family, subunits 1, 2, 8 and 10; mu family, subunits 3, 4, 6 and 9 and pi family, subunit 7 only. Subunit 5 has yet to be assigned (MEYER et al., 1984, 1985a; KETTERER et al., 1985, 1986; SCOTT et al., 1987; COLES, MEYER, LALOR, KETTERER, unpublished information).

Within a multigene family, subunits may form heterodimers. GSH trans-ferases $1-2$, $3-4$ and $6-9$ and GSH transferases $3-6$, $3-9$, $4-6$ and $4-9$ have been identified (KETTERER et al., 1986).

## 4.2. Enzymic activity

### 4.2.1. Model substrates

Much of the work done with GSH transferases involves substrates chosen, not because of their biological significance, but because their GSH conjugation results in an optical density change which can be made the basis of a con-venient spectrophotometric assay. The most commonly used substrate is 1-chloro-2,4-dinitrobenzene (see reaction 8) which is utilized well by all rat subunits except 5 and 9 (see Table 2). It is often treated as a "universal" substrate for unfractionated GSH transferases and, since subunits 5 and 9 are usually present in small amounts, this is a reasonable practice. Other substrates are chosen because they are relatively specific for a particular subunit and may be used to detect and quantify it (see Table 2). These sub-strates undergo one or other of the four types of reaction with GSH described above. The following undergo GSH conjugation: 1-chloro-2,4-dinitro-benzene, 1,2-dichloro-4-dinitrobenzene (reaction 1), trans-4-phenyl-3-buten-2-one and 4-hydroxy non-2-enal (reaction 4) and 1,2-epoxy-3-(p-nitrophenoxy) propane (reaction 3). Cumene hydroperoxide on the other hand undergoes reduction (reaction 14) and $\triangle^5$-androstene-3,17-dione which undergoes GSH dependent isomerization (reaction 9).

Table 2. Rat GSH transferases

| Substrate | Enzymes by class | | | | | | | |
|---|---|---|---|---|---|---|---|---|
| | alpha | | | mu | | | pi | un-assigned |
| | 1—1 | 2—2 | 8—8 | 3—3 | 4—4 | 6—9 | 7—7 | 5—5 |
| 1-Chloro-2,4-dinitrobenzene | 40.0 | 38.0 | 10.0 | 50.0 | 20.0 | 190.0 | 20.0 | < 0.15 |
| 1,2-Dichloro-4-nitrobenzene | 0.15 | 0.15 | 0.12 | 8.4 | 0.7 | 2.4 | < 0.05 | nil |
| Trans-4-phenyl-3-buten-2-one | 0.1 | 0.1 | 0.1 | 0.1 | 1.2 | 0.2 | 0.02 | < 0.001 |
| 1,2-Epoxy-3-(p-nitrophenoxy)-propane | 0.7 | 0.9 | nd | 0.2 | 0.9 | < 0.5 | 1.0 | 25.5 |
| Ethacrynic acid | 0.3 | 2.1 | 7.0 | 0.4 | 1.0 | (< 0.5) | 4.0 | nil |
| 4-Hydroxynon-2-enal* | 2.6 | 0.7 | 170.0 | 2.7 | 6.9 | nd | nd | nd |
| Cumene hydroperoxide | 1.4 | 3.0 | 1.1 | 0.1 | 0.4 | 0.04 | 0.01 | 12.5 |
| Linoleate hydroperoxide | 3.0 | 1.6 | 0.2 | 0.2 | 0.02 | 0.06 | 1.5 | 5.3 |
| $\triangle^5$-Androstene-3,17-dione | 0.23 | 0.07 | nd | 0.02 | 0.002 | nd | < 0.001 | nd |

Activities ($\mu$mol $\cdot$ min$^{-1}$ $\cdot$ mg$^{-1}$) towards model substrates some of which are used to specify subunits
* Data are from MANNERVIK et al., 1985a; JENSSON et al., 1986. Otherwise they are from the Cancer Research Campaign Molecular Toxicology Research Group.

## 4.2.2. Selected xenobiotic substrates

Biologically significant substrates long associated with the action of GSH transferases are certain of those xenobiotics which are either electrophilic per se or give rise to electrophiles upon metabolism. Some examples are: 1) industrial chemicals such as ethylene dibromide, vinyl chloride, styrene oxide and aromatic amines; 2) pesticides and herbicides such as alachlor, diazinon and atrazine; 3) pharmaceutical products such as paracetamol, melphalan, carmustine and misonidazole; 4) environmental pollutants derived

B. KETTERER; J. B. TAYLOR

Table 3. Rat GSH transferases

| Substrate | Enzyme by class | | | | | | |
|---|---|---|---|---|---|---|---|
| | alpha | | | mu | | pi | unassigned |
| | 1−1 | 2−2 | 8−8 | 3−3 | 4−4 | 7−7 | 5−5 |
| Cholesterol-5,6-oxide | 0.14 | nd | nd | nil | nil | nd | nil |
| 4-Hydroxynon-2-enal[a] | 2.6 | 0.7 | 170 | 2.7 | 6.9 | nd | nd |
| 10,11-Epoxyeicosatrienoic acid[b] | 0.003 | 0.002 | nd | 0.022 | 0.001 | nd | 0.136 |
| Leukotriene $A_4$ methyl ester[a] | 0.008 | 0.008 | nd | 0.003 | 0.102 | nd | nd |
| N-Acetyl-p-benzoquinone imine | 24.0 | 48.0 | nd | 6.0 | 3.0 | 60.0 | nd |
| α-Bromoisovaleryl urea | 0.01 | 0.1 | nd | 0.08 | 0.06 | nd | nd |
| Aflatoxin $B_1$-8,9-oxide* | 0.001 | 0.001 | nd | nil | nil | nil | nd |
| Benzo(a)pyrene-4,5-oxide[c] | 0.011 | 0.004 | nd | 0.087 | nd | nd | 0.069 |
| anti-Benzo(a)pyrene-7,8-diol-9,10-oxide[d] | 0.01 | 0.08 | nd | 0.03 | 0.33 | 0.33 | nd |
| 1-Nitropyrene-4,5-oxide | 0.01 | 0.03 | nd | 0.30 | 0.30 | 0.02 | nd |

Activities ($\mu$mol · min$^{-1}$ · mg$^{-1}$) towards some biologically relevant electrophilic substrates
Data are from [a]MANNERVIK et al., 1985a; JENSSON et al., 1986; [b]SPEARMAN et al., 1985; [c]NEMOTO et al., 1975; [d]JERNSTROM et al., 1985. Otherwise form the Cancer Research Campaign Molecular Toxicology Research Group.
* These results were obtained with aflatoxin $B_1$-8,9-oxide generated by mouse microsomes and arc only approximate (COLES et al., 1985).

from the combustion of petrochemicals and other organic material such as benzo(a)pyrene and 1-nitropyrene; 5) the mould product aflatoxin B and 6) the amino acid pyrolysis product 3-hydroxyamino-1-methyl-5H-pyrido-[4,3-b]indole (Trp-P-2) (SAITO et al., 1983). However, in recent times the importance of substrates of endogenous origin has also become evident, namely lipid and nucleic acid hydroperoxides, the lipid peroxidation byproducts such as hydroxyalkenals and epoxides of cholesterol and polyunsaturated fatty acids.

Table 3 shows the activity of rat GSH transferase isoenzymes towards various xenobiotic electrophiles. Some of these are selected for discussion below.

(i)  Aflatoxin-8,9-oxide
Aflatoxin B$_1$ (AFB$_1$) is a powerful hepatocarcinogen, in the rat, the carcinogenic metabolite being AFB$_1$-8,9-oxide. As shown in Table 3,

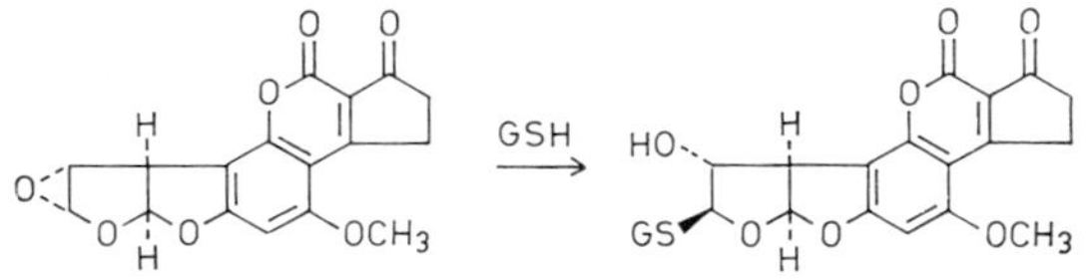

(10)

AFB$_1$-8,9-oxide is utilized by subunits 1 and 2 only and is a poor substrate
in both cases. In the rat, induction of subunit 1 by, for example, ethoxyquin
results in a 4 to 5 fold increase in the biliary excretion of AFB$_1$-GSH conjugate
and a marked decrease in genotoxicity as demonstrated by a 90% reduction
in DNA adducts and a greater than 95% reduction in the percentage of liver
occupied by preneoplastic foci (KENSLER et al., 1986).

Ethoxyquin is one of a wide range of chemicals, both naturally occurring
and synthetic, which induce GSH transferases in the rat including barbiturates,
polyaromatic hydrocarbons (IGARASHI et al., 1987), other carcinogens such as
N,N-dimethyl-4-aminoazobenzene, certain antioxidants such as butylated
hydroxy-toluene (BHT), butylated hydroxy-anisole (BHA) and ethoxyquin
(BENSON et al., 1978, 1979; SATO et al., 1984); the GSH transferase substrate
trans-stilbene oxide (DI SIMPLICIO et al., 1983) and natural products such as
dithiolthiones isolated from the Brassica family (PEARSON et al., 1983). Most
of these inducers are selective for subunits 1 and 3 and can assist in the detoxication of substrates for these enzymes.

(ii)  N-acetyl benzoquinone imine
Metabolism of paracetamol (P) otherwise known as acetaminophen gives N-
acetyl benzoquinone imine (NABQI), an electrophile, which reacts with GSH
spontaneously very rapidly to give two products namely 3-(glutathion-S-yl)
paracetamol (PSG) see reaction 11) and paracetamol in the ratio 3:2 (ALBANO
et al., 1985). Table 3 shows that the reaction of NABQI with GSH is very
effectively catalysed by GSH transferases $7-7$, $2-2$ and $1-1$ and much

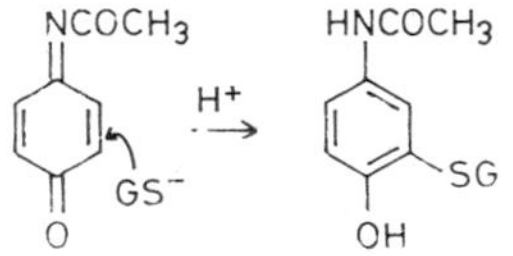

(11)

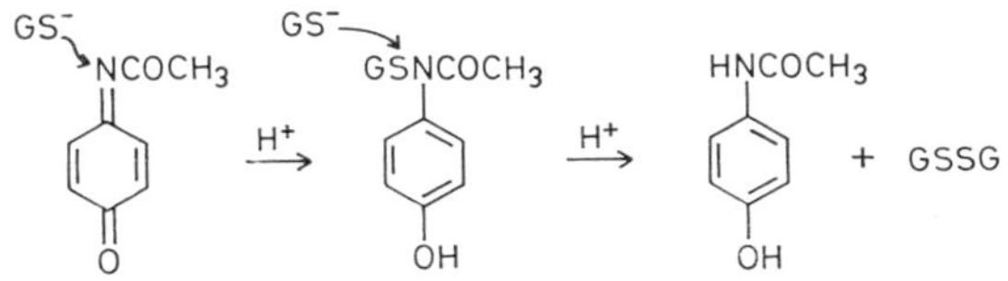

(12)

less so by GSH transferases 3−3 and 4−4. When the products of the enzyme catalysed reactions are analysed, subunits 1 and 2 are seen to produce P and PSG in a similar ratio to that produced by the spontaneous reaction, indicating that these two subunits are good catalysts of both reactions. Subunit 7, on the other hand, is selective and produces predominantly PSG. Thus, at first sight, GSH transferases containing subunits 1 and 2 appear to behave, not only as a GSH transferase, but also as a quinone reductase. However, the mechanism of the reaction, which results in the reduction of NABQI to para-cetamol, may also involve GSH conjugation. It is postulated that an **ipso** adduct with GSH is formed, which is reduced by further GSH (see reactions 12).

Since the spontaneous reaction is so rapid ($k_2 = 3.1 \times 10^4 \, M^{-1} \, s^{-1}$), the need for catalysis for effective detoxication in vivo, at therapeutic doses of paracetamol and physiological levels of GSH, might be questioned (COLES et al., 1988) however, it is probable that both routes are involved. When very high doses of paracetamol are given, hepatic GSH is consumed, first by the spontaneous and enzymic reactions of GSH with NABQI and subsequent-ly, predominantly by the enzymic reaction. This proceeds rapidly until GSH levels fall well below the $K_m$ for GSH, which, in the case of subunits 1 and 2, is $10^{-4}$ M. After such profound GSH depletion the hepatocyte becomes susceptible to both oxygen toxicity and the attack of NABQI on critical protein thiols etc. with perhaps lethal consequences. Therefore, in paracetamol overdoses, GSH transferases may assist toxicity rather than protect against it.

(iii) α-Bromoisovaleryl urea
α-Bromoisovaleryl urea, a narcotic drug, is an electrophile, with such low reactivity, that its toxicity is negligible. It is a substrate for GSH transferases, but the enzymic reaction has such a high $K_m$ and low

$$(13)$$

$V_{max}$ that, in terms of catalytic efficiency it is very poor indeed (Table 5). Nevertheless the enzymic reaction is enantioselective with all isoenzymes so far examined, the selectivity differing from one family to the other, GSH trans-ferases 1−1 and 2−2 giving the (R) enantiomer and GSH transferases 3−3 and 4−4 the (S) enantiomer (TE KOPPELE et al., 1988).

### 4.2.3. Endogenous electrophilic substrates

Electrophiles which are endogenous are attracting increasing attention. Some endogenous electrophiles are cytotoxic or genotoxic by-products of lipid per-oxidation (SLATER et al., 1984). Examples are cholesterol oxide (MEYER and

KETTERER, 1982) and hydroxyalkenals (e.g. 4-hydroxy non-2-enal) which are both formed as a result of the metal-catalysed decomposition

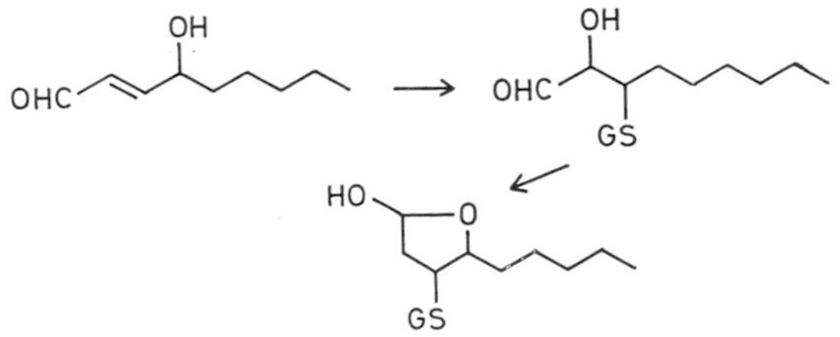

(14)

of polyunsaturated fatty acyl hydroperoxides are another. While cholesterol oxide is a poor substrate, hydroxyalkenals are among the best substrates for GSH transferases yet measured (DANIELSON et al., 1987) (Table 3).

The potential for lipid peroxidation, and the formation of its electrophilic by-products, exists in all cells. However, the melanocytes and cells of the substantia nigra region in the brain produce a specific electrophile, 3,4-phenyl-alanyl quinone (DOPA quinone).

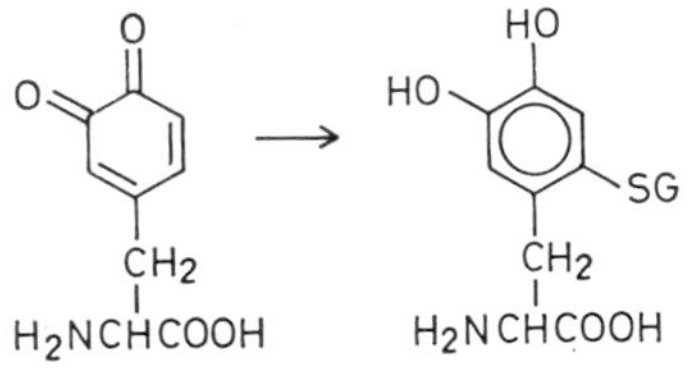

(15)

A glutathione adduct, 3,4-dihydroxy-5(glutathion-S-yl) phenylalanine, has been identified and its cysteinyl metabolite, when detected in urine, has been shown to be a good marker for metastatic melanoma. Like the quinone imine, NABQI, DOPA quinone undergoes rapid spontaneous conjugation with GSH in addition to enzymic conjugation (MIRANDA et al., 1987).

The above endogenous electrophiles are toxic and like the xenobiotic electro-philes described above, give GSH conjugates the only fate of which is to be excreted. There are other endogenous electrophiles which are precursors of GSH conjugates which are biologically active and may function as local hormones. Thus, in leukocytes and perhaps other cell types, arachidonic acid is oxidized via lipoxygenase to the epoxide leukotriene A$_4$, which is metabolized to its GSH conjugate, leukotriene C$_4$, a local hormone with profound effects on inflammation. At its site of biosynthesis leukotriene A$_4$ is probably a substrate for membrane-bound GSH transferases (BACH et al., 1984),

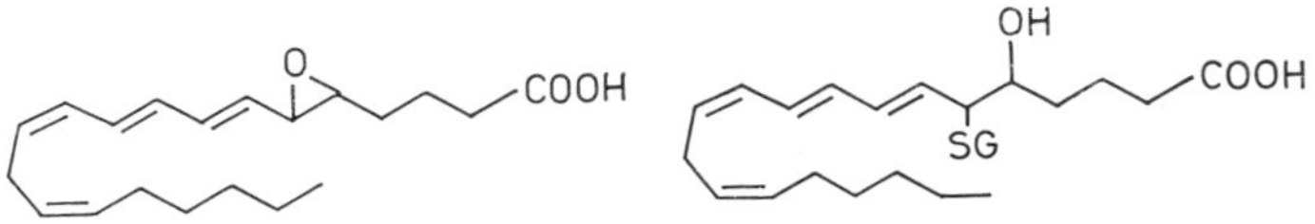

(16)

Table 4. Rat GSH transferases

| Substrate | Enzyme by class | | | | | | |
|-----------|-------|-----|-----|-----|-----|-----|------------|
| | alpha | | | mu | | pi | unassigned |
| | 1—1 | 2—2 | 3—3 | 4—4 | 6—9 | 7—7 | 5—5 |
| Cumene hydroperoxide | 1.4 | 3.0 | 0.1 | 0.4 | 0.04 | 0.01 | 12.5 |
| Linoleate hydroperoxide | 3.0 | 1.6 | 0.2 | 0.2 | 0.06 | 1.5 | 5.3 |
| Arachidonate hydroperoxide | 2.6 | 1.7 | 0.2 | 0.2 | nd | 1.5 | nd |
| 5-Hydroperoxymethyl uracil | 0.1 | 1.00 | 0.5 | 0.3 | 1.3 | 1.3 | nd |
| DNA hydroperoxide | nil | nil | 0.02 | 0.03 | 0.01 | 0.01 | 0.5 |

Activities ($\mu$mol $\cdot$ min$^{-1}$ $\cdot$ mg$^{-1}$) towards hydroperoxide substrates
Data are from the Cancer Research Campaign Molecular Toxicology Research Group.

but soluble enzymes are also active in vitro (see Table 4). Arachidonate can also be metabolized by cytochrome P-450 in vitro to four regioisomeric eicosatrienoic epoxides (EET), namely cis 5,6-, 8,9-, 11,12- and 14,15-EET all of which are substrates for GSH transferases (SPEARMAN et al., 1985) (see Table 3.). EET-GSH conjugates

$$ \text{(17)} $$

may also be biologically active (FALCK et al., 1983). Whether or not EET-GSH conjugates, like leukotrienes have a physiological role, is not known.

### 4.2.4. Lipid and nucleic acid hydroperoxide substrates

(i) Lipid hydroperoxides
Although the GSH conjugating activity of GSH transferases has attracted most of the attention in this field, the Se-independent GSH peroxidase activity of these enzymes may also be important, because lipid and nucleic acid hydroperoxides are good substrates. Although these hydroperoxides form in appreciable amounts during exposure to high energy irradiation and under oxidative stress, they may also form at low levels, during normal metabolism, as by-products of oxygen utilization. The first indication that GSH transferases might detoxify lipid hydroperoxides came from observations that GSH

transferases can inhibit microsomal lipid peroxidation in vitro (Burk et al., 1980; Tan et al., 1984). This inhibition was shown to require microsomal phospholipase in order to release the free fatty acids hydroperoxides, which, rather than their glyceryl esters, are substrates for GSH transferases (see Table 4, which shows that subunits 1, 2 and 5 are best enzymes for these substrates).

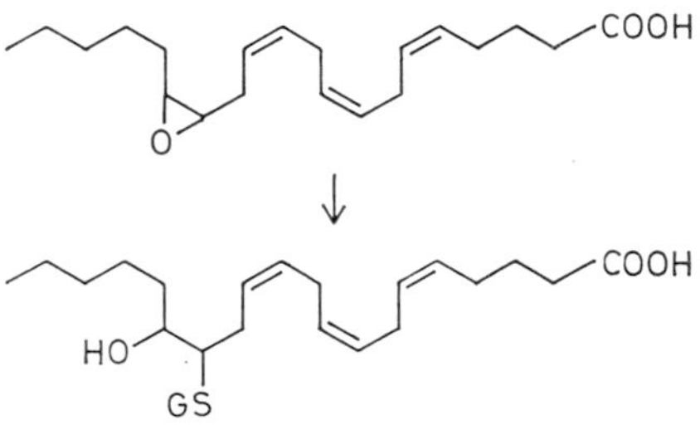

(18)

In vivo it has been proposed that GSH peroxidases (either Se-dependent or Se-independent) are part of a system for the repair of free radical damage to membrane phospholipids involving the successive action of phopholipase, GSH peroxidase and lysophospatide fatty acyl CoA transferase (van Kuijk et al., 1987).

(ii) Nucleic acid hydroperoxides
The thymine hydroperoxide, 5-hydroperoxymethyl-uracil, has also been shown to be a substrate for GSH peroxidases (Tan et al., 1986).

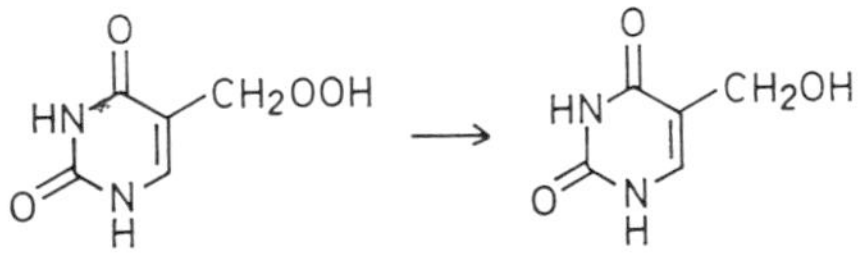

(19)

This has raised the possibility that peroxidized thymine residues in the DNA macromolecule might also be substrates. This appears to be the case since DNA, exposed to high energy irradiation in the presence of oxygen, is also a substrate (Table 4.) (Tan et al., 1987a). The isoenzymes, for which peroxidized DNA is a good substrate, are quite different from those which utilize 5-hydroperoxymethyl-uracil. This may be due to a quite different interaction between substrate and enzyme, which presumably occurs when substrate moieties are not free to diffuse independently, but are linked together in a double stranded polynucleotide of high viscosity. Alternatively, it may be due to the presence of hydroperoxide moieties, other than 5-hydroperoxy-methyl-uracil, with a different isoenzyme specificity, such as 5-hydroperoxy-6-hydroxy-5,6-dihydrothymine and 5-hydroxy-6-hydroperoxy-5,6-dihydro-thymine.

It has been proposed that GSH transferases may be part of a repair system for thymine hydroperoxide residues resulting from radical damage to DNA, involving the successive actions of GSH transferase, 5-hydroxymethyl-uracil or thymine glycol DNA glycosylase, endonuclease, DNA polymerase and DNA ligase (KETTERER et al., 1987). Both hydroxymethyl-uracil and thymine glycol are normal products of excretion and perhaps arise from the operation of the above mechanism (AMES and SAUL, 1985). Whereas cytoplasmic enzymes are involved in the protection against lipid peroxidation, nuclear enzymes are involved in DNA repair. It has been shown that the rat nucleus contains a chromatin-associated GSH transferase isoenzyme with a high activity towards DNA hydroperoxide (TAN et al., 1988) which is referred to as GSH transferase 5*−5*, because it is similar but not identical to GSH transferase 5−5. Its concentration in the rat nucleus, 140,000 molecules per nucleus, is similar to that of the well studied repair enzyme $O^6$-alkylguanine alkyl transferase (LINDAHL, 1982).

(iii) The relative importance of Se-dependent and Se-independent GSH per-oxidases
The relative contribution of these two types of GSH peroxidase to overall activity in any one organ varies with the species. For example, rat liver is relatively rich in the Se-dependent activity, while guinea pig and human liver are relatively rich in the Se-independent activity (LAWRENCE and BURK, 1978). In any one species it varies with the organ (for example, in the rat, there is relatively more Se-independent activity in the testis than in the liver) and, within a cell, it varies with the organelle (for example, there is relatively more Se-independent activity in the nucleus than in the cytoplasm of liver) (TAN, MEYER, and KETTERER, unpublished information).

## 4.3. GSH transferases and the pharmacological activity of nitroglycerin

Nitroglycerin and similar polyhydric nitrate esters such as erythrityl tetra-nitrate are substrates of GSH transferases (see reaction 13) (KEEN et al., 1976). In the case of nitroglycerin the products are nitrite, 1,3-dinitroglycerol and GSSG. Many of these compounds are vasodilators and it is believed that their pharmacological effect is due to their ability to release nitrite intra-cellularly, giving rise to nitrosothiols (for example, S-nitrosoglutathione) which activate guanylate synthetase in appropriate vascular endothelial tissues (LOSCALZO, 1985). In the case of nitroglycerin that subunits are active in the order 4 > 2 > 1 > 3 (KEEN et al., 1976).

## 4.4. The enzyme active site

A GSH binding site of relatively high specificity is demonstrated by the fact that relatively few compounds can substitute for GSH as a substrate or act as inhibitors by binding at this site. Gamma-glutamyl cysteine can replace GSH, but with an order of magnitude increase in $K_m$, suggesting that, while the gamma-glutamyl moiety is of prime importance, the C-terminal glycine residue is essential for full substrate activity (SUGIMOTO et al., 1985). Analogues which are inhibitors are gamma-glutamyl seryl glycine ($\gamma$ESG) and gamma-glutamyl alanyl glycine ($\gamma$EAG, norophthalmic acid). Whereas $\gamma$ESG has a similar $K_I$ to the Km for GSH, that of $\gamma$EAG is an order of magnitude greater, suggesting that the presence of a heteroatom is essential for maximal binding (CHEN et al., 1985).

In general GSH conjugates are inhibitors of GSH transferases. In a homologous series, the more hydrophobic the S-substituent, the greater the inhibition, for example, with S-n-alkyl substituents, the inhibitory effect increases with chain length. Leukotriene $C_4$, an S-n-alkyl substituent, has an affinity constant with GSH transferase $1-1$ as high as $8 \times 10^8 \, M^{-1}$ (SUN et al., 1986). Similar effects are seen with polyaromatic hydrocarbon GSH conjugates (CHEN et al., 1986). This suggests that the GSH binding site is associated with a hydrophobic site, which it is assumed is related to the electrophilic substrate binding site. Because many hydrophobic ligands are non-competitive inhibitors, it is not possible to make a simple statement about the nature of the electrophilic substrate binding site.

It has been said that the enzymic reaction of GSH transferases may only be due to a propinquity effect, i.e. the bringing of the two substrates into close proximity (JAKOBY, 1978). However, one activation mechanism which has often been suggested is enzyme-induced reduction of the $pK_a$ of the GSH thiol. Spectroscopic studies indicate that the thiol of GSH bound to rat GSH transferase $4-4$, is approximately 6.8, that is, at least two pH units below that of GSH in aqueous solution (GRAMINSKI et al., 1988). This effect could be brought about by an appropriately placed basic amino acid residue with a $pK_a$ about neutrality at the active site. Activation of the electrophilic substrate might also occur: another charged group at the active site could polarize the electrophilic centre, at least in those substrates which take up a suitable orientation there (MANNERVIK et al., 1978).

Single subunits, alone, are inactive and require to be in a dimeric structure to express activity. When GSH transferase subunits are separated after partial denaturation and then allowed to renature, they regain activity according to second order reaction kinetics, implying that dimer formation is rate limiting, and therefore, essential for activity (MEYER, COLES, LALOR, TAYLOR, and KETTERER, unpublished information). In the dimer, each active site appears to function independently of the other (DANIELSON and MANNERVIK, 1985). The activity of a GSH transferase dimer corresponds therefore to the sum of its component monomers.

## 4.5. Inhibition

GSH transferase activity is highly sensitive to inhibition by numerous compounds both exogenous and endogenous. Among exogenous inhibitors are trialkyl tin salts (HENRY and BYINGTON, 1976; TIPPING et al., 1979), tetra-bromosulphophthalein (CLARK et al., 1967) and Cibacron blue F3GA (TAHIR et al., 1985). Endogenous metabolites, which are strong inhibitors of GSH transferase activity and may modulate enzyme activity in vivo, include bilirubin (SIMONS and VANDER JAGT, 1980; VANDER JAGT et al., 1985), haematin (YALCIN et al., 1983; VANDER JAGT et al., 1985), bile acids (HAYES and CHALMERS, 1983; HAYES and MANTLE, 1986), unsaturated fatty acids (MEYER and KETTERER, 1987) and a number of products of GSH transferase activity e.g. leukotriene C (SUN et al., 1986).

## 4.6. Specificity constant and rate enhancement

The ratio of $k_{cat}$ to $K_m$ is an apparent second order rate constant. It determines the specificity of an enzyme for competing substances. It cannot be greater than any second order rate constant on the forward reaction pathway and therefore sets a lower limit on the rate constant for the association of enzyme and substrates. In certain well known efficient enzymes $k_{cat}/K_m$ approaches the diffusion controlled rate of reaction between enzyme and substrate. For

**Table 5.** Some kinetic constants

| Substrate | Isoenzyme | $k_{cat}$ (s$^{-1}$) | $K_m$ (M $\times$ 10$^{-6}$) | $k_{cat}/K_m$ (M$^{-1} \cdot$ s$^{-1}$) |
|---|---|---|---|---|
| N-Acetyl benzoquinone imine[1] | 2—2 | 35 | 0.7 | $5 \times 10^7$ |
| 4-Hydroxynon-2-enal[2] | 8—8 | — | — | $4.2 \times 10^6$ |
| 1-Nitropyrene-4,5-oxide[3] | 4—4 | 0.3 | 1.0 | $3 \times 10^5$ |
| Linoleate hydroperoxide[4] | 1—1 | 3.2 | 20.0 | $1.6 \times 10^5$ |
| 5-Hydroperoxymethyl-uracil[4] | 3—3 | 0.6 | 3.0 | $5 \times 10^4$ |
| α-Isobromovaleryl urea[5] | 2—2 | 0.1 | 600 | 167 |

Data are from [1]COLES et al., 1988; [2]DANIELSON and MANNERVIK, 1987; [3]DJURIC et al., 1987; [4]MEYER, unpublished information; [5]TE KOPPELE et al., 1988.

---

example, the $k_{cat}/K_m$ for crotonase, fumarase and catalase are $2.8 \times 10^8$, $1.6 \times 10^8$ and $4 \times 10^7$ $M^{-1}\,s^{-1}$ respectively (FERSHT, 1984). Similar high values are obtained with N-acetylbenzoquinone imine and 4-hydroxynon-2-enal; intermediate values with other substrates such as 1-nitropyrene-4,5-oxide, linoleic acid hydroperoxide and 5-hydroperoxymethyl-uracil while a very low value is obtained with $\alpha$-bromoisovaleryl urea (Table 5.).

The ratio of $k_{cat}/K_m$ to $k_2$, the second order rate constant for the spontaneous reaction with GSH, has been mentioned as a possible indicator of rate enhancement (DOUGLAS, 1986) brought about by an enzyme. In the case of NABQI and GSH transferase 2 –2 this value is $1.7 \times 10^3$ but in the case of 1-nitropyrene-4,5-oxide and GSH transferase 4 –4 it is $6.5 \times 10^5$ which is more than two orders of magnitude greater. Thus, although the specificity and rate of the transferase 2 – 2 catalysed reaction of N-acetylbenzoquinone imine with GSH, is very high, these advantages are offset by the very high second order rate constant for the spontaneous reaction.

### 4.7. Cell and tissue distribution in differentiation, carcinogenesis and in cells in vitro

a) Normal adult tissue
So far GSH transferases have been found in almost every rat tissue examined although not necessarily in every cell type. CDNB-transferase activity varies considerably from high values in the liver and testis to very low values in the non-lactating mammary gland and negligible levels in epididymal sperm (MEYER et al., 1983). In addition, striking differences in isoenzyme distribution occur from one tissue to another, a phenomenon the significance of which is not yet understood (KETTERER et al., 1985, 1986, and unpublished information).

The content of GSH transferase subunits in a tissue may be determined by reverse phase HPLC-analysis of the GSH transferase fraction obtained from tissue homogenates by GSH agarose affinity chromatography (OSTLUND FARRANTS et al., 1987). Analyses of the liver, kidney, lung, interstitial cells and spermatogenic tubules of the testis are shown in Figure 1. In the liver, subunits 1, 2, 3, 4, 6 and 8 are clearly distinguished but, since the retention time of subunits 4 and 9 are very similar, these two subunits are not always resolved. Subunits 1 and 8 on the other hand, both resolve into two forms. An appreciable amount of subunit 7 is noteworthy for its absence in the normal liver.

In some tissues one subunit predominates, for example, in the red blood cell it is subunit 8 (LALOR, COLES, MEYER, ALIN, MANNERVIK, and KETTERER, unpublished information), in the small intestine it is subunit 7 (TAHIR et al., 1988), while in the lactating mammary gland and the adrenal gland it is sub-

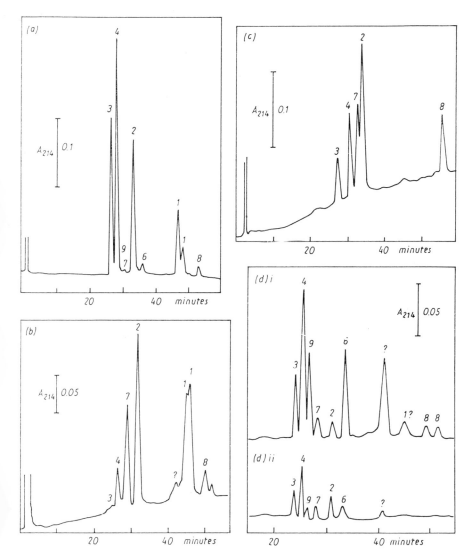

**Fig. 1.** The distribution of GSH transferase subunits in a) liver, b) kidney, c) lung di) seminiferous tubules and dii) testicular interstitial cells of the rat. Analyses were by reverse phase h.p.l.c. according to OSTLUND FARRANTS et al., 1987.

unit 2. In some tissues a subunit usually encountered in small amounts may be abundant. Thus, the testis is rich in subunits 6 and 9 and the brain in subunit 6.

Immunohistochemistry has revealed interesting differences within tissues; it has shown that in the normal liver, subunit 7 is present in all bile duct cells and in occasional hepatocytes (TATEMATSU et al., 1985), and that, although subunits 1, 2, 3 and 4 are present in all hepatocytes, they are more abundant in the area around the central vein than in the periportal region (REDICK et al., 1982; TATEMATSU et al., 1985). In the brain, GSH transferases have been detected only in the astroglial and ependymal cells and not at all in the neuronal stroma (ABRAMOWITZ et al., 1988).

b) Development changes
Changes in GSH transferase gene expression during development may occur throughout the body, but they have been studied in most detail in the liver. During the perinatal period overall activity is low and only two subunits are seen on reverse phase HPLC, namely subunits 2 and 10. Subunit 10 is a representative of the alpha family and a particularly active Se-independent GSH peroxidase (MEYER et al., 1985a). GSH transferase levees increase after birth so that, at about two weeks, subunits associated with the adult have appeared and increase in quantity with time. Subunit 10 is no longer obvious, although it may be present at relatively low concentrations. At maturity an interesting difference between the sexes develops, the alpha family being the more abundant in the female and the mu family more abundant in the male (IGARASHI et al., 1986, 1987; JOHNSON, MEYER, COLES, and KETTERER, unpublished information).

c) Carcinogenesis
An early effect of feeding a hepatocarcinogen, such as 3'-methyl-N,N-dimethyl-4-aminoazobenzene or aflatoxin $B_1$ is an overall induction in subunits 1 and 3 (KITAHARA et al., 1984). Subsequently foci of hepatocytes rich in subunits 1, 3 and 7 appear (SATO et al., 1984; POWER et al., 1987) many developing into nodules where, in addition to the above GSH transferase subunits, epoxide hydrolase, NAD(P)H-quinone oxido-reductase and UDP-glucuronyl transferase are also seen to be increased (ERIKSSON et al., 1983). Most foci and nodules redifferentiate into apparently normal liver tissue, but some nodules persist and become neoplastic (TATEMATSU et al., 1983). The primary hepatomas which result have a similar isoenzyme content to that in the nodules (MEYER et al., 1985b). During the progression of hepatoma to greater malignancy all but subunit 7 may be repressed (JOHNSON, MEYER, COLES, and KETTERER, unpublished information). Subunit 7 is therefore a very useful marker for both hepatocellular preneoplasia and hepatocarcinoma.

d) Cells in culture
Hepatocytes in primary culture undergo a remarkable change in gene expression within 48 h. Subunits 1, 2 and 8, that is, the alpha family, fall to very low levels, subunit 3 is increased, subunit 4 remains unchanged and subunit 7

B. KETTERER; J. B. TAYLOR

is expressed de novo (VANDERBERGHE et al., 1988). These changes in enzyme levels, which are reflected by similar changes in mRNA (VANDERBERGHE, GUILLOUZO, GUILLOUZO, PEMBLE, TAYLOR, and KETTERER, unpublished information) can be partially prevented by adding nicotinamide or dimethylsulphoxide to the medium.

## 5. Soluble GSH transferases in the human

### 5.1. Isoenzymes, nomenclature and genetic relationship

The first separation of human liver GSH transferases involved a tissue sample from a single individual. Five related "cationic" forms named $\alpha$-$\varepsilon$ were separated (KAMISAKA et al., 1975) and are now known to be related to the alpha class of rat GSH transferases. "Near neutral" and "anionic" forms were found in later studies of human liver from various sources and shown to be homologous with the mu and pi families in the rat respectively (MANNERVIK et al., 1985a; HUSSEY et al., 1986; SOMA et al., 1986; TU and QIAN, 1986; RHOADS et al., 1987; KANO et al., 1987; COWELL et al., 1988).

Using chromatofocusing FPLC, it is now possible to separate up to 15 forms of the alpha class (VANDER JAGT et al., 1985; OSTLUND FARRANTS et al., 1987). Isoenzymes at each extreme of the range of isoelectric points have been shown to be homodimers while those with intermediate isoelectric points are heterodimers. Homodimers and heterodimers are most readily identified and quantified by reverse phase HPLC (see Fig. 2). The monomeric forms have

**Table 6.** Nomenclature and physicochemical properties of human glutathione transferases

| Dimer | Subunit | Family | App Mol Wt on SDS PAGE $\times 10^{-3}$ | pI | Subunit retention time on reverse phase HPLC (min)[d] |
|---|---|---|---|---|---|
| $\alpha$ | $\alpha_x\alpha_x$ | alpha | 25 | 7.9[a] | 42 |
| $\beta, \gamma, \delta$ | $\alpha_x\alpha_y$ | alpha | 25 | 8.35, 8.55, 8.75[a] | 42 + 37 |
| $\varepsilon$ | $\alpha_y\alpha_y$ | alpha | 25 | 8.9[a] | 37 |
| Skin 9—9 | | alpha | 27 | 9.9[b] | nd |
| $\mu$ | | mu | 26 | 6.6[c] | 28 |
| $\psi$ | | mu | 26 | 5.5[c] | nd |
| $\pi$ | | pi | 23 | 4.8[c] | 25 |

Data are from [a]KAMISAKA et al., 1975; [b]DEL BOCCIO et al., 1987; [c]JOHNSON, HAYES, MEYER, unpublished information; [d]OSTLUND FARRANTS et al., 1987.

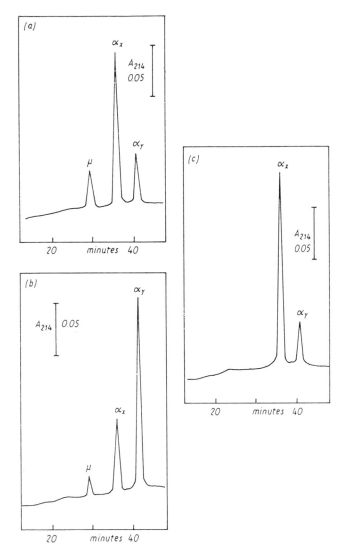

**Fig. 2.** Interindividual variation in GSH transferase subunit composition of human liver. The GSH transferase content of samples of three human livers (a — c) were determined by reverse phase HPLC according to OSTLUND FARRANTS et al., 1987.

been named either $B_1$ and $B_2$ (STOCKMAN et al., 1987), $Y_1$ and $Y_4$ (SOMA et al., 1986) or $\alpha_x$ and $\alpha_y$ (OSTLUND FARRANTS et al., 1987). $B_1$, $Y_1$ and $\alpha_x$ being the more basic of the two subunits in each nomenclature. Two homogeneous monomers would be expected to give two homodimers and one heterodimer. The

B. KETTERER; J. B. TAYLOR

multiplicity of forms separated by chromatofocussing FPLC is yet to be explained.

Two full length cDNA clones for alpha family subunits have been described (Tu and Qian, 1986; Rhoads et al., 1987) having 96% identity in the nucleotide sequence of the coding region, but greater differences in the non-coding regions. These clones are referred to as $Ha_1$ and $Ha_2$. Deduced amino acid sequences show that, at two points, glutamine residues in $Ha_1$ are represented by lysine in $Ha_2$. This suggests that the polypeptide equivalent to $Ha_1$ has an isoelectric point lower than that equivalent to $Ha_2$ and raises the possibility that $Ha_1$ and $Ha_2$ might be equated with $\alpha_y$ and $\alpha_x$ respectively. Comparisons of cDNA clones show that, between species, human and rat alpha class GSH transferases are approximately 75% identical in amino acid sequence.

Mu class enzymes are not present in all individuals. From genetic studies of 179 liver samples from Chinese, Indian and Caucasian subjects it was concluded the mu locus (GST1) has two expressing genes and a null allele (Board, 1981). This conclusion has been supported by other population studies (Laisney et al., 1984; Strange et al., 1984; Harada et al., 1987). It has been suggested that the null allele either gives no product, or an inactive product, incapable of forming dimers, with the result that the mu family is either absent or represented by either one or other of two active homodimers or an active heterodimer (Board, 1981). Comparison between sequences at present available for human and rat GSH transferases of the mu family show considerable homology between the species.

The only pi class enzyme so far found in humans is GSH transferase $\pi$ which shows very close homology with rat subunit 7 cDNA (Kano et al., 1987).

## 5.2. Enzymic activity

Table 7 shows the activity of the above mentioned GSH transferases towards a number of substrates.

It is seen that, within the same family, human and rat isoenzymes have similar enzymic activities. Subunits $\alpha_x$ and $\alpha_y$ are associated with Se-independent GSH peroxidase activity, subunit $\alpha_x$ being more active than $\alpha_y$. GSH transferase $\mu$, like rat subunit 4, has activity towards **trans**-4-phenyl-3-butene-2-one and is a good enzyme for trans-stilbene oxide and epoxide derivatives of hydrocarbons, so much so, that trans-stilbene oxide can be used to detect and measure small amounts of GSH transferase $\mu$ in tissue samples (Seidergard et al., 1987). GSH transferase $\pi$, like rat subunit 7, is a good enzyme for ethacrynic acid and a better Se-independent GSH peroxidase with linoleate hydroperoxide than with cumene hydroperoxide. Unlike rat subunit 7, GSH transferase $\pi$ is a good enzyme with 1,2-epoxy-3(p-nitrophenoxy)propane.

**Table 7.** Human GSH transferases

| Substrate | Enzyme by class | | | |
|---|---|---|---|---|
| | alpha $\alpha_x\alpha_x$ | $\alpha_y\alpha_y$ | mu $\mu$ | pi $\pi$ |
| 1-Chloro-2,4-dinitrobenzene | 55 | 60 | 90 | 72 |
| 1,2-Dichlorobenzene-4-nitrobenzene | 0.3 | 0.8 | 0.03 | 0.6 |
| Trans-4-phenyl-3-buten-2-one | nil | nil | 0.36 | 0.01 |
| 1,2-Epoxy-3-(p-nitrophenoxy)propane | nil | nil | 0.11 | 1.8 |
| Ethacrynic acid | 0.11 | 0.14 | 0.081 | 1.9 |
| Cumene hydroperoxide | 6.2 | 2.1 | 0.63 | 0.04 |
| Linoleate hydroperoxide | 4.7 | 1.6 | nd | 0.44 |

Activities ($\mu$mol · min$^{-1}$ · mg$^{-1}$) towards substrates used to specify subunits.
Data are from the Cancer Research Campaign Molecular Toxicology Research Group.

## 5.3. Tissue distribution

A survey of a number of tissues shows that, in most cases, humans have more 1-chloro-2.4-dinitrobenzene-GSH transferase activity than the rat (BAARS et al., 1981). Many data are now available concerning isoenzyme content of the liver and inter-individual variation in that organ (Fig. 2). Information in the same detail is rarely available for other tissues, where in most cases the enzymes present are usually identified as "cationic", "near neutral" or "anionic" and, although these broad designations may sometimes equate with alpha, mu and pi, this equivalence cannot be taken for granted. In an early study all three forms of GSH transferase namely "anionic", "near neutral" and "acidic" were found in the adrenal, testis and ovary; two only, namely "cationic" and "anionic" were found in the kidney; and "anionic" alone occurred in the lung (SHERMAN et al., 1983). In certain tissues such as erythrocyte, platelet (LOSCALZO and FREEDMAN, 1986); thyroid (DEL BOCCIO et al., 1987); placenta, lung, heart, spleen, kidney and pancreas (TAKEOKA et al., 1987) the acidic transferase has been identified as GSH transferase $\pi$. It is usually abundant, sometimes predominant (TATEOKA et al., 1987). Subsequently the kidney has been analysed more precisely and three classes of GSH transferases were found to be present in significant amounts. Multiple forms of the alpha class together with the pi class are well represented and small amounts of the mu class are present (SINGH et al., 1987; TAKEOKA et al., 1987). This pattern is comparable to that found in rat kidney.

Multiple "cationic" enzymes were also found in five adult hearts, where, as Se-independent. GSH peroxidases, they may have an important role in protecting against oxygen toxicity (DI ILIO et al., 1986). Information concerning the occurrence in the heart of "near neutral" and "anionic" enzymes is less precise but a muscle specific isoenzyme GST4 (LAISNEY et al., 1984; SUZUKI et al., 1987) is present.

Two further enzymes associated with extrahepatic tissues are GST5, which has so far only been found in brain (SUZUKI et al., 1987), and the alpha class GST transferase 9−9, which has been isolated from skin and which differs from hepatic alpha enzymes in its N-terminal sequence, which is identical to that of rat GSH transferase subunit 2 and its higher apparent molecular weight (DEL BOCCIO et al., 1987).

Relatively little is known of the occurrence of GSH transferase mu in extrahepatic tissues. The availability of the substrate trans-stilbene oxide, as a sensitive means for detecting it, could be used to gather information about this enzyme, since the considerable variation in isoenzyme content between individuals should result in corresponding variations in their susceptibility to drugs and toxins. An interesting study with mononuclear leukocytes measured activity towards trans-stilbene oxide in a survey of smokers, who were either control subjects without disease or lung cancer patients. It was found that a greater proportion of control smokers (59%) had trans-stilbene oxide-GSH transferase activity than those with cancer (35%), which led to the conclusion that the gene expressing mu may be a host determinant of susceptibility to lung cancer (SEIDERGARD et al., 1985, 1987).

## 5.4. Human tumours

In carcinogenesis, GSH transferases are of interest for several reasons. They may confer an advantage on preneoplastic and neoplastic cells over their normal neighbours, they may confer resistance to chemotherapeutic drugs and also they may have diagnostic value as tumour markers.

In the case of liver tumours, present evidence suggests that the expression of alpha enzymes indicates a hepatocellular carcinoma, and pi enzymes, either a cholangiocarcinoma or a metastasis from the colon. Unlike subunit 7 in rat hepatoma, GSH transferase $\pi$ is not a generally recognized marker for human hepatocellular carcinoma (HAYES et al., 1987; SOMA et al., 1986; KODATE et al., 1986).

Many other human tumours show a common pattern, namely enhanced expression of GSH transferase $\pi$. This has been demonstrated immunohistochemically in cancer of the colon, stomach, pancreas and uterine cervix (KODATE et al., 1986 and SATO et al., 1987) and by immunoblotting techniques in adenocarcinoma of the breast and lung, nodular small cell lymphoma and mesothelioma (SHEA and HENNER, 1987) and metastatic melanoma (MANNERVIK et al., 1987). It has also been demonstrated by isoelectric focussing in renal cortex tumours (DI ILIO et al., 1987).

Many tumour cell lines also have elevated GSH transferase $\pi$ e.g. both small cell (AWATHI et al., 1988) and non-small cell lung carcinoma cell lines, ovarian carcinoma and the EJ6 bladder carcinoma (WOLF et al., 1987). There are however, human tumour cell lines with very low levels of GSH transferases, for example, the mammary carcinoma cell line MCF7. Induction of resistance to adriamycin in MCF7 cell lines, however, has resulted in a 45-fold increase in an acid isoenzyme believed to be GSH transferase $\pi$ (BATIST et al., 1986). Adriamycin is not itself a substrate for the GSH transferase $\pi$, which it induces in MCF7 cells, however, endogenous hydroperoxides resulting from its redox cycling may be.

# 6. Soluble GSH transferases in other organisms

Multiple GSH transferase isoenzymes have been observed in the livers of all mammalian species studied. Information enabling the placing of these GSH transferases into the multigene families has been obtained in a few cases. Thus in the mouse all three families are well represented, the pi enzyme being the major enzyme in the female, but not the male (WARHOLM et al., 1985; HATA-YAMA et al., 1986). We have seen that the rat lacks the pi enzyme, but alpha and mu enzymes are well represented and that man has an even more limited distribution in that alpha enzymes predominate, mu enzymes are present in only 50% of the population and the pi enzyme is poorly expressed or absent. The hamster also has a limited distribution, but in this case the mu enzyme predominates (FOLIOT, MEYER, COLES, KETTERER, unpublished information).

In the guinea pig liver basic GSH transferase with GSH peroxidase activity, and presumably of the alpha family, are abundant. In this respect the guinea pig resembles man, hepatic GSH peroxidase activity being largely Se-independent (LAWRENCE and BURK, 1978). The presence of such large differences in the interspecies distribution of multigene families species in the same organ is yet to be explained.

In general GSH transferase have been detected in members of most animal phyla including protozoa, sponges, coelenterates, molluscs, helminths, annelids, arthropods, tunicates and vertebrates (STENERSEN et al., 1987) and, although studies of plants have been few, for example the fungus *Fusarium oxysporum* (CHEN et al., 1986), the angiosperms *Hevea brasiliensis* (BALABAS-KARAN and MURIANDY, 1984), *Pisum sativum* (FREAR and SWANSON, 1973; DIES-PERGER and SANDERMAN, 1979) and *Zea mays* (MOZER et al., 1983; MOORE et al., 1986), these enzymes are presumably as widespread in plants as in animals.

Studies in the rat have been pursued largely as a model for man, but other species have their own intrinsic interest. Thus, insecticide and herbicide resistance has occasionally been traced to GSH transferases in insects (CLARK and DAUTERMAN, 1982; CLARK and SHAMAAN, 1984; CLARK et al., 1984) and plants (FREAR and SWANSON, 1970; LAY and CASSIDA, 1976) while in schistosomes the GSH transferases have proved interesting because of their antigenicity in mammals and therefore their potential as components of vaccines against schistosomiasis (TAYLOR et al., 1988).

## 7.  Protein structure, gene structure and gene expression

The structures of GSH transferases are known from organisms as remote in phylogeny as maize, schistosomes, rat and man. They have in common a dimeric structure with subunits of molecular weight conserved between 23,000 and 26,000. Within the amino acid sequences of the subunits occurs a similar heptadecapeptide beginning about 70 residues from the N-terminus (TAYLOR et al.,.1987).

So far the structures of six genes are known, namely a rat class α gene (subunit 1) (ROTHKOPF et al., 1986), a rat class mu gene (subunit 3 or 4) (TU et al., 1987), a rat class pi gene (subunit 7) (OKUDA et al., 1987), a human class pi gene (π) (COWELL et al., 1988), a mouse class α gene (DANIEL et al., 1987) and a maize gene encoding GSH transferase I (SHAH et al., 1986). The rat and mouse class α genes are highly homologous in that they are both composed of seven exons interspaced with six introns at equivalent positions in the coding sequences. A high degree of homology in the coding sequences results in only ten amino acid differences over their length of 222, and this high homology extends at the nucleotide level into the intron sequences (70−80%, for 50−200 bases depending on the intron). Comparisons of the cDNA sequences of rat subunit 1 and subunit 2 suggest that exons 2 and 4 encode sequences which are conserved and exons 3 and 5 encode divergent sequences.

The class pi genes encoding rat subunit 7 and human π also have an intron-exon structure of seven exons interrupted by six introns at precisely equivalent positions in their sequences, but these boundary positions differ from those of the α class genes. Furthermore, the high sequence homology of the coding regions for subunit 7 and π is not sustained beyond five nucleotides into the introns.

The upstream regulatory regions of the class α and π genes also show substantial differences. Thus both pi class genes have sequence motifs associated with "house keeping" genes, such as a high G and C and CpG content around their promoters and GC boxes matching the consensus sequence for their binding site of the transcription factor SP1. In addition both genes have a phorbol ester responsive element (TRE) (OKUDA et al., 1987; COWELL et al., 1988). It has been shown that c-Ha-ras and phorbol esters, both of which can activate the polyoma virus enhancer, act through an enhancer element closely related in sequence to the TRE (IMLER et al., 1988). The possibility that ras acts through TRE elements of cellular genes is highly relevant to tumour specific induction of GSH transferase $\pi$ expression, as amplified or activated ras genes are frequently found in human tumours. GSH transferase 7−7 appears to be ras responsive, since it can be induced de novo in rat liver epithelial cells by transformation with an N-ras gene (POWER et al., 1987).

The 5' flanking region of the rat (TELAKOWSKI-HOPKINS et al., 1987) and mouse (DANIEL et al., 1988) class $\alpha$ genes do not appear to contain either the motif associated with "housekeeping genes" or the phorbol ester responsive element. The rat gene contains an element required for maximal basal promoter activity and also a $\beta$-naphthoflavone responsive element. The latter is presumably associated with inducibility of GSH transferase subunit 1: this gene is only expressed in cells which possess a TCDD receptor. As yet it is not known whether the $\beta$-naphthoflavone responsive element is activated directly or indirectly by a trans-acting protein. Both genes contain tissue-specific regulatory elements, for example a liver-specific element is seen in the mouse gene.

Another distinguishing feature of the class pi genes is that in both rat and human, there is only one transcribed gene. In the case of GSH transferase $\pi$ the Southern blot is in accord with the existence of a single gene (COWELL et al., 1988) while Southern blots for GSH transferase 7−7 are complex due to hybridization to four processed pseudogenes in addition to the single expressing gene (OKUDA et al., 1987). In contrast $\alpha$ genes are members of expressing multigene families.

Little else is known of the gene structure of these enzymes beyond the fact that although the gene for rat GSH transferase subunits 1 and 7 contain seven exons, that for subunit 3 contains eight and the gene for maize GSH transferase I contains only three.

# 8. Microsomal GSH transferases

Although soluble enzymes are extensively studied membrane-bound GSH transferases also exist. Of these, the microsomal GSH transferase has received much attention. When isolated from rat liver, microsomes contain both strongly adsorbed soluble GSH transferases and an intrinsic enzyme (MORGEN-STERN et al., 1982; MORGENSTERN et al., 1983). The intrinsic enzyme referred to as microsomal GSH transferase has a mol wt of 17,200 and a primary structure which has no apparent homology with the amino acid sequences of soluble GSH transferase subunits 1, 2, 3, 4 and 7 (MORGENSTERN et al., 1985). This enzyme is activated by reagents which react with thiols including N-ethylmaleimide, disulphides, and perhaps also quinones and products of $CCl_4$ metabolism (MORGENSTERN and DEPIERRE, 1985; MORGENSTERN et al., 1987). When activated, this enzyme has appreciable activity towards CDNB, CuOOH, methyl iodide and p-nitrobenzyl chloride with levels of activity similar to those of GSH transferase $1-1$. In the rat the microsomal enzyme is found principally in the liver although it has been detected in extrahepatic tissues and a similar enzyme has been detected in the liver of other mammals (MORGEN-STERN and DEPIERRE, 1985).

A specific role of microsomal GSH transferases has yet to be revealed. Other membrane-bound GSH transferases exist, notably membrane-bound leukotriene $C_4$ synthetase, but this enzyme is distinct from the microsomal GSH transferase (BACH et al., 1984).

# 9.   Some prospects

Traditionally GSH transferases have been regarded as enzymes which detoxify xenobiotics and enable man to resist the toxicity of many of the drugs and carcinogens which give rise to electrophilic metabolites. In recent years it has been shown that they are not always beneficial to man. For example, treatment with anticancer drugs sometimes results in the emergence of resistant cells with an elevated expression of GSH transferases, in particular GSH transferase $\pi$. This finding has given an impetus to studies of the regulation of these genes, since, if a means of supressing their expression in tumours could be found, it might be possible to reverse their resistance to drugs.

Development of resistance to insecticides and herbicides is also seen in insects and plants respectively and sometimes this too is associated with the acquisition by resistant strains of GSH transferases which utilize the insecticide or herbicide as substrate. By the use of genetic manipulation this sort of resistance could be made beneficial to man. It should be possible to design agricultural plants (e.g. maize) equipped with a GSH transferase which confers resistance to a specific herbicide thereby giving these resistant plants an advantage.

GSH transferases also appear to be an important part of the armoury of the organism which enables it to deal with oxygen toxicity. If we were looking for substrates common to all organisms, which, early in aerobic life, might have provided the evolutionary pressure for development of GSH transferases, the toxicity of lipid and DNA hydroperoxides or their decomposition products such as hydroxy alkenals might provide the answer. Subsequent gene duplication and divergence might then have given rise to multigene families with the added advantage of the ability to detoxify a wide range of xenobiotics.

A totally different function of GSH transferase activity is the production of leukotriene $C_4$ which, rather than a detoxication product is a local hormone, or perhaps the precursor of a local hormone (leukotriene $D_4$). This is so potent that one assumes that its formation is not a general phenomenon but the function of specialized cells.

---

One of the more remarkable findings at present is the variation in GSH transferase distribution from tissue to tissue and within a tissue from species to species. At the moment this phenomenon is difficult to interpret and suggests that we still have much to learn about the function of these enzymes.

## Acknowledgements

The authors wish to thank the Cancer Research Campaign for their generous support, David J. MEYER for valuable discussion and Lucia CHRISTODOULIDES for the preparation of the manuscript.

## 10. References

ABRAMOWITZ, M., HOMMA, H., ISHIGAKI, S., TANCEY, F., CAMMER, W., and LISTOWSKY, I. (1988), J. Neurochem. **50**, 50—57.

ALBANO, E., RUNDGREN, M., HARVISON, P. J., NELSON, S. D., and MOLDEUS, P. (1985), Mol. Pharmacol. **28**, 306—311.

AMES, B. N. (1983), Science **221**, 1256—1264.

AMES, B. N. and SAUL, R. (1985), in: 'Genetic Toxicology of Environmental Chemicals', Alan Liss, New York, pp. 1—16.

AWASTHI, Y. C., BHATNAGAR, A., and SINGH, S. V. (1987), Biochem. Biophys. Res. Commun. **143**, 965—970.

AWASTHI, Y. C., SINGH, S. V., AHMAD, H., MOLLER, P. C., and GUPTA, V. (1988), Carcinogenesis **9**, 89—93.

BAARS, A. J., MUKHTAR, H., ZOETEMELK, C. E. M., JANSEN, M., and BREIMER, D. D. (1981), Comp. Biochem. Physiol. **70C**, 285—288.

BACH, M. K., BRASHLER, J. R., PECK, R. E., and MORTON, D. R. (1984), J. Allergy & Clin. Immunol. **74**, 353—357.

BALABASKARAN, S. and MUNIANDY, N. (1984), Phytochemistry **23**, 251—256.

BATIST, G., TULPULE, A., SINHA, B. K., KATKI, A., MYERS, E., and COWAN, K. H. (1986), J. Biol. Chem. **261**, 15544—15549.

BENSON, A. M., BATZINGER, R. P., OU, S.-Y. L., BUEDING, E., CHA, Y.-N., and TALALAY, P. (1978), Cancer Res. **38**, 4487—4495.

BENSON, A. M., CHA, Y.-N., BUEDING, E., HEINE, H. S., and TALALAY, P. (1979), Cancer Res. **39**, 2971—2977.

BOARD, P. G. (1981), Ann. Hum. Genet. **33**, 36—43.

BURK, R. F., TRUMBLE, M. J., and LAWRENCE, R. A. (1980), Biochim. Biophys. Acta **618**, 35—41.

CHASSEAUD, L. F. (1979), Adv. Cancer Res. **29**, 175—274.

CHEN, W.-J., BOEHLERT, C. C., RIDER, K., and ARMSTRONG, R. N. (1985), Biochem. Biophys. Res. Commun. **128**, 233—240.

CHEN, W.-J., DE SMIDT, P. C., and ARMSTRONG, R. N. (1986), Biochem. Biophys. Res. Commun. **141**, 892—897.

CLARK, A. G., DARBY, F. G., and SMITH, J. N. (1967), Biochem. J. **103**, 49—54.

CLARK, A. G. and DAUTERMAN, W. C. (1982), Pestic. Biochem. Physiol. **17**, 307—314.

CLARK, A. G. and SHAMAAN, N. A. (1984), Pestic. Biochem. Physiol. **22**, 249—261.

CLARK, A. G., SHAMAAN, N. A., DAUTERMAN, W. C., and HAYAOKA, T. (1984), Pestic. Biochem. Physiol. **22**, 51—59.

COHEN, E., GAMLIEL, A., and KATAN, J. (1986), Pestic. Biochem. Physiol. **26**, 1—9.
COLES, B., MEYER, D. J., KETTERER, B., STANTON, C. A., and GARNER, R. C. (1985), Carcinogenesis **6**, 693—697.
COLES, B., WILSON, I., WARDMAN, P., HINSON, J. A., NELSON, S. D., and KETTERER, B. (1988), Archives Biochem. Biophys. **264**, 253—260.
COWELL, I. G., DIXON, K. H., PEMBLE, S. E., KETTERER, B., and TAYLOR, J. B. (1988), Biochem. J. **255**, 79—83.
DANIEL, V., SHARON, R., TICHAUER, Y., and SARID, S. (1987), DNA **6**, 317—324.
DANIEL, V., TICHAUER, Y., and SHARON, R. (1988), Nucleic Acids Research **16**, 351.
DANIELSON, U. H. and MANNERVIK, B. (1985), Biochem. J. **231**, 263—267.
DANIELSON, U. H., ESTERBAUER, H., and MANNERVIK, B. (1987), Biochem. J. **247**, 707—713.
DEL BOCCIO, G., DI ILIO, C., CASALONE, E., PENNELLI, A., ACETO, A., SACCHETTA, P., and FEDERICI, G. (1987), Italian J. Biochem. **36**, 8—17.
DIESPERGER, H. and SANDERMAN, H. (1979), Plant **146**, 643—648.
DI ILIO, C., DEL BOCCIO, G., ACETO, A., and FEDERICI, G. (1987), Carcinogenesis **8**, 861—864.
DING, G. J.-F., LU, A. Y. H., and PICKETT, C. B. (1985), J. Biol. Chem. **137**, 125—129.
DI SIMPLICIO, P., JENSSON, H., and MANNERVIK, B. (1983), Acta Chem. Scand. **1337**, 255—257.
DJURIC, Z., COLES, B., FIFER, E. K., KETTERER, B., and BELAND, F. A. (1987), Carcinogenesis **8**, 1781—1786.
DOUGLAS, K. T. (1986), Adv. Enzymol. Related Areas Mol. Biol. **59**, 103—167.
ERIKSSON, L. C., SHARMA, R. N., ROOMI, M. W., HO, R. K., FARBER, E., and MURRAY, R. K. (1983), Biochem. Biophys. Res. Commun. **117**, 740—745.
FALCK, J. R., MANNA, S., MOLZ, J., CHACOS, N., and CAPDEVILA, J. (1983), Biochem. Biophys. Res. Commun. **114**, 743—749.
FERSHT, A. (1984), Enzyme Structure and Mechanism, W. H. Freeman and Co. New York.
FREAR, D. S. and SWANSON, H. R. (1970), Phytochem. **9**, 2123—2132.
FREAR, D. S. and SWANSON, H. R. (1973), Pestic. Biochem. Physiol. **3**, 473—482.
GUTHENBERG, C., WARHOLM, M., RANE, A., and MANNERVIK, B. (1986), Biochem. J. **235**, 714—745.
GRAMINSKI, G. F., CHEN, W.-J., and ARMSTRONG, R. N. (1987), Fed. Proc. **46**, 1934.
HARADA, S., ABEI, M., TANAKA, N., AGARWAL, D. P., and GOEDDE, H. E. (1987), Hum. Genet. **75**, 322—325.
HATAYMA, I., SATOH, K., and SATO, K. (1986), Biochem. Biophys. Res. Commun. **140**, 581—588.
HAYES, J. D. and CHALMERS, J. (1983), Biochem. J. **215**, 581—588.
HAYES, J. D. and MANTLE, T. J. (1986), Biochem. J. **233**, 407—415.
HAYES, P. C., PORTMANN, B., ALDOIS, P. M., WILLIAMS, R., and HAYES, J. D. (1987), in: 'Glutathione S-Transferases and Carcinogenesis' (T. J. MANTLE, C. B. PICKETT, and J. D. HAYES, eds.), Taylor and Francis, London, pp. 175—187.
HENRY, R. A., and BYINGTON, K. H. (1976), Biochem. Pharmacol. **25**, 2291—2295.
HUSSEY, A. J., STOCKMAN, P. K., BECKETT, G. J., and HAYES, J. D. (1986), Biochim. Biophys. Acta **874**, 1—12.
IGARASHI, T., TOMIHARI, N., OHMORI, S., UENO, K., KITAGAWA, H., and SATOH, T. (1986), Biochem. Internat. **13**, 641—648.
IGARASHI, T., IROKAWA, N., ONO, S., OHMORI, S., UENO, K., and KITAGAWA, H. (1987), Xenobiotica **17**, 127—137.
IMLER, J. L., SCHATZ, C., WASYLYK, C., CHATTEN, B., and WASYLYK, B. (1988), Nature **322**, 275—278.

---

JAKOBY, W. B. (1978), in: 'Functions of Glutathione in Liver and Kidney' (H. SIES and A. WENDEL, eds.), Springer Verlag, Berlin, pp. 157—163.

JAKOBY, W. B., KETTERER, B., and MANNERVIK, B. (1984), Biochem. Pharmacol. 33, 2539—2540.

JENSSON, H., GUTHENBERG, C., ALIN, P., and MANNERVIK, B. (1986), FEBS Lett. 203, 207—209.

JERNSTROM, B., MARTINEZ, M., MEYER, D. J., and KETTERER, B. (1985), Carcinogenesis 6, 85—89.

KAMISAKA, K., HABIG, W. H., KETLEY, J. M., ARIAS, I. H., and JAKOBY, W. B. (1975), Eur. J. Biochem. 60, 153—161.

KANO, T., SAKAI, M., and MURAMATSU, M. (1987), Cancer Res. 47, 5626—5630.

KEEN, J. H., HABIG, W. H., and JAKOBY, W. B. (1976), J. Biol. Chem. 251, 6183—6188.

KENSLER, T. W., EGNER, P. A., DAVIDSON, N. E., ROEBUCK, B. D., PIKUL, A., and GROOPMAN, J. D. (1986), Cancer Res. 46, 3924—3931.

KETTERER, B., MEYER, D. J., COLES, B., and TAYLOR, J. B. (1985), in: 'Microsomes and Drug Oxidations' (A. R. BOOBIS, J. CALDWELL, F. DE MATTEIS, and C. R. ELCOMBE, eds.), Taylor and Francis, London, pp. 166—167.

KETTERER, B., MEYER, D. J., COLES, B., TAYLOR, J. B., and PEMBLE, S. E. (1986), in: 'Antimutagenesis and Anticarcinogenesis: Mechanisms' (D. M. SHANKEL, P. HARTMAN, T. KADA, and A. HOLLAENDER, eds.), Plenum Press, New York, pp. 103 to to 126.

KETTERER, B., TAN, K. H., MEYER, D. J., and COLES, B. (1987), in: 'Glutathione S-Transferases and Carcinogenesis' (T. J. MANTLE, C. B. PICKETT, and J. D. HAYES, eds.), Taylor and Francis, London, pp. 149—163.

KITAHARA, A., SATOH, K., NISHIMURA, K., ISHIKAWA, T., RUIKE, K., SATO, K., and TSUDA, H. (1984), Cancer Res. 44, 2698—2703.

KODATE, C., FUKUSHI, A., NARITA, T., and KUDO, H. (1986), Jpn. J. Cancer Res. 77, 226—229.

LAI, H.-C. J., LI, N.-Q., WEISS, M. J., REDDY, C. C., and TU, C.-P. D. (1984), J. Biol. Chem. 259, 5536—5542.

LAISNEY, V., VAN CONG, N., CROSS, M. S., and FREZAL, J. (1984), Hum. Genet. 68, 221—227.

LAWRENCE, R. A. and BURK, R. F. (1978), J. Nutr. 108, 211—215.

LAY, M. M. and CASIDA, J. E. (1976), Pestic. Biochem. Physiol. 6, 442—456.

LINDAHL, T. (1982), Ann. Rev. Biochem. 51, 61—87.

LOSCALZO, J. (1985), J. Clin. Invest. 76, 703—708.

LOSCALZO, J. and FREEDMAN, J. (1986), Blood 67, 1595—1599.

MANNERVIK, B., GUTHENBERG, C., JAKOBSEN, I., and WARHOLM, M. (1978), in: 'Conjugation Reactions in Drug Biotransformation' (A. ALIO, ed.), Elsevier, Holland, pp. 101—110.

MANNERVIK, B., ALIN, P., GUTHENBERG, C., JENSSON, H., TAHIR, M. K., WARHOLM, M., and JORNVALL, H. (1985a), Proc. Natl. Acad. Sci. U.S.A. 82, 7202—7206.

MANNERVIK, B., ALIN, P., GUTHENBERG, C., JENSSON, H., and WARHOLM, M. (1985b), in: 'Microsomes and Drug Oxidations' (A. R. BOOBIS, J. CALDWELL, F. DE MATTEIS, and C. R. ELCOMBE, eds.), Taylor and Francis, London, pp. 221—228.

MANNERVIK, B., CASTRO, V. M., DANIELSON, U. H., TAHIR, M. K., HANSSON, J., and RINGBORG, U. (1987), Carcinogenesis, 12, 1929—1932.

MEYER, D. J. and KETTERER, B. (1982), FEBS Lett. 150, 499—502.

MEYER, D. J., CHRISTODOULIDES, L. G., NYAN, O., SCHUSTER BRUCE, R., and KETTERER, B. (1983), in: 'Extrahepatic Drug Metabolism and Chemical Carcinogenesis' (J. RYDSTROM, J. MONTELIUS, and M. BENGTSSON, eds.), Elsevier Science Publishers B.V., pp. 189—190.

---

MEYER, D. J., CHRISTODOULIDES, L. G., TAN, K. H., and KETTERER, B. (1984), FEBS Lett. **173**, 327—330.

MEYER, D. J., BEALE, D., TAN, K. H., COLES, B., and KETTERER, B. (1985a), FEBS Lett. **184**, 139—143.

MEYER, D. J., TAN, K. H., CHRISTODOULIDES, L. G., and KETTERER, B. (1985b), in: 'Free Radicals in Liver Injury' (G. POLI, K. H. CHEESEMAN, M. V. DIANZANI and T. F. SLATER, eds.), IRL Press, Oxford, pp. 221—224.

MEYER, D. J. and KETTERER, B. (1987), in: 'Glutathione S-Transferases and Carcinogenesis' (T. J. MANTLE, C. B. PICKETT, and J. D. HAYES, eds.), Taylor and Francis, London, pp. 53—55.

MIRANDA, M., DI ILIO, C., BONFIGLI, A., ARCADIA, A., PITARI, G., DUPRE, S., FEDERICI, G., and DEL BOCCIO, G. (1987), Biochim. Biophys. Acta **913**, 386—394.

MOORE, R. E., DAVIES, M. S., O'CONNELL, K. M., HARDING, E. I., WIEGAND, R. C., and TIEMEIER, D. C. (1986), Nucleic Acids Res. **14**, 7227—7235.

MORGENSTERN, R., GUTHENBERG, C., and DE PIERRE, J. W. (1982), Eur. J. Biochem. **128**, 243—248.

MORGENSTERN, R., GUTHENBERG, C., MANNERVIK, B., and DE PIERRE, J. W. (1983), FEBS Lett. **160**, 264—268.

MORGENSTERN, R. and DE PIERRE, J. W. (1985), in: 'Reviews in Biochemical Toxicology' (E. HODGSON, J. R. BEND, and R. M. PHILPOT, eds.), Vol **7**, Elsevier/North Holland, New York, pp. 67—104.

MORGENSTERN, R., DE PIERRE, J. W., and JORNVALL, H. (1985), J. Biol. Chem. **260**, 13976—13983.

MORGENSTERN, R., WALLIN, H., and DE PIERRE, J. W. (1987), in: 'Glutathione S-Transferases and Carcinogenesis' (T. J. MANTLE, C. B. PICKETT, and J. D. HAYES, eds.), Taylor and Francis, London, pp. 29—38.

MOZER, T. J., TIEMEIER, D. C., and JAWORSKI, E. G. (1983), Biochemistry **22**, 1068 to 1072.

NEMOTO, N., GELBOIN, H. V., PABST, W. H., and JAKOBY, W. B. (1975), Nature **252**, 512.

OKUDA, A., SAKAI, M., and MURAMATSU, M. (1987), J. Biol. Chem. **262**, 3858—3863.

OSTLUND FARRANTS, A.-K., MEYER, D. J., COLES, B., SOUTHAN, C., AITKEN, A., JOHNSON, P. J., and KETTERER, B. (1987), Biochem. J. **245**, 423—428.

PEARSON, W. R., WINDLE, J. J., MORROW, J. F., BENSON, A. M., and TALALAY, P. (1983), J. Biol. Chem. **258**, 2052—2062.

PICKETT, C. B., TELAKOWSKI-HOPKINS, A., DING, G. J.-F., ARGENTBRIGHT, L., and LU, A. Y. H. (1984), J. Biol. Chem. **259**, 5182—5188.

POWER, C., SINHA, S., WEBBER, C., MANSON, M. M., and NEAL, G. E. (1987), Carcinogenesis 8, 797—801.

REDICK, J. A., JAKOBY, W. B., and BARON, J. (1982), J. Biol. Chem. **257**, 15200—15203.

RHOADS, D. M., ZARLENGO, R. P., and TU, C. P. (1987), Biochem. Biophys. Res. Commun. **145**, 474—481.

ROTHKOPF, G. S., TELAKOWSKI-HOPKINS, C. A., STOTISH, R. L., and PICKETT, C. B. (1986), Biochemistry **225**, 993—1002.

SAITO, K., YAMAZOE, Y., TETSUYA, K., and KATO, R. (1983), Carcinogenesis 4, 1551 to 1557.

SATO, K., KITAHARA, A., YIN, Z., WARAGAI, F., NISHIMURA, K., HATAYAMA, I., EBINA, T., YAMAZAKI, T., TSUDA, H., and ITO, N. (1984), Carcinogenesis 5, 473—477.

SCOTT, T. R. and KIRSCH, R. E. (1987), Biochimica et Biophysica Acta. **926**, 264—269.

SEIDERGARD, J., DE PIERRE, J. W., and PERO, R. W. (1985), Carcinogenesis 6, 1211 to 1216.

SEIDERGARD, J., GUTHENBERG, C., PERO, R. W., and MANNERVIK, B. (1987), Biochem. J. **246**, 783—785.

SHAH, D. M., HIRONAKA, C. M., WIEGAND, R. C., HARDING, E. I., KIRIVI, G. G., and TIEMEIER, D. C. (1986), Plant Mol. Biol. 6, 203—211.

SHEA, T. C. and HENNER, W. D. (1987), in: 'Glutathione S-Transferases and Carcinogenesis' (T. J. MANTLE, C. B. PICKETT, and J. D. HAYES, eds.), Taylor and Francis, London, pp. 227—230.

SHERMAN, M., TITMUSS, S., and KIRSCH, R. E. (1983), Biochem. Int. 6, 109—118.

SIMONS, P. C., and VANDER JAGT, D. L. (1980), J. Biol. Chem. 255, 4740—4744.

SINGH, S. V., LEAL, T., ANSARI, A. S., and AWASTHI, Y. C. (1987), Biochem. J. 246, 179—186.

SLATER, T. F. (1984), Biochem. J. 222, 1—15.

SOMA, Y., SATOH, K., and SATO, K. (1986), Biochim. Biophys. Acta 869, 247—258.

SPEARMAN, M. E., PROUGH, R. A., ESTABROOK, R. W., FALCK, J. R., MANNA, S., LEIBMAN, K. C., MURPHY, R. C., and CAPDEVILA, J. (1985), Arch. Biochem. Biophys. 242, 225—230.

STENERSEN, J., KOBRO, S., BJERKE, M., and AREND, U. (1987), Comp. Biochem. Physiol. 86C, 73—82.

STOCKMAN, P. K., MCLELLAN, L. I., and HAYES, J. D. (1987), Biochem. J. 244, 55—61.

STRANGE, R. C., FAULDER, C. G., DAVIS, B. A., HUME, R., BROWN, J. A., COTTON, W., and HOPKINSON, D. A. (1984), Ann. Hum. Genet. 48, 11—20.

SUGIMOTO, M., KUHLENKAMP, J., OOKHTENS, M., AW, T. Y., REEVE, J., and KAPLOWITZ, N. (1985), Biochem. Pharmacol. 34, 3643—3647.

SUGUOKA, Y., KANO, T., OKUDA, A., SAKAI, M., KITAGAWA, T., and MURUMATSU, M. (1985), Nucleic Acids Res. 13, 6049—6057.

SUN, F. F., CHAU, L.-Y., SPUR, B., COREY, E. J., LEWIS, R. A., and AUSTEN, K. F. (1986), J. Biol. Chem. 261, 8540—8546.

SUZUKI, T., COGGAN, M., SHAW, D. C., and BOARD, P. G. (1987), Ann. Human Genet. 51, 95—106.

TAHIR, M. K., GUTHENBERG, C., and MANNERVIK, B. (1985), FEBS Lett. 181, 249—252.

TAHIR, M. K., OZER. N., and MANNERVIK, B. (1988), Biochem. J., 253, 759—766.

TAN, K. H., MEYER, D. J., and KETTERER, B. (1984), Biochem. J. 220, 243—252.

TAN, K. H., MEYER, D. J., COLES, B., and KETTERER, B. (1986), FEBS Lett. 207, 231—233.

TAN, K. H., MEYER, D. J., COLES, B., GILLIES, N., and KETTERER, B. (1987a), Biochem. Soc. Trans. 15, 628—629.

TAN, K. H., MEYER, D. J., and KETTERER, B. (1987b), Free Rad. Res. Commun. 3, 273—278.

TAN, K. H., MEYER, D. J., GILLIES, N., and KETTERER, B. (1988), Biochem. J. 245, 841—845.

TATEOKA, N., TSUCHIDA, S., SOMA, Y., and SATO, K. (1987), Clin. Chim. Acta 166, 207—218.

TATEMATSU, M., NAGAMINE, Y., and FARBER, E. (1983), Cancer Res. 43, 5049—5058.

TAYLOR, J. B., CRAIG, R. K., BEALE, D., and KETTERER, B. (1984), Biochem. J. 219, 223—231.

TAYLOR, J. B., PEMBLE, S. E., COWELL, I. G., DIXON, K., and KETTERER, B. (1987), Biochem. Soc. Trans. 15, 578—581.

TAYLOR, J. B., VIDAL, A., TORPIER, G., MEYER, D. J., ROITSCH, C., BALLOUL, J. M., SOUTHAN, C., SONDERMEYER, P., PEMBLE, S., LECOCQ, J. P., CAPRON, A., and KETTERER, B. (1988), EMBO J. 7, 465—472.

TE KOPPELE, J. M., COLES, B., KETTERER, B., and MOLDER, G. J. (1988), Biochem. J. 252, 137—142.

TELAKOWSKI-HOPKINS, C. A., ROTHKOPF, G. S., and PICKETT, C. B. (1985), Proc. Natl. Acad. Sci. U.S.A. 83, 9393—9397.

TELAKOWSKI-HOPKINS, C. A., KING, R. G., and PICKETT, C. B. (1987), Proc. Natl. Acad. Sci. (USA) **83**, 9393−9397.

TIPPING, E., KETTERER, B., CHRISTODOULIDES, L., ELLIOTT, B. M., ALDRIDGE, W. N., and BRIDGES, J. W. (1979), Chem.-Biol. Interact. **24**, 317−327.

TU, C. P.-D. and QIAN, B. (1986), Biochem. Biophys. Res. Commun. **141**, 1170−1176.

TU, C. P.-D., LAI, H.-C. J., and REDDY, C. C. (1987), in: 'Glutathione S-Transferases and Carcinogenesis' (T. S. MANTLE, C. B. PICKETT, and J. D. HAYES, eds.), Taylor and Francis, London, pp. 87−110.

VANDERBERGHE, Y., GLAISE, D., MEYER, D. J., GUILLOUZO, A., and KETTERER, B. (1988), Biochem. Pharmacol. **37**, 2482−2485.

VANDER JAGT, D. L., HUNSAKER, L. A., GARCIA, K. B., and ROYER, R. E. (1985), J. Biol. Chem. **260**, 11603−11610.

VAN KUIJK, F. J. G. M., SEVANIAN, A., HANDELMAN, G. J., and DRATZ, E. A. (1987), TIBS **12**, 31−34.

WARHOLM, M., JENSSON, H., TAHIR, M. K., and MANNERVIK, B. (1986), Biochemistry **25**, 4119−4125.

WOLF, C. R., LEWIS, A. D., CARMICHAEL, J., ANSELL, J., ADAMS, HICKSON, I. J., HARRIS, A., BALKWILL, F. R., GRIFFEN, D. B., and HAYES, J. D. (1987), in: 'Glutathione S-Transferases and Carcinogenesis' (T. J. MANTLE, C. B. PICKETT, and J. D. HAYES, eds.), Taylor and Francis, London, pp. 199−212.

YALCIN, S., JENSSON, H., and MANNERVIK, B. (1983), Biochem. Biophys. Res. Commun. **114**, 829−834.

# Chapter 9

# Epoxide Hydrolases: Molecular Properties, Induction, Polymorphisms and Function

H. Thomas, C. W. Timms, and F. Oesch

# 1. Introduction

Epoxides are generated mainly by the action of microsomal cytochrome P-450 dependent monooxygenases from a wide variety of naturally occurring compounds as well as man made chemicals and drugs containing an aliphatic or aromatic double bond (JERINA and DALY, 1974; SUGIMURA et al., 1980; SNYDER et al., 1982). These intermediates have gained considerable importance in toxicology and chemical carcinogenesis due to their reactivity towards essential nucleophilic cellular constituents (JERINA et al., 1977, 1978; OESCH, 1982). In addition to their reaction with proteins and nucleic acids (DOERJER et al., 1984; HODGSON et al., 1983), epoxides may undergo degradation by several alternative pathways, including cytochrome P-450 mediated reduction to the parent compound, rearrangement to phenols and corresponding aliphatic aldehydes or ketones, conjugation to glutathione (HABIG et al., 1974; MANNERVIK, 1985) and hydration to form trans-1,2-dihydrodiols. The latter reaction may proceed non enzymatically in some cases but was shown to be catalyzed by epoxide hydrolases in the majority of instances (OESCH, 1973; LU and MIWA, 1980; GUENGERICH, 1982). Enzymatic hydration is essentially irreversible and produces metabolites of lower reactivity which, in addition, can be readily conjugated and excreted. Therefore the action of epoxide hydrolases is generally regarded detoxifying. However, the resulting metabolites may provide in some cases precursors for further activation to ultimate carcinogens as demonstrated for pre-bay dihydrodiols of polycyclic aromatic hydrocarbons (KAPITULNIK et al., 1978; LEVIN et al., 1978).

Epoxide hydrolase, in older papers often referred to as epoxide hydrase or hydratase, has been assigned the IUPAC number EC 3.3.2.3. Also, with the discovery of multiple forms of the enzyme there is, as yet, no official nomenclature for the separate forms: therefore, the following classification is used during this review.

Microsomal epoxide hydrolase (mEH$_b$) has a broad substrate specificity and is present in all species investigated. A diagnostic substrate for mEH$_b$ is benzo[a]pyrene 4,5-oxide. mEH$_{ch}$ is another microsomal epoxide hydrolase with very limited substrate specificity for certain steroids only, and is present in all species investigated. The standard substrate is cholesterol 5,6-oxide. Cytosolic epoxide hydrolase (cEH) is an enzyme with complementary substrate specificity to mEH$_b$ and is present in all species investigated. The standard substrate is trans-stilbene oxide. A distinct soluble epoxide hydrolase (LTA$_4$ hydrolase) catalyzes the conversion of leukotriene A$_4$ into leukotriene B$_4$ and seems to be ubiquitously distributed in all species investigated.

Excellent reviews on epoxide hydrolase have been published (OESCH, 1973; OESCH, 1979; LU and MIWA, 1980; GUENGERICH, 1982; SEIDEGÅRD and DEPIERRE, 1983; WIXTROM and HAMMOCK, 1985; MEIJER and DEPIERRE,

H. THOMAS; C. W. TIMMS; F. OESCH

1988), the latter being the most recent. The aim of this chapter is to focus on areas of recent research including the molecular biology of epoxide hydrolases which were not covered by those reviews.

## 2.  Historical background

In 1950, almost 200 years after Sir Percival Pott drew the causal relationship between the incidence of certain human cancers and exposure to chemicals in soot (POTT, 1763). BOYLAND (1950) proposed that the hydroxylation of polycyclic aromatic hydrocarbons proceeds via an epoxide intermediate, and that the hydration of the epoxide proceeds enzymatically. Soon afterwards, the formation of epoxides from insecticides (DAVIDOW and RADOMSKI, 1953) and steroids (BLOOM and SHULL, 1955) was shown. However, it was not until 1961 that BREUER and KNUPPEN (1961) demonstrated that the hydration of certain epoxides, in this case an estrogen, is enzymatic. Five years later, SIMS (1966) showed that the hydration of epoxides of polycyclic aromatic hydrocarbons is also mediated enzymatically, and at the same time BROOKS (1966) reported the enzymatic hydration of chlordene epoxide. Further work in the next five years confirmed the formation of epoxide intermediates in the hydroxylation of naphthalene (JERINA et al., 1970a), styrene (LEIBMANN and ORTIZ, 1968) and aliphatic olefins (WATABE and MAYNERT, 1968). Further work into the 1970's showed the broad range of epoxidic compounds which undergo hydration in the microsomal cell fraction (OESCH et al., 1971c; GROVER et al., 1971). The first enrichment of microsomal epoxide hydrolase activity was carried out in 1970 (WATABE and KANEHIRA, 1970), and the first purification to apparent homogeneity in 1975 (BENTLEY and OESCH, 1975a; LU et al., 1975), from rat liver, and in 1979, from human liver (LU et al., 1979; GUENGERICH et al., 1979b).

In 1974, experiments investigating the metabolism of a juvenile hormone (phenylgeranyl ether epoxide) indicated for the first time that hydrolase activity exists in the cytosolic fraction (GILL et al., 1974). However, it was not until the late 1970's that greater interest was shown in this fraction (HAMMOCK et al., 1980b) and the first purification schemes were published in 1982 for rabbit and human liver cytosolic epoxide hydrolase respectively (WAECHTER et al., 1982b; WANG et al., 1982).

In 1981, WATABE et al. presented evidence for a distinct form of microsomal epoxide hydrolase specific for cholesterol 5,6-oxide. Although a few more substrates have been found for this enzyme, the substrate specificity remains very narrow, and the enzyme has yet to be purified.

# 3. Assay methods

Examples of the various methods used for assaying epoxide hydrolase are shown in Table 1: each method will be briefly described in this section. For a thorough discussion of the merits and dangers inherent in the various assay systems, see WIXTROM and HAMMOCK (1985).

**Radiometric extraction** involves the differential extraction of radiolabelled substrate and product from the incubation medium. Due to the ease and speed with which this method is carried out, it is mainly used for routine assays, and has the additional advantage that dual-enzyme measurements are possible. For example, cEH and glutathione-S-transferase activities have been determined concomitantly with trans-ethyl styrene oxide (TESO) and trans-stilbene oxide (TSO) (MULLIN and HAMMOCK, 1980; GILL et al., 1983a), while our laboratory recently developed an assay method allowing concomitant determination of the same enzyme activities with various methylated styrene oxides as substrates (SCHLADT et al., 1988; MILBERT et al., 1986). A partition assay for the simultaneous determination of insect juvenile hormone esterase and epoxide hydrolase activity was reported by SHARE and ROE (1988). However, precaution is sometimes necessary to ensure that only the desired enzymes are metabolizing the substrate (MOODY et al., 1986), and that the finally extracted radioactivity consists solely of the anticipated product.

**Thin-layer chromatography** allows separation of almost any mixture of epoxide substrate and diol product under appropriate conditions, but occasionally has the disadvantage that the radiolabelled substances must be scraped off the plate for quantification. Possible solutions to this problem are either a sensitive radio-scanner, or an automated high-resolution zonal scraper (SNYDER, 1964; WIXTROM and HAMMOCK, 1985).

**Gas-liquid chromatography.** The high sensitivity of this method, especially with an electron-capture detector, has made it a very useful technique in the analysis of the metabolism of pesticides and their derivatives (BROOKS et al., 1970). Also, in combination with a mass spectrometer immediate characterization of each metabolite is possible (WATABE et al., 1981). The only disadvantage is that in some cases a time-consuming derivatization of the metabolites is necessary to optimize the chromatographic separation, e.g. silylation (BROOKS et al., 1970), reaction with n-butyl boronic acid (HAMMOCK and HASE-GAWA, 1983).

**Spectrophotometry** is despite the somewhat reduced sensitivity of this method, still very popular due to the advantages that no radiolabel is necessary and that the measurement is continuous, and therefore useful for kinetic work

---

H. THOMAS; C. W. TIMMS; F. OESCH

**Table 1.** Assay methods for measuring epoxide hydrolase activity

| | Substrate | Reference |
|---|---|---|
| 1. Radiometric extraction | STO<br>BPO<br>OE<br>TESO<br>methylated styrene oxides<br><br>TSO<br>AE<br>juvenile hormones<br>farnesoic acid epoxides | OESCH and DALY (1971a)<br>SCHMASSMANN et al. (1976)<br>BINDEL et al. (1979)<br>MULLIN and HAMMOCK (1980)<br>SCHLADT et al. (1986)<br>MILBERT et al. (1986)<br>OESCH and GOLAN (1980)<br>OESCH et al. (1981)<br>MUMBY and HAMMOCK (1979a)<br>SHARE and ROE (1988) |
| Modifications | STO<br><br><br><br><br><br>BPO<br>TSO | OESCH et al. (1971b)<br>NEBERT et al. (1972)<br>OESCH, et al. (1973)<br>GANU et al. (1977)<br>SEIDEGARD et al. (1977)<br>GANU and ALWORTH (1978)<br>GLATT et al. (1983)<br>GILL et al. (1983a) |
| 2. Thin-layer chromatography | estratetraenol<br>16α,17α-epoxide<br>STO, BPO and eight other poly-<br>cylic aromatic hydrocarbons<br>TSO, TESO and p-nitro<br>styrene oxide<br>2,3-epoxypropyl ethers<br><br>hepoxilin A₃<br>cholesterol 5,6-oxide<br>N-methyl 2β, 3β-imino-5α-<br>cholestane | WATABE et al. (1979)<br>JERINA et al. (1976)<br><br><br>WANG et al. (1982)<br><br>VAN DEN EECKHOUT et al. (1985)<br>PACE-ASCIAK et al. (1986)<br>BLACK and LENGER (1979)<br><br>WATABE and SUZUKI (1972) |
| Modifications | cholesterol 5,6-oxide | SEVANIAN et al. (1980)<br>LEVIN et al. (1983) |
| 3. Gas-liquid chromatography | HEOM and four derivatives<br>TESO and TSO<br><br>cholesterol 5,6-oxide | BROOKS et al. (1970)<br>HAMMOCK and HASEGAWA (1983)<br>WATABE et al. (1981) |
| Modifications | HEOM | CRAVEN et al. (1982) |
| 4. Spectrophoto-metry | cyclohexene oxide<br>+ 15 epoxides<br>BPO + nine polycyclic<br>aromatic hydrocarbons | GUENGERICH and MASON (1980)<br>ARMSTRONG et al. (1980) |

**Table 1.** (continued)

| | Substrate | Reference |
|---|---|---|
| 4. Spectrophoto-metry | TSO<br>p-nitrostyrene oxide | MULLIN and HAMMOCK (1982)<br>WESTKAEMPER and HANZLIK (1981) |
| 5. Fluorometry | BPO + three polycyclic aromatic hydrocarbons | DANSETTE et al. (1979) |
| Modifications | BPO | HUKKELHOVEN et al. (1982) |
| 6. High-pressure liquid chromatography | BPO + four polycyclic aromatic hydrocarbons<br>phenylglycidyl ethers<br>1,3,5(10)-estratriene 3,16α,17α-epoxide<br>3-(p-nitrophenoxy) 1,2-propene oxide<br>Arachidonic acid epoxides<br>5α-pregnan-20-one 3β,5α,6α-epoxide<br>cholesterol 5,6-oxide | THAKKER et al. (1977 b)<br>SINSHEIMER et al. (1987)<br>WATABE et al. (1979)<br>GIULIANO et al. (1980)<br>CHACOS et al. (1983)<br>WATABE et al. (1981)<br>WATABE et al. (1981) |
| 7. Immunological | radial immuno-diffusion<br><br><br>immunoblotting<br>ELISA (enzyme-linked immuno-sorbent assay) | THOMAS et al. (1981)<br>THOMAS et al. (1982)<br>PARKINSON et al. (1983)<br>HUNT et al. (1987)<br>GILL et al. (1982)<br>GRIFFIN and GENGOZIAN (1984)<br>MOODY et al. (1987)<br>ZHIRI et al. (1987) |

(ARMSTRONG et al., 1980; WESTKAEMPER and HANZLIK, 1981). Interestingly, GUENGERICH and MASON (1980), have developed a continuous assay by coupling alcohol dehydrogenase to the epoxide hydrolase reaction, and measuring the transfer of electrons to $NAD^+$ at 340 nm.

**Fluorometry** for polycyclic aromatic hydrocarbons in particular, is a very sensitive method and, like spectrophotometry, continuous. However, care must be taken that at higher substrate concentrations ($> 10 \, \mu M$) the absorbance is still sufficiently low to avoid quenching, and additionally that no signi-

H. THOMAS; C. W. TIMMS; F. OESCH

ficant photodecomposition of the substrate occurs (DANSETTE et al., 1979). Indeed, HUKKELHOVEN et al. (1982) have developed an endpoint assay which solves some of the above-mentioned problems.

**High-pressure liquid chromatography** although essentially expensive, tedious and time-consuming, does allow simultaneous qualitative and quantitative analysis of metabolites, using either radiolabelled compounds (WATABE et al., 1981) or spectrophotometric detection (GIULIANO et al., 1980).

**Immunological methods.** Although the catalytic activity of the enzyme cannot be measured, these methods are attractive due to their high sensitivity in measuring enzyme concentrations in complex mixtures, i.e. cell fractions. The only disadvantage is that the pure protein must first be isolated and antibodies raised against it: however, when this condition has been fulfilled, sensitivity limits of $2-5$ ng $mEH_b$ have been reported in an ELISA (enzyme-linked immunosorbent assay) (GILL et al., 1982). This has also proved to be a very useful method in the detection of $mEH_b$ in serum of rats and humans (GRIFFIN and GENGOZIAN, 1984).

# 4.  Properties of epoxide hydrolases

## 4.1.  Substrate specificity

All the epoxide hydrolases catalyze the addition of water to an epoxide, to form the vicinal dihydrodiol, which almost always has the trans configuration. Tables 2 and 3 show the most active substrates for $mEH_b$ and cEH respectively, and Table 4 compares the differential substrate-specificity of the two forms. Whilst $mEH_b$ exhibits a broad substrate-specificity, ranging from simple aliphatics e.g. octene oxide, to large polycyclic aromatic hydrocarbons (HODGSON et al., 1985), e.g. benzo[a]pyrene-4,5-oxide, cEH appears to be more selective, not being able, for example, to metabolize the bulky steroids and polycyclic aromatic hydrocarbons.

Systematic studies have shown that monosubstituted oxiranes with a large hydrophobic substituent tend to be excellent substrates of $mEH_b$ (OESCH et al., 1971c), whilst other monosubstituted epoxides are hydrated by both systems (HAMMOCK and HASEGAWA, 1983). The latter study has also confirmed that 1,2-disubstituted and tri- and tetra-substituted oxiranes are excellent, selective substrates for cEH, which on the other hand can only poorly hydrate 1,2-cis-epoxides, when one or more substituents is a phenyl moiety. A good example is stilbene oxide (Table 3), where the cis-isomer is a good

---

**Table 2.** Substrates for mEH$_b$ from rat liver microsomes

| Substrate | Structure | Specific activity (nmol product/ mg protein/min) | Reference |
|---|---|---|---|
| Styrene 7,8-oxide (STO) | | 8.9 | Bentley and Oesch (1975a) |
| Octene 1,2-oxide | | 13.1 | Jerina et al. (1976) |
| Phenanthrene 9,10-oxide | | 27.3 | Bentley et al. (1976) |
| Benz(a)anthracene 5,6-oxide | | 6.8 | Bentley et al. (1976) |
| Dibenz(a,h)-anthracene 5,6-oxide | | 0.4 | Bentley et al. (1976) |
| Benzo(a)pyrene 4,5-oxide (BPO) | | 7.2 | Jerina et al. (1976) |
| Oestroxide (OE) | | 12.5 | Bindel et al. (1979) |
| Androstene-oxide (AE) | | 55.0 | Oesch et al. (1981) |
| HEOM | | 3.0 | Walker et al. (1978) Walker et al. (1986) |
| Epichlorohydrin | | 92.0 | Guengerich and Mason (1980) |
| cis-Stilbene-oxide | | 11.1 | Gill et al. (1983a) |

**Table 3.** Substrates for cEH from mouse liver cytosol

| Substrate | Structure | Specific activity (nmol product/ mg protein/min) | Reference |
|---|---|---|---|
| trans-$\beta$-Propyl-styrene-oxide | | 267.9 | GILL et al. (1983a) |
| trans-$\beta$-Ethyl-styrene-oxide (TESO) | | 136.4 | GILL et al. (1983a) |
| trans-Stilbene-oxide (TSO) | | 8.4 | GILL et al. (1983a) |
| trans-$\beta$-Methyl-styrene-oxide | | 2.3 | GILL et al. (1983a) |
| Phenylgeranyl-monoethyl-epoxide (trans) | | 12.6 | MUMBY and HAMMOCK (1979) |
| Phenylgeranyl-monopropyl-epoxide (trans) | | 12.6 | MUMBY and HAMMOCK (1979) |
| Phenylgeranyl-monopropyl-epoxide (cis) | | 25.8 | MUMBY and HAMMOCK (1979) |
| Squalene 2,3-23,24-dioxide | | no figure given | HAMMOCK et al. (1980b) |
| cis-9,10-Epoxy-methyl-stearate | | 21.1 | GILL and HAMMOCK (1979) |
| 14,(15)Oxido-5,8,11-eicosa-trienoic-acid | | 1260 (diol and S-lactone) Partially purified cEH | CHACOS et al. (1983) |

**Table 4.** Differential substrates for microsomal and cytosolic fractions of mouse liver

| Substrate | Structure | Specific activity (nmol product/ mg protein/min) | | Reference |
|---|---|---|---|---|
| | | Micro- somes | Cytosol | |
| trans-Methyl- styrene-oxide | | 0.4 | 136 | GILL et al. (1983a) |
| cis-Methyl- styrene-oxide | | 0.3 | 0.2 | GILL et al. (1983a) |
| trans-Stilbene- oxide | | N.D. | 8.4 | GILL et al. (1983a) |
| cis-Stilbene- oxide | | 4.1 | 0.9 | GILL et al. (1983a) |
| 1,2-Limonene- oxide | | 10.6 | N.D. | GILL et al. (1983a) |
| 7,8-Limonene- oxide | | 16.6 | 0.3 | GILL et al. (1983a) |
| Benzo(a)pyrene 4,5-oxide | | 1.6 | N.D. | OESCH and GOLAN (1980) |
| Benz(a)anthra- cene 5,6-oxide | | 1.4 | 0.2 | OESCH and GOLAN (1980) |

N.D.: not detectable

$mEH_b$ substrate and the trans-isomer is rapidly hydrolyzed by cEH (HAM-MOCK et al., 1980a). An exception, however, is the class of cis-disubstituted phenylgeranyl ether epoxides, which is metabolized by cEH more rapidly than the trans-isomers (MUMBY and HAMMOCK, 1979). Further work by OESCH (1974) indicated that epoxides with small side-chains are probably poor sub-strates for $mEH_b$, although a few were metabolized at significant rates, especially when a lipophilic group is present (GUENGERICH and MASON, 1980). In contrast to cEH which rapidly metabolizes trans-stilbene oxide, this sub-strate is not metabolized at all by the cytosolic $LTA_4$ hydrolase. The latter appears to exclusively and specifically hydrate the endogenous substrate leukotriene $A_4$ to form the 12-hydroxy metabolite 5(S),12(R)-dihydroxy-6,14-cis-8,10-trans-icosatetraenoic acid ($LTB_4$), while cEH opens the epoxide to yield erythro-5(S),6(R)-dihydroxy-7,9-trans-11,14-cis-icosatetraenoic acid (HAEGGSTRÖM et al., 1986).

As already described, epoxides of cyclic systems including BPO are hydrated almost exclusively by $mEH_b$, and BROOKS et al. (1970) found that a range of dieldrin derivatives were almost all metabolized by hepatic micro-somal epoxide hydrolase from various species. The authors suggested that these differences might be explained by the relative ease of approach to the epoxide ring; for example, in dieldrin the insertion of a methylene bridge in the epoxycyclohexane ring of HEOM enforces a "boat"-conformation which is known to be remarkably resistant to enzymatic hydrolysis. Also, the intro-duction of a halogen atom adjacent to the epoxide ring decreases the turnover rate, probably due to a redirection of the attack of the hydroxyl ion to the more hindered side of the molecule. Similarly, various steroid epoxides have also been shown to be exclusively metabolized by $mEH_b$: $16\alpha,17\alpha$-epoxy-androst-4-en-3-one (AE) (OESCH et al., 1981), $16\alpha,17\alpha$-epoxy-1,3,5(10)-estratrien-3-ol (OE) (BINDEL et al., 1979) (Table 2), and norethisterone-$4\beta,5\beta$-oxide (WHITE, 1980; PETER et al., 1981).

Of particular interest is the discovery of a number of lipophilic fatty acid epoxides which are good substrates for cEH: cis- and trans-epoxymethyl stearates (GILL and HAMMOCK, 1979; HALARNKAR et al., 1989), arachidonic acid epoxides (CHACOS et al., 1983) and 18-hydroxy-9,10-epoxyoctadecanoic acid (KOLATTUKUDY and BROWN, 1975). In addition, squalene oxide and dioxide, and lanosterol oxide (HAMMOCK et al., 1980b) 1,2-epoxy-cycloalkanes (MAGDA-LOU and HAMMOCK, 1988) and leukotriene $A_4$ (HAEGGSTRÖM et al., 1986; MIKI et al., 1989) are also hydrated by cEH.

Recently, a microsomal epoxide hydrolase isozyme, $mEH_{ch}$, (WATABE et al., 1981) has aroused much interest, after it was discovered in rat liver using the substrate, cholesterol 5,6-oxide. Table 5 shows the substrates which are known to be metabolized by $mEH_{ch}$, and it is clear that no substrate has been found which is metabolized by both this form and $mEH_b$, i.e. the substrate-specificities are mutually exclusive. Also, it has been observed that $2\alpha,3\alpha$- and

**Table 5.** Substrates for $mEH_{ch}$ determined with mouse (A) and rat (B) liver microsomes

| Substrate | Structure | Specific activity (nmol product/ g liver/min) | Reference |
|---|---|---|---|
| **A** | | | |
| 5,6α-Cholesterol-epoxide | | 7.4 | Watabe et al. (1983) |
| 5,6α-Epoxy-5α-cholestane | | 3.8 | Watabe et al. (1983) |
| 5,6α-Epoxy-20-methyl-5α-pregnan-3β-ol | | 2.7 | Watabe et al. (1983) |
| 5.6α-Epoxy-5α-pregnan-20-on-3β-ol | | 0.4 | Watabe et al. (1983) |
| 2α,3α-Epoxy-5α-cholestane | | 0.1 | Watabe et al. (1983) |
| 2β,3β-Imino-5α-cholestane | | ? | Watabe and Suzuki (1972) |
| **B** | | | |
| 5,6α-Cholesterol-epoxide | | 0.480 nmol/mg/ min | Sevanian and McLeod (1986) |
| 5,6β-Cholesterol-epoxide | | 0.451 nmol/mg/ min | Sevanian and McLeod (1986) |

?: no quantitative data given

H. Thomas; C. W. Timms; F. Oesch

$2\beta,3\beta$-epoxy-$5\alpha$-cholestane are metabolized by mouse liver microsomes (WATABE et al., 1983), but it is not clear which isozyme is responsible. It is even possible that another form exists. Similarly, it was shown that rabbit liver microsomes are capable of converting $2\beta,3\beta$-imino-$5\alpha$-cholestane to $2\beta$-amino-$3\alpha$-hydroxy-$5\alpha$-cholestane, and the enzyme catalyzing the reaction was named "aziridine hydrolase" (WATABE and SUZUKI, 1972; WAZABE et al., 1971 b). However, WATABE and AKAMATSU (1974) were able to inhibit the hepatic microsomal metabolism of cis-stilbene oxide to threo-stilbene glycol, with trans-stilbenimine indicating similarity of this hydrolase with $mEH_b$. Therefore, the identity of "aziridine hydrolase" and its relationship to the other microsomal forms remains to be established.

## 4.2. Regio- and stereoselectivity

As already mentioned, in animals $mEH_b$ initiates a trans-opening of the epoxide ring (JERINA et al., 1970 b; WATABE and AKAMATSU, 1972), as do cEH (GILL and HAMMOCK, 1979) and $mEH_{ch}$ (WATABE and SAWAHATA, 1979; WATABE et al., 1981).

Table 6 shows the various substrates used for investigating the regio- and stereoselectivity of $mEH_b$ and cEH. For $mEH_b$, hydroxide ion attack occurs at the least sterically hindered epoxide carbon atom, and regiospecific discrimination is not abolished until almost complete symmetry exists around the ring, as is the case with benzo[a]pyrene 4,5-oxide (BPO). Furthermore, whilst the metabolism of benzo[a]pyrene 4,5-oxide exhibits low regioselectivity and high enantioselectivity (THAKKER et al., 1977 b), that of styrene oxide (STO) and naphthalene oxide (JERINA et al., 1970a; HANZLIK et al., 1978) and benzo[a]pyrene 7,8-oxide (THAKKER et al., 1977 b) demonstrate high regio- but low enantioselectivity, and the diol is formed from 9,10-epoxy-stearic acid with both low regio- and stereoselectivity (WATABE and AKAMATSU, 1972). Similarly, whilst the (S) carbon atom is the preferred site of attack for substrates such as benzene oxide, naphthalene oxide (JERINA et al., 1970a), the stilbene oxides (DANSETTE et al., 1978), benzo[a]pyrene 4,5-oxides and only the (+)benzo[a]pyrene 7,8-oxide (ARMSTRONG et al., 1981), phenanthrene 9,10-oxide (JERINA et al., 1970b) and (−)benzo[a]pyrene 7,8-oxide (ARMSTRONG et al., 1981) are preferentially attacked at the (R) carbon. Therefore, it is difficult to make appropriate generalizations concerning the enantioselectivity, regio- and stereoselectivity of $mEH_b$, and each substrate must be investigated in itself. However, SAYER et al. (1985), have recently investigated the stereoselective metabolism of diol epoxides and tetrahydroepoxides derived from benz[a]anthracene. When the tetrahydroepoxides are only fair or poor substrates of $mEH_b$ (i.e. 1,2-, 8,9- or 10,11-epoxides) the additional retardation introduced by adjacent hydroxyl groups, e.g. in bay-region diol epoxides,

---

**Table 6.** Regio- and stereospecificity of rat liver mEH$_b$ and cEH

| Substrate | Position of attack | % Attack | % Optical enantiomers | Reference |
|---|---|---|---|---|
| Styrene 7,8-oxide | 8 | 90 | (−) 50 (+) 50 | JERINA et al. (1970b) DANSETTE et al. (1978) HANZLIK et al. (1978) |
| cis-β-Methylstyrene 7,8-oxide | 8 | 89 | — | HANZLIK et al. (1978) |
| cis-Octene 2,3-oxide | 2 | 85 | — | HANZLIK et al. (1978) |
| Tetradecene 1,2-oxide | 1 | 98 | — | HANZLIK et al. (1976) |
| cis-Stilbene-oxide | — | — | (+) 98 | WATABE et al. (1971a) |
| trans-Stilbene-oxide | | | meso form | WATABE and AKAMATSU (1972) DANSETTE et al. (1978) |
| (2-Naphthyl)-ethene 1,2-oxide | 1 | 97 | — | HANZLIK et al. (1976) |
| Naphthalene 1,2-oxide | 2 | 90 | (−) in small excess | JERINA et al. (1970b) JERINA et al. (1970a) |
| Benzo(a)pyrene 4,5-oxide (±) | 4 | 54 | (−) 89 (+) 11 | THAKKER et al. (1977b) |
| | 5 | 46 | | |
| | 4 | 95 | (−) 71 (+) 29 | YANG et al. (1977b) |
| (4S,5R) | 4 | 96 | (−) 99 (+) 1 | ARMSTRONG et al. (1981) |
| (4R,5S) | 5 | 80 | (−) 85 (+) 15 | |
| From benzo[a]pyrene metabolism | — | — | (−) 96 (+) 4 | THAKKER et al. (1977a) |
| | 4 | 95 | (−) 100 | YANG et al. (1977a) |
| Benzo(a)pyrene 7,8-oxide (±) | 8 | 98 | (−) 54 (+) 46 | THAKKER et al. (1977a) |
| | 8 | 95 | (−) 86 (+) 14 | YANG et al. (1977b) |
| (+) | — | — | (−) 100 (+) 0 | ARMSTRONG et al. (1981) |
| (−) | — | — | (−) 0 (+) 100 | |

Table 6. (continued)

| Substrate | Position of attack | % Attack | % Optical enantiomers | Reference |
|---|---|---|---|---|
| From benzo(a)pyrene metabolism | | | (−) 96 (+) 4 | THAKKER et al. (1977 a) |
| | 8 | 95 | (−) 100 (+) 0 | YANG et al. (1977 a) |
| Benzo(a)pyrene 9,10-oxide (±) | − | − | (−) 61 (+) 39 | THAKKER et al. (1977 b) |
| From benzo(a)pyrene metabolism | 9 | 95 | (−) 100 (+) 0 | YANG et al. (1977 b) |
| | | | (−) 96 (+) 4 | THAKKER et al. (1977 a) |
| 1,2,3,4-Tetrahydro-benz(a)anthracene 1,2-diol, 3,4-epoxide | 4 | 95 | − | SAYER et al. (1985) |
| 1,2,3,4-Tetrahydro-benz(a)anthracene 3,4-diol, 1,2-epoxide | 1 | 95 | − | SAYER et al. (1985) |
| 1,2,3,4-Tetrahydro-benz(a)anthracene 3,4-epoxide | | | | |
| (3S,4R) | 4 | 80 | − | SAYER et al. (1985) |
| (3R,4S) | 4 | 95 | − | |
| 1,2,3,4-Tetrahydro-benz(a)anthracene 1,2-epoxide | | | | |
| (1R,2S) | 2 | 86 | − | SAYER et al. (1985) |
| (1S,2R) | 2 | 88 | − | |
| Regiospecificity of cEH 1-(4′-ethylphenoxy)-3,7-dimethyl-cis-octane 6,7-oxide | 6 | 96 | − | HAMMOCK et al. (1980 b) |

causes the enzyme-catalyzed hydrolysis to be insignificantly slow. In contrast, (−)-tetrahydrobenz[a]anthracene (3R,4S)-epoxide has the largest turnover rate with rat liver mEH$_b$ of any substrate investigated (V$_{max}$ = 6800 nmol/min/mg protein), and the two diastereomeric diol epoxides (1,2-diol-3,4-epoxides), unlike other known diol epoxides, are moderately good substrates. The authors claim that this is due to the very high reactivity of the 3,4-epoxide system which overcomes the kinetically unfavourable effect of hydroxyl substitution.

Also, although the more reactive benzylic carbon is usually the kinetically preferred site of attack, only 20% attack occurs in this position (C-3), indicating the strong tendency of $mEH_b$ to attack at a centre of (S) chirality, even when this is kinetically unfavourable. The authors then extended their studies to naphthalene, and have constructed a mathematical model for predicting the enantiomeric composition of the dihydrodiols formed in reconstituted systems, containing specific $mEH_b$ concentrations (VAN BLADEREN et al., 1985). Similarly YANG (1988) investigated the stereoselectivity of $mEH_b$ with a number of planar and nonplanar polycyclic aromatic hydrocarbons. From the enantiomeric compositions of epoxides formed in the metabolism of planar benz[a]-anthracene (MUSHTAG et al., 1986), benzo[a]-pyrene, and chrysene, and nonplanar benzo[c]phenanthrene (YANG et al., 1987a), 12-methylbenz[a]anthracene (YANG et al., 1987b) and 7,12-dimethylbenz[a]anthracene by liver microsomes as well as their dihydrodiol composition following $mEH_b$-catalyzed hydration can be concluded: regardless of the absolute configuration, non K-region epoxides are converted to trans-dihydrodiols by epoxide hydrolase-catalyzed water attack at the allylic carbon. While the S-center of K-region S,R-epoxide enantiomers derived from planar benz[a]anthracene, benzo[a]pyrene and chrysene is the major site of water attack, is in contrast the R-center of K-region S,R-epoxide enantiomers from nonplanar polycyclic hydrocarbons the preferred site of attack. However, the K-region R,S-epoxides of all six studied polycyclic aromatic hydrocarbons are hydrated by $mEH_b$ with varying degrees of regioselectivity.

The enantioselectivity of $mEH_b$ also has biological consequences, especially in its interaction with cytochrome P-450. For example, Table 6 shows clearly the high optical purity of dihydrodiols formed from the microsomal metabolism of benzo[a]-pyrene, as compared to the action of purified $mEH_b$ in the absence of cytochrome P-450, indicating the high degree of enantioselectivity of the latter. Indeed, further metabolism of (−)benzo[a]pyrene 7,8-dihydrodiol, in the presence of either liver microsomes from 3-methylcholanthrene-treated rats or a purified cytochrome P-448-containing monooxygenase system, forms predominantly (±)-7$\beta$,8$\alpha$-dihydroxy-9$\alpha$,10$\alpha$-epoxy-7,8,9,10-tetrahydrobenzo-[a]-pyrene. This diastereomer proved to be both a more potent mutagen in Chinese hamster V-79 cells and 5−10 fold more tumorigenic, on application to mouse skin, than the corresponding 9$\beta$,10$\beta$-epoxide, indicating its identity as a potent carcinogen derived from benzo[a]pyrene (THAKKER et al., 1977a). Therefore, although most evidence presented until now argues against a functional coupling of cytochrome P-450 and $mEH_b$, the stereo- and enantioselectivity of both enzyme systems plays a central role in determining the susceptibility of a particular tissue to the toxicity and carcinogenicity of polycyclic aromatic hydrocarbons.

## 4.3. Mechanism of action

Although both acid- (WATABE et al., 1971a) and base-catalysis (HANZLIK et al., 1976) have been proposed as a mechanism of action of mEH$_b$, evidence for the latter is more convincing (HANZLIK et al., 1976; HANZLIK et al., 1978; DANSETTE et al., 1978). It is proposed that the enzymatic hydration consists of mEH$_b$-specific activation of water to a hydroxide ion, through a general base-catalysis mechanism involving a single histidine-imidazole residue (DU BOIS et al., 1978), followed by a "backsite" nucleophilic addition at the least hindered epoxide carbon atom of the incoming hydroxide ion. This results in the ring opening away from this ion with the subsequent formation of

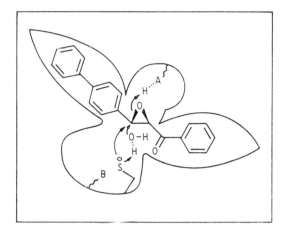

**Fig. 1.** Proposed active-site topography for the interaction of chalcone oxide inhibitors with cEH.

The representative inhibitor is depicted interacting with the two hydrophobic regions extending from the active core of the enzyme. An essential cysteine in its thiolate form and a possible specific acid catalyst A and a general base catalyst B are also included (from MULLIN and HAMMOCK, 1982).

trans-dihydrodiol (BELLUCCI et al., 1981). In addition to this, ARMSTRONG et al. (1980), have proposed the existence of a transition state during the hydration, involving a covalent enzyme-substrate intermediate: this proposal, however, must be carefully examined as other workers have presented evidence against it (BELLUCCI et al., 1981; YANG et al., 1977b; THAKKER et al., 1977b). Furthermore, two areas in the enzyme situated behind the epoxide have been described as important for substrate-binding, enantioselectivity and regio-selectivity of mEH$_b$ (ARMSTRONG et al., 1981; BELLUCCI et al., 1981; SAYER et al., 1985).

---

Epoxide Hydrolases

As is evident from Table 6, the mechanism of action of cEH has been less intensively investigated, but apparently cEH-catalyzed hydration also proceeds through a "backside" nucleophilic addition of the hydroxide ion, similar to that of mEH$_b$, to give trans-diols (MULLIN and HAMMOCK, 1982). These authors also described the differential characteristics of the active site of cEH, using a series of chalcone oxide derivatives, which is illustrated in Figure 1. Further work showed that the activation of the oxirane ring to nucleophilic attack and consequently hydration of the substrate is substantially reduced with epoxyketones because these compounds can apparently bind to the hydrolytic site electrophile through hydrogen bonding to the carbonyl group (PRESTWICH et al., 1985).

## 4.4. Physical and chemical properties

The molecular weight of mEH$_b$ in all species has been shown to be between 46,000 and 50,000 daltons (BENTLEY et al., 1975; HALPERT et al., 1979; KNOWLES and BURCHELL, 1977a; GUENGERICH et al., 1979b; LU et al., 1979) with minor differences being explained by the various gel-electrophoresis systems used. In the absence of detergent, rat mEH$_b$ aggregates forming an oligomer of molecular weight, 600,000—700,00 daltons (BENTLEY et al., 1975); LU et al., 1975; GUENGERICH and DAVIDSON, 1982). The absorption spectrum indicated a high tryptophan content (BENTLEY et al., 1975) and an absence of heme or flavin as prosthetic group (LU et al., 1975).

Amino acid analyses for mEH$_b$ demonstrated a high content of tryptophan, tyrosine, methionine and phenylalanine, endowing a hydrophobic character, which might explain the aggregation in absence of detergent (BENTLEY et al., 1975; GUENGERICH et al., 1979b; DU BOIS et al., 1982). HALPERT et al. (1979) calculated that 45% of the total amino acid residues are hydrophobic. Additionally, automated Edman degradation analysis of human and rat mEH$_b$ has identified methionine as the NH$_2$-terminal amino acid (DU BOIS et al., 1979; DU BOIS et al., 1982), whilst BENTLEY et al., (1975) concluded that the C-terminal amino acid of rat mEH$_b$ is either asparagine or glutamine. In 1984, HEINEMANN and OZOLS (1984a, b) determined the complete amino acid sequence of mEH$_b$ purified from phenobarbital-treated New Zealand White rabbits: 455 amino acid residues were found to constitute a single polypeptide chain, with a molecular weight of 52,691 daltons: this compares to 468 as calculated by LU et al. (1975) and 455 amino acids (PORTER et al., 1986) deduced for rat mEH$_b$ from the coding cDNA sequence.

Reagents specific to sulphydryl groups had little effect on the catalytic activity (LU et al., 1975), although buried cysteine residues have been reported in rat (DU BOIS et al., 1978) and rabbit enzyme preparations (HALPERT et al., 1979). Whilst lysine and histidine appear to be essential for catalytic activity

(DU BOIS et al., 1978; THOMAS et al., 1988), metal chelating agents had no effect (LU et al., 1975).

The isoelectric point of pure $mEH_b$ is 5.6 from rabbit (HALPERT et al., 1979) and 5.7—6.5 for multiple forms purified from rat and human (GUENGE-RICH et al., 1979b). Also, whilst the pH optima are substrate-dependent (LU et al., 1977), values differed considerably between purified preparations and microsomal fractions. In addition the lipid concentration appears to play a significant role here as well as the enzyme kinetic characteristics have also been altered according to lipid concentration in the medium (OESCH and DALY, 1971; LU et al., 1975; HALPERT et al., 1979). This is particularly true of lipophilic substrates, e.g. benz[a]anthracene 5,6-oxide, and LU et al. (1977) have suggested that when the lipid concentration exceeds the critical micelle concentration, the micelles function as a substrate reservoir and dispersing agent. Also, the addition of microsomal phospholipids to purified preparations has been shown to cause primarily a stabilization of the enzyme activity, but also a slight increase in activity towards STO (BULLEID et al., 1986), BPO (LU et al., 1977) and HEOM (WALKER et al., 1986). However, this was not the case with the more hydrophilic substrate, octene 1,2-oxide (LU et al., 1977) thereby emphasizing the importance of substrate solubility.

In contrast to $mEH_b$, cEH exists as a dimer with a monomer molecular weight of 55,000 in rabbit (WAECHTER et al., 1982b), 59,000 in mouse (GILL, 1983; MEIJER and DE PIERRE, 1985a, b) 61,000 in rat (SCHLADT et al., 1988b,) 60,00 in guinea pig (MIKI et al., 1989) and 58,000 in human liver (WANG et al., 1982; SCHLADT et al., 1988a). Amino acid analysis of the rabbit preparation showed 76% serine, 17% glycine and 7% threonine, with very similar or identical monomers, both containing an N-terminal serine (WAECHTER et al., 1982b). Similarly, in mouse no significant differences were observed between the monomers (MEIJER and DE PIERRE, 1985a, b).

Isoelectric points between 5.0 and 7.3 have been reported for cEH from the various species' (WAECHTER et al., 1982b; MEIJER and DE PIERRE, 1985a, b; WANG et al., 1982; MIKI et al., 1989; SCHLADT et al., 1988b): also, pH optima and enzyme kinetics have been described which are significantly different from those of $mEH_b$ (WAECHTER et al., 1982b; WANG et al., 1982; SCHLADT et al., 1988b).

From human liver ($cEH_{PNSO}$: WANG et al., 1982; $cEH_{cSO}$: SCHLADT et al., 1988a) as well as lung (GUENTHNER and KARNEZIS, 1986a, b) forms of cytosolic epoxide hydrolase with properties different from the classical cEH, and substrate specificities for paranitro styrene oxide, cis-stilbene oxide or benzo[a]pyrene-4,5-oxide have been reported. The liver forms are characterized by an isoelectric point of 9.2 compared to 5.7 and 7.0 for human liver cEH and $mEH_b$ respectively (SCHLADT et al., 1988a). Subunit molecular weights of 49,000 daltons resemble closely those of rat and human $mEH_b$. Immunoprecipitation experiments with antibodies against rat liver cEH and $mEH_b$

---

revealed a close immunological relationship of human liver cytosolic cis-stilbene oxide hydrolase (SCHLADT et al., 1988a) as well as lung cytosolic benzo[a]pyrene-4,5-oxide hydrolase (GUENTHNER and KARNEZIES, 1986a) with mEH$_b$, however not with cEH.

Leukotriene A$_4$ hydrolase, also a cytosolic enzyme, was shown by SDS-gel electrophoresis to have a molecular weight of 68,000—70,000 daltons in human and rat leukocytes (RÅDMARK et al., 1984; EVANS et al., 1985) and an apparent molecular weight of 54,000 daltons in human erythrocytes (McGEE and FITZ-PATRICK, 1985). Maximum enzyme activity was observed between pH 7.4 and 8.6 (IZUMI et al., 1986) which ranges between the pH optima for cEH of 7.4 (SCHLADT et al., 1988b) and mEH$_b$ of about 9 (GUENTHNER and OESCH, 1983). There appears to be no immunological relationship between LTA$_4$ hydrolase and any other so far characterized epoxide hydrolase.

## 4.5. Modulation of epoxide hydrolase activity

For a definitive list of modulators of microsomal epoxide hydrolase activity, the reader is referred to PARKKI (1982).

### 4.5.1. Modulation of mEH$_b$ activity

**Induction of mEH$_b$:** The most commonly used inducers of mEH$_b$ activity are shown in Table 7, with the corresponding activities towards various substrates. Besides these, a number of structurally little related aliphatic and aromatic compounds (MEIJER and DE PIERRE, 1987), substituted urea and dichloro-phenyl-trichloroethyl-oxirane herbicides (SCHOKET and VINCZE, 1986; MOODY and HAMMOCK, 1987), aliphatic nitrosamines (CRAFT et al., 1988), 2-ethyl-hexanoic acid (LUNDGREN et al., 1987), diethylhexylphthalate (GILL and KAUR, 1987) and 1-benzylimidazole (SCHLADT et al., 1987) have been shown to cause an 1.5- to 4-fold induction of hepatic mEH$_b$ in different rodent species, the inducing potency depending in part on strains and sexes tested. HgCl$_2$ has been identified as an organ-specific mEH$_b$-inducer for rat kidney (KROLL et al., 1988). Although induction studies have provided clear evidence for the differentiation of cEH and mEH$_b$ activities, the evidence for multiple forms of mEH$_b$ is limited. For example, differential induction has been demonstrated for microsomal and nuclear epoxide hydrolase (BORNSTEIN et al., 1979; GONTOVNICK and BELLWARD, 1981; GONZALEZ and KASPER, 1981a; GUENTHNER, 1986b; MOODY et al., 1988), and for mEH$_b$ from various rat organs (OESCH et al., 1977; SCHMASSMANN et al., 1978; BENSON et al., 1979; GUENTHNER, 1986a; DI BIASIO et al., 1989). In some cases, ambiguous results have been obtained using 3-methylcholanthrene as inducer particularly with respect to a failure of this compound to induce mEH$_b$ in female rats

H. THOMAS; C. W. TIMMS; F. OESCH

**Table 7.** In vivo inducers of rat and mouse liver $mEH_b$ activity

| Inducer | Substrate | % Control activity | Reference |
|---|---|---|---|
| **In rat:** | | | |
| Phenobarbital | STO | 234 | BRESNICK et al. (1977a) |
| | octene 1,2-oxide | 252 | |
| | 3-MC 11,12-oxide | 233 | |
| | naphthalene 1,2-oxide | 271 | |
| | STO | 263 | HASANI and BURCHELL (1979) |
| | STO | 200 | GONTOVNICK and BELLWARD (1980) |
| | BPO | 230 | CRESTEIL et al. (1980) |
| | CSO | 160 | DI BIASIO et al. (1989) |
| 3-Methyl-cholanthrene | octene 1,2-oxide | 150 | BRESNICK et al. (1977a) |
| | 3-MC,11,12-oxide | 150 | |
| | STO | 109 | HASANI and BURCHELL (1979) |
| | STO | 100 | GONTOVNICK and BELLWARD (1980) |
| trans-Stilbene-oxide | BPO | 295 | SCHMASSMANN et al. (1978) |
| | STO | 321 | |
| | STO | 513 | SEIDEGÅRD et al. (1981) |
| | STO | 350 | GONTOVNICK and BELLWARD (1980) |
| 2-Acetylamino-fluorene | STO | 762 | ÅSTRÖM and DE PIERRE (1981) |
| | STO | 650 | |
| | BPO | 800 | DENT and GRAICHEN (1982) |
| | STO | 500 | ÅSTRÖM et al. (1986a) |
| 2(3)-tert-Butyl-4-hydroxyanisole | STO | 280 | CHA et al. (1978) |
| | CSO | 230 | DI BIASIO et al. (1989) |
| **In mouse:** | | | |
| Phenobarbital | CSO | 93 | DI BIASIO et al. (1989) |
| 2(3)-tert-Butyl-4-hydroxyanisole | STO | 275 | BENSON et al. (1979) |
| | BPO | 650 | HAMMOCK and OTA (1983) |
| | CSO | 240 | DI BIASIO et al. (1989) |

(ÅSTRÖM et al., 1987). Also, no specific inducer for $mEH_b$ has been found which at the same time does not induce other enzymes (GONTOVNICK and BELLWARD, 1980; SEIDEGÅRD et al., 1981). Indeed, using structural analogous of stilbene, the latter authors proposed the following structural requirements for $mEH_b$ induction by aromatic compounds: the presence of two phenyl rings, with a reduction in the inductive effect if the rings are substituted or if one of the ring carbon atoms is replaced by a nitrogen. Alternatively, induction can be increased by a carbon-bridge between the phenyl groups, especially if the bridge contains an epoxy group or one or two keto groups. Furthermore, in a systematic study of $mEH_b$ induction by polychlorinated biphenyls, PARKINSON et al. (1983) have proposed that an asymmetrical distribution of chlorine atoms increases the inductive effect. This would agree with KOHLI et al. (1979) who suggested that the total stereoelectronic properties of the biphenyl rings are of more importance for induction than the specific positions of the halogens.

The mechanism of induction has been shown to be similar for various compounds, including ethoxyquin (KAHL, 1980), phenobarbital (GONZALEZ and KASPER, 1980), 2-acetylaminofluorene (GONZALEZ et al., 1982; GRAICHEN and DENT, 1984) and aliphatic nitrosamines (CRAFT et al., 1988): an increased rate of transcription of the respective genes followed by an increase in epoxide hydrolase mRNA and enzyme protein (WAECHTER et al., 1982a). The primary translation product appears to have the same molecular weight as the mature enzyme, thereby ruling out the possibility of precursor forms. Although in previous studies both induction and immunoquantitation of $mEH_b$ after induction (THOMAS et al., 1981) have failed to provide convincing evidence for a multiplicity of mEH forms in rat liver, GUENTHNER (1986a, b) presents conclusive data for a selective induction of different mEH forms in mouse liver, in that compounds such as benzofuran, butylated hydroxyanisole and phenobarbital induce microsomal benzo[a]pyrene 4,5-oxide hydrolase significantly, while the microsomal TSO hydrolase activity remains unchanged. Further, characterization of $mEH_b$-induction in rats by 2-acetylaminofluorene and a number of structurally related compounds indicated that there was no correlation between this induction and that of five isozymes of cytochrome P-450 (ÅSTRÖM et al., 1986a), despite earlier work which had indicated an association between $mEH_b$ induction and the induction of certain cytochrome P-450 isoenzymes (GRAICHEN and DENT, 1984). Therefore, it would appear that although both $mEH_b$ and various isozymes of cytochrome P-450 are induced by the same compounds, there is no functional coupling between the two enzyme systems.

**In vitro modulation of $mEH_b$ activity:** Table 8 illustrates the range of activators of microsomal $mEH_b$ activity. Uncompetitive activation is claimed as the mechanism for most substances, indicating that they bind to a site distinct from the catalytic site where the substrate binds. This led GANU and ALWORTH (1978) to suggest a single, general scheme of equilibria of nonessential activation, involving an enzyme-substrate-activator complex. However, an important factor in this theory is substrate-inhibition of STO-activity at high concentrations, and because other groups could not confirm this experimentally (OESCH et al., 1971c), the theory must be treated with caution.

---

H. THOMAS; C. W. TIMMS; F. OESCH

**Table 8.** In vitro activators of liver microsomal $mEH_b$ activity

Activity measured with STO

| Species | Activator | (mM) | Activation (%) | Conclusion | Reference |
|---|---|---|---|---|---|
| Male Hartley guinea pigs | metyrapone | 2.00 | 116 | | OESCH et al. (1971 c) |
| | cyclohexanol | 2.00 | 20 | | |
| | α-tetralone | 2.00 | 15 | | |
| | glycidol | 2.00 | 12 | | |
| | 2-cyclohexen-1-ol | 2.00 | 9 | | |
| Male Sprague-dawley rats | chalcone | 1.87 | 310 | compounds containing an aryl carbonyl, substituted with additional hydrophobic groups, are more effective stimulators | GANU and ALWORTH (1978) |
| | 9-fluorenone | 0.49 | 128 | | |
| | benzophenone | 0.49 | 93 | | |
| | metyrapone | 0.49 | 79 | | |
| | diphenylcyclopropenone | 0.49 | 79 | | |
| | propylphenylketone | 0.12 | 68 | | |
| | Benzylphenylketone | 1.87 | 60 | | |
| Male Sprague-dawley rats | isoquinoline | 1.00 | 485 | compounds with a bicyclic structure incorporating a pyridine nitrogen β to ring fusion, are most potent | VAZ et al. (1981) |
| | harman | 1.00 | 390 | | |
| | norharman | 1.00 | 304 | | |
| | ellipticine | 0.05 | 265 | | |
| | 4-azafluorene | 1.00 | 141 | | |
| | 4-phenylpyridine | 1.00 | 71 | | |
| | pyridine | 1.00 | 47 | | |
| | quinoline | 1.00 | 14 | | |
| Male Sprague-dawley rats | benzil | 1.60 | 539 | both phenyl and keto groups are required for activation, and substitution of nitrogen in the ring abolishes activation | SEIDEGÅRD and DE PIERRE (1980) |
| | chalcone oxide | 1.60 | 294 | | |
| | metyrapone | 1.60 | 244 | | |
| | chalcone | 1.60 | 201 | | |
| | desoxybenzoin | 0.80 | 172 | | |
| | benzylacetophenone | 1.60 | 157 | | |
| | dibenzoylmethane | 1.60 | 133 | | |
| Male Sprague-dawley rats | clotrimazole | 1.60 | 690 | | SEIDEGÅRD et al. (1986) |
| | benzil | 1.60 | 451 | | |
| | chalcone oxide | 1.60 | 272 | | |
| | metyrapone | 1.60 | 226 | | |

Epoxide Hydrolases                                                              301

**Table 9.** In vitro activation of purified fractions of liver mEH$_b$

| Species | Activator | mM | Substrate | Activation (%) Rat | Human | Guinea-pig | Reference |
|---|---|---|---|---|---|---|---|
| Male Sprague-Dawley rats, male guinea pigs (undetermined strain). Human liver autopsy samples | metyrapone 1-(2-Isopropylphenyl)-imidazole 1-(2-Cyanophenyl)-imidazole | 2.0 2.0 2.0 | STO STO STO | 87 76 76 | 87 76 57 | 74 162 61 | Oesch (1974) |
| PB-treated, male Long evans rats | metyrapone 1-(2-Isopropylphenyl)-imidazole | 1.5 1.5 | STO STO | 220 220 | | | Lu et al. (1975) |
| PB-treated, male Long evans rats | metyrapone | 6.0 | octene 1,2-oxide BA 5,6-oxide BPO BP 11,12-oxide DBA 5,6-oxide | 210 40 0 −60 −60 | | | Levin et al. (1978b) |
| Male Sprague-dawley rats: untreated, PB-, 3 MC- and TSO-treated, various human liver autopsy samples | metyrapone | 6.0 | STO BP 7,8-oxide DBA 5,6-oxide 3-MC 11,12-oxide BA 5,6-oxide | 0 to 150 −50 to +70 −30 to +50 −80 to +20 −40 to −10 | | | Guengerich et al. (1979b) |
| PB-treated, male Long evans rats, one human liver autopsy sample | metyrapone | 15.0 | STO octene 1,2-oxide BPO BP 11,12-oxide DBA 5,6-oxide | 230 189 4 −80 −83 | 161 82 48 −77 −79 | | Lu et al. (1979) |

| male sprague-dawley rats | | | | | SEIDEGÅRD et al. (1986) |
|---|---|---|---|---|---|
| metyrapone | 1.60 | STO | 97 | | |
| | 0.04 | BPO | 8 | | |
| | 0.11 | OE | −12 | | |
| | 0.28 | AE | −1 | | |
| chalcone oxide | 1.60 | STO | 107 | | |
| | 0.04 | BPO | 4 | | |
| | 0.11 | OE | −5 | | |
| | 0.28 | AE | −9 | | |
| benzil | 1.60 | STO | 166 | | |
| | 0.04 | BPO | 3 | | |
| | 0.11 | OE | 5 | | |
| | 0.28 | AE | −14 | | |
| clotrimazole | 1.60 | STO | 233 | | |
| | 0.04 | BPO | −17 | | |
| | 0.11 | OE | −56 | | |
| | 0.28 | AE | −85 | | |

SEIDEGÅRD and DE PIERRE, (1980) have shown that when microsomes from male Sprague-Dawley rats pretreated with various inducers, are activated by benzil, metyrapone and chalcone oxide, only small differences can be observed between the resultant activities towards STO. In contrast to this, whilst STO activity is stimulated (500—600%) in rat, mouse, rabbit and human liver microsomes, clotrimazole reduces androstene oxide (AE) activity to 10—40% of the control activity (SEIDEGÅRD et al., 1986): a similar effect, although not so extreme, is also observed with oestroxide (OE) activity. This illustrates, once again, the substrate-dependence of various phenomena and may in contrast to in vivo induction studies indicate multiple forms of $mEH_b$.

Table 9 illustrates the effects of various activators on purified $mEH_b$ and it is immediately apparent that the degree of activation is substrate-dependent. With STO and octene 1,2-oxide the enzyme is consistently activated, whilst the activities with steroids and polycyclic aromatic hydrocarbons have a reduced tendency towards activation and are in some cases inhibited. Possibl, this reflects a reduced affinity of the bulkier molecules, due to the binding of the modulator substance to a neighbouring site, whilst the smaller substrates, e.g. STO, remain unaffected: alternatively, the existence of multiple forms, which are differentially modulated by the activators, is also possible. Also, the relatively small species differences indicate similarities between the purified enzyme from the various sources.

Whilst sonication, freeze-thawing and KCl concentrations up to 100 mM, have little or no effect on $mEH_b$ activity (BURCHELL et al., 1976), the effect of detergents is unclear. In contrast to BURCHELL et al. (1976), who report a 30—60% activation of STO activity from male Sprague-Dawley rats, with neutral or anionic detergents (1—2 g: 1 g protein), BULLEID and CRAFT (1984) report an inhibition of STO activity from male Wistar rats pretreated with diethylnitrosamine, with both Lubrol and sodium cholate, as do HALPERT et al., (1979) using Renex 690 and phenobarbital-induced rabbit liver microsomes. Additionally, BULLEID et al. (1986) observed almost complete inhibition of purified rat liver microsomal STO-activity by 0.02% (w/v) Lubrol PX. Furthermore, VOGEL-BINDEL et al. (1982) observed inhibition of the activity towards AE and OE with both ionic and non-ionic detergents (1%, v/v), with only sodium cholate preserving the activity. Also, whilst activation by various substances in vitro is unaffected by concentrations of detergents less than 1%, VAZ FIORICA and GRIFFIN (1981) have reported a reduction in ellipticine-activation of $mEH_b$, in the presence of 2% sodium cholate. Therefore, the effects of detergents are complex, being dependent on the type and concentration of detergent used and the substrate used for activity determination. Generally however, ionic detergents such as sodium deoxycholate tend to preserve epoxide hydrolase activity better than non-ionic detergents such as Cutscum (TIMMS, 1987). For a discussion of the reason for this and of the physico-chemical characteristics of detergent-solubilization of membranes, the reader is referred to HELNIUS and SIMONS (1975) and LICHTENBERG et al. (1983).

Care must also be taken in the choice of solvents for both substrates and effectors: for example, both tetrahydrofuran (HALPERT et al., 1979; SEIDEGÅRD and DE PIERRE, 1980) and ethanol (OESCH, 1974; HALPERT et al., 1979) activate $mEH_b$, although LU et al. (1975) claimed that ethanol inhibited the reaction rate of purified rat liver $mEH_b$. In contrast, the enzyme activity is relatively insensitive to acetone, acetonitrile and dimethylsulphoxide (OESCH, 1974; SEIDEGÅRD and DE PIERRE, 1980).

Inhibition of $mEH_b$: The systematic studies investigating the inhibition of $mEH_b$ activity in guinea-pig (OESCH et al., 1971c), humans (OESCH, 1974) and insects (BROOKS, 1973), all gave similar results. Competitive inhibitors include oxiranes with a 1-aryl or 1-alkyl substituent, juvenile hormone epoxides, and recently, the antiepileptic drug valpromide (PACIFIC et al., 1986; ROBBINS et al., 1990) and a series of cyclopropyl oxiranes which have been shown to be both competitive and reversible inhibitors of $mEH_b$ activity (PRESTWICH et al., 1985). Non-competitive inhibitors include 1,2,3,4-tetrahydronaphthalene-1,2-epoxide and cyclohexene oxide, whilst 3,3,3-trichloropropane-1,2-epoxide inhibits uncompetitively (OESCH et al., 1971c). However, as observed with in vitro activators, substrates can play an important role and LEVIN et al. (1978b) observed differential inhibition of $mEH_b$ activity using cyclohexene oxide. So did GUENTHNER (1986a), who found a potent noncompetitive inhibition of mouse liver microsomal trans-stilbene oxide hydrolase activity by 4-phenylchalcone oxide and 4'-phenylchalcone oxide, while microsomal benzo[a]pyrene 4,5-oxide hydrolase activity remained essentially unchanged. Conversely, compounds that are potent inhibitors of benzo[a]pyrene 4,5-oxide hydrolase, including cyclohexene oxide and TCPO, inhibited microsomal trans-stilbene oxide hydrolase only at very high concentrations.

Mercury and zinc are non-competitive inhibitors of $mEH_b$ (PARKKI, 1980) and although nickel, cobalt, chromium and lead salts are inactive, cadmium iodide is a potent inhibitor, competitively inhibiting STO activity (AITIO et al., 1978).

### 4.5.2. Modulation of cEH activity

Induction: All inducers of cEH discovered until now, are also peroxisome, proliferators, including hypolipidemic agents, e.g. clofibrate, plasticizers-e.g. di-(2-ethylhexyl)-phthalate (HAMMOCK and OTA, 1983: WAECHTER et al., 1984) and pesticides (MOODY and HAMMOCK, 1987). Indeed, a recent study showed that strong peroxisome proliferators e.g. fenofibrate and tiadenol are good inducers of rat liver cEH and weak peroxisome-proliferators, e.g. acetylsalicylic acid, 1-benzylimidazole, cause only a mild induction of cEH,

---

leading the authors to suggest that cEH induction may be related to peroxisome-proliferation (SCHLADT et al., 1987; STEINBERG et al., 1988). Additional evidence was provided in the same studies for the concomitant induction of both cEH and peroxisomal $\beta$-oxidation.

Although there is a clear differentiation in the types of compounds which induce $mEH_b$ and cEH activities, the induction is not quite mutually exclusive as evidenced by the weak induction of rat liver cEH by phenobarbital (HAM-MOCK and OTA, 1983). Additionally, cEH inducers, e.g. clofibrate, simultane-ously induce $mEH_b$ activity, significantly in mice (LOURY et al., 1985), and at a very much reduced level in rats (SCHLADT et al., 1987). Induction of cEH activity, measured using trans-stilbene oxide as substrate, has been observed in the liver and kidneys, but not testis, of rat (MOODY et al., 1986; SCHLADT et al., 1987) and mouse (LOURY et al., 1985) and characterization of this in-creased activity has shown it to be identical with the enzyme from control animals (PRESTWICH and HAMMOCK, 1985; LOURY et al., 1985; MEIJER and DE PIERRE, 1985a).

**In vitro modulation of cEH activity:** the first inhibition characteristics of cEH were investigated by MUMBY and HAMMOCK (1979a) using alkyl epoxides of 6,7-epoxygeranylphenyl ether and in 1982, chalcone oxide (trans-1-benzoyl-2-phenyloxirane) was shown to be an optimal inhibitory structure which, it was proposed, reacted covalently with an essential cysteine residue (MULLIN and HAMMOCK, 1982). Further work showed various chalcone oxide derivatives to be selective and potent inhibitors of mouse liver cytosolic epoxide hydrolase (MIYAMOTO et al., 1987), while cyclopropyl oxiranes are merely moderate inhibitors not only of cEH but also of $mEH_b$ (PRESTWICH et al., 1985a); indeed, recent work has shown the importance of the $\alpha,\beta$-epoxyketone moiety for the inhibitory potency of these compounds (PRESTWICH et al., 1985). An inhibition of cEH activity has also been observed in the presence of non-ionic detergents such as Lubrol PX and Emulgen 911 (SCHLADT, et al., 1987).

cEH also exhibits greater sensitivity to solvents than $mEH_b$, with aceto-nitrile, isopropanol, dimethylformamide, dioxane, tetrahydrofuran and ethyl acetate being inhibitory, but ethanol, methanol and dimethyl sulphoxide being inactive (MULLIN and HAMMOCK, 1982; MEIJER and DE PIERRE, 1985b; SCHLADT et al., 1987, 1988b). So far there has been no report of an in vitro activation of cEH.

### 4.5.3. Modulation of $mEH_{ch}$ activity

At present, neither in vivo inducers nor in vitro activators have been found for this activity and indeed some of the $mEH_b$ inducers have been shown to inhibit $mEH_{ch}$ activity in rats (LEVIN et al., 1983). The classical $mEH_b$ inhibitor 3,3,3-trichloropropane-1,2-epoxide has no effect on $mEH_{ch}$

H. THOMAS; C. W. TIMMS; F. OESCH

whilst slightly inhibiting cEH. The sterole imine 5,6α-imino-5α-cholestan-3β-ol alternatively completely inhibits mEH$_{ch}$, whilst having no effect on either mEH$_b$ or cEH activities and therefore has been used diagnostically in the identification of mEH$_{ch}$ (Watabe et al., 1983; Oesch et al., 1984). Also of interest, are an apparent product inhibition by cholestane triol, (Sevanian and McLeod, 1985), as well as inhibitory cholesterol oxidation products such as 7-ketocholesterol, 6-ketocholestanol, 7-ketocholestanol (Sevanian and McLeod, 1986) and 7-dehydrocholesterol 5,6β-oxide (Nashed et al., 1985). Furthermore, mEH$_{ch}$ activity in male mouse liver microsomes was inhibited 30—40% with sodium cholate and Triton WR-1339 and 100% with Emulgen 108 and 911, and Lubrol PX (Watabe et al., 1986b).

Although mEH$_{ch}$ activity is stable against repeated freezing and thawing, it is inhibited by acetone, methanol, tetrahydrofuran, dimethylsulphoxide and various concentrations of acetonitrile; however, the best reaction rates are obtained with 6.25% acetonitrile (Levin et al., 1983).

There have been no reports so far of an induction, inhibition or activation of leukotriene A$_4$ hydrolase.

# 5. Distribution of epoxide hydrolases

## 5.1. Subcellular localization

### 5.1.1. Subcellular localization of mEH$_b$

One of the most informative cell subfractionations of rat liver with respect to mEH$_b$ was reported by Stasiecki et al. (1979), using BPO and STO as substrates. They found the highest specific activity in smooth endoplasmic reticulum (ER), which they designated 100%, with activities in the other fractions ranging from 80% in rough ER, to approximately 30% in total nuclei and 18—23% in nuclear membranes. mEH$_b$ activity in Golgi apparatus and plasma membranes ranged from 18—25%, with mitochondrial fractions showing negligible activity (Stasiecki et al., 1980). Similar results have been obtained by Waechter et al. (1982a), using immuno-electron-microscopy, as well as changes in the extent of labelling and enzyme activity of the corresponding subcellular fractions upon induction of the animals with phenobarbital and 2-acet-aminofluorene. However, in contrast to these results, Galteau et al. (1985) reported that purified antibodies, raised against mEH$_b$, exclusively stained smooth ER a well-developed reticulated network in various cell lines.

One of the most striking features from the above subfractionation, is the relatively large amount of mEH$_b$ activity associated with the nuclear fraction, which Fahl et al. (1978) and Bresnick et al. (1977b) as well as Moody et al. (1988) showed, using marker enzymes, could not be explained as microsomal contamination. The initial observation that an epoxide hydrolase activity exists in the nuclear fraction had been made by JERNSTRÖM et al. (1976) and much work was then invested in further characterization of this activity. Several workers have compared the induction and inhibition characteristics of this nuclear activity to those of mEH$_b$ (BORNSTEIN et al., 1979; GONTOVNICK and BELLWARD, 1981; GONTALEZ and KASPER, 1981; WAECHTER et al., 1982a; MUKHTAR et al., 1980; GAZOTTI et al., 1981a): however, although quantitative differences were observed qualitative differences have not been reported.

Indeed, GONTOVNICK and BELLWARD (1981) have proposed that the quantitative differences may be explained by a decreased rate of incorporation of newly synthesized enzyme molecules into the nuclear membrane, as compared to the endoplasmic reticulum. Also, MUKHTAR et al. (1980) have drawn attention to the differential membrane environments in the subcellular compartments and the consequences that this may have on the enzyme kinetics. Despite this, BORNSTEIN et al. (1979) observed less than 10% differences in the relative rates of metabolism of 10 polycyclic aromatic hydrocarbons between rat nuclei and microsomes. Immunoaffinity purified epoxide hydrolase preparations from both fractions gave similar molecular weights (approx. 49,000 daltons), and immunological identity has been reported by THOMAS et al. (1979), BENTLEY et al. (1979) and WAECHTER et al. (1982a). Also, GONZALEZ and KASPER (1981a) showed that after phenobarbital administration the induced levels of the epoxide hydrolase messenger RNAs associated with the nuclear envelope rapidly decline concomitantly with their accumulation in rapidly sedimenting endoplasmic reticulum.

Therefore, the available evidence indicates that a form of epoxide hydrolase, either identical or very similar to mEH$_b$, exists in the nuclear fraction of rat liver. The existence of such an enzyme is understandable, considering the proximity of DNA, and the favourable substrate-specificity for the epoxides of lipophilic polycyclic aromatic hydrocarbons. Indeed, the covalent binding of such metabolites to DNA is probably a crucial step in the carcinogenic process (JERNSTRÖM et al., 1976). However, the ratio of epoxide hydrolase to monooxygenase activity must also be considered and GONTOVNICK and BELLWARD (1981), have calculated a ratio of 4.5 in nuclei (compared to 1.5 in microsomes), indicating that ample hydrolase is present to metabolize any epoxides, although the danger of secondary activation may not be ignored. Also of toxicological interest is the late ontogenic development of nuclear epoxide hydrolase (ROMANO et al., 1983), and the characteristic of biphasic kinetics in human adult liver samples (PACIFICI and RANE, 1982). The recent discovery by GUENTHNER (1986b) of a nuclear envelope-associated epoxide hydrolase in mouse liver, with similar characteristics to cEH, is perhaps of relevance to this latter observation.

H. THOMAS; C. W. TIMMS; F. OESCH

### 5.1.2. mEH$_b$ in cytosol

In 1982, GILL et al. made the interesting observation that low levels of mEH$_b$ could be immunologically detected in rat liver cytosol, which constituted approximately 0.02% of total cytosolic protein. The amount of antigen in cytosol is inducible by typical mEH$_b$ inducers (KIZER et al., 1985), and has also been found in rhesus monkey, as well as both normal and neoplastic human liver samples (GILL et al., 1983b), and human lung samples (GUENTH-NER and KARNEZIS, 1986a) predominantly in the bronchial epithelium (GUENTH-NER and KARNEZIS, 1986b). In each case, the specific activity was between 1 and 5% of the microsomal-bound activity and exhibited similar substrate-specificity, inhibition and induction characteristics as the native mEH$_b$ in the microsomes.

Therefore, it would seem that three possible explanations can be suggested for the existence of this mEH$_b$-form in the cytosolic fraction. Firstly, mEH$_b$ is not totally membrane-bound and a certain amount exists under 'normal' conditions in the soluble fraction. Secondly, mEH$_b$ is totally membrane-bound but can be released under certain conditions (unknown at the moment), apparently particularly so in human fractions where relatively high levels have been found (GILL et al., 1983b). Thirdly, the enzyme is only immuno-logically and biochemically similar to mEH$_b$, and actually constitutes a new form. However, although no decisive evidence has been presented for either of the three proposals, as more similarities in the two mEH$_b$-forms character-istics are discovered the third proposition becomes more likely.

### 5.1.3. Particulate activity similar to cEH

Various reports have presented evidence for a mitochondrial epoxide hydrolase activity in male Swiss-Webster mouse liver towards trans-stilbene oxide and trans-ethyl styrene oxide, with similar substrate-specificity, molecular weight (GILL and HAMMOCK, 1981; GILL, 1985; KAUR and GILL, 1986) isoelectric point and immunological characteristics (GILL, 1984), to cEH. However, using somewhat more sensitive techniques, e.g. isopycnic centrifugation, WAECHTER et al. (1983) proposed that some, and perhaps all, of this "mitochondrial" activity is indeed peroxisomal in origin: also, similar evidence has been provided for exclusively peroxisomal activity in mouse liver (PATEL et al., 1986). In the meantime there seems to be substantial evidence for a very close relationship if not identity of mouse liver peroximal epoxide hydrolase with cEH (MEIJER and DE PIERRE 1988a; JOSTE and MEIJER, 1989). Our laboratory has recently concentrated on this activity in rat liver and also found exclusive localization in peroxisomes with negligible activity in the mitochondria, using differential gradient centrifugation in sucrose and Nycodenz (TIMMS et al., 1988). In addi-tion, when the purified peroxisomal fractions were investigated for immunolo-

---

gical reactivity with antibodies raised against purified rat liver mEH$_t$ (Fig. 2), additional minor bands were observed in the proximity of the main band A, which has exactly the same molecular weight as rat liver purified mEH$_b$. Also, a significant second band, B, with a higher molecular weight was observed, the intensity of which correlated well with the peroxisomal marker enzymes activities, indicating the possible presence of multiple epoxide hydrolases in peroxisomes.

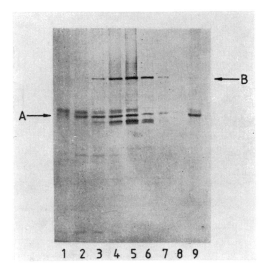

**Fig. 2.** Western blot analysis of peroxisomal fractions with antibodies against purified rat liver mEH$_b$.
Following zonal centrifugation in a nycodenz gradient consecutive peroxisomal fractions (lanes 1—9) with a maximum in lane 5 were analysed by Western blotting using polyclonal antibodies against rat liver mEH$_b$. Besides the band of molecular weight 49,000 Da corresponding to purified rat liver mEH$_b$ (A) minor immunoreactive bands with molecular weights slightly higher and lower than 49,000 Da were observed as well as a band of a molecular weight around 80,000 Da (B) indicating a possiblem ultiplicity of epoxide hydrolases in peroxisomes.

As described in earlier sections, an activity towards trans-stilbene oxide was found in the mouse liver microsomal fraction which could not be removed by multiple washings (GUENTHNER and OESCH, 1983), and has been shown to have similar substrate-specificity, induction and inhibition characteristics as cEH (GUENTHNER, 1986a). This activity has also been found in the microsomal fraction of human lung (GUENTHNER and KARNEZIS, 1986a), and it is reasonable to assume that this activity can be assigned to peroxisomal and/or mitochondrial contamination, as described above.

### 5.1.4. Localization of $mEH_{ch}$ activity

Exclusive compartmentation of this activity in the microsomes has been proposed by WATABE et al. (1983), LEVIN et al. (1983), and OESCH et al. (1984), whilst ASTRÖM et al. (1986b) have specifically localized it in both rough and smooth ER. In contrast, a small but significant amount of activity in the cytosol has been reported by WATABE and SAWAHATA (1979) and SEVANIAN et al. (1980, 1981).

### 5.2. Localization within liver regions and distinct liver cell populations

Immunohistochemical studies of the distribution of epoxide hydrolases within the liver, have been limited so far to $mEH_b$. BENTLEY et al. (1979) showed more extensive staining in parenchymal, as compared to non-parenchymal cells in rat liver, and this has been confirmed recently by measuring $mEH_b$ enzyme activity (BPO and CSO as substrates) in rat (LAFRANCONI et al., 1986; STEINBERG et al., 1988) and mouse liver (STEINBERG et al., 1987). Localization within the rat liver has been found to follow the order: Centrilobular > Midzonal $\geq$ Periportal (BENTLEY et al., 1979; BARON et al., 1980; REDICK et al., 1980).

Further work from KAWABATA et al. (1981) showed no sex-differences and uniform staining patterns within the left, right, median and caudate liver lobes. Conflicting evidence has been presented about the effects of various inducers, the general tendency being that the distribution of $mEH_b$ within the lobules becomes more uniform (BENTLEY et al., 1979; KAWABATA et al., 1981; WOLF et al., 1984).

The localization of enzymes is of toxicological importance for activation and deactivation of xenobiotics. For example, when epoxide hydrolases and glutathione-S-transferases are increased upon induction in the same areas as P-450 isozymes, e.g. P-450IA1, a favourable situation exists for deactivation. In contrast, differential localization of the deactivating enzymes and a P-450 isozymes, e.g. P-450IA2, creates a situation where toxic metabolites are produced but cannot be locally deactivated (WOLF et al., 1984).

### 5.3. Organ distribution

The organ-distribution of $mEH_b$ has been reviewed by OESCH (1979), and so only a summary will be given here with emphasis on multiple forms. OESCH et al. (1977) have demonstrated the ubiquity of $mEH_b$ in rat organs and interestingly an exceedingly high activity in the testis and ovary, as compared to liver, of NMRI mice. In contrast, the hamster has 100-fold lower activity in testis than in liver and 30-fold lower than in kidney and lung: in addition, the kidney activity is 50-fold higher than in the mouse. The significance of

---

**Table 10 a.** Species differentiation in organ distribution of mEH$_b$ activity

| Species | Organ distribution | Substrate | Reference |
|---|---|---|---|
| Rabbit: | | | |
| New Zealand | liver>testis>kidney>lung | BPO, HEOM | WALKER et al. (1978) |
| White | liver>testis>lung>kidney | STO | |
| Rat: | | | |
| Sprague-Dawley | liver>testis>kidney>lung | BPO | OESCH et al. (1977) |
| | liver>testis>kidney>lung | CSO | MOODY et al. (1986) |
| | liver>testis>kidney>lung | CSO | ÅSTRÖM et al. (1986a) |
| | liver>kidney>testis | BPO | MOODY et al. (1986) |
| Mouse: | | | |
| Swiss-Webster | testis>liver>lung>kidney | CSO, BPO | LOURY et al. (1985) |
| C57 BL | testis>liver>lung>kidney | CSO | DE PIERRE et al. (1983) |
| NMRI | testis>liver>lung>kidney | BPO | OESCH et al. (1977) |
| DBA/2 | liver > testis > lung > kidney | STO | WAECHTER et al. (1988) |

these results with respect to possible endogenous roles will be discussed later. Also, the use of several substrates has shown differential organ distribution in various species (Table 10a).

Differential induction has also been observed, particularly in rat. For example, phenobarbital (DE PIERRE et al., 1983), Aroclor 1254 (OESCH et al., 1977) and TCDD (2,3,7,8-tetrachlorodibenzo-p-dioxin) (PARKKI and AITIO, 1978) induce only liver mEH$_b$ activity: in contrast, trans-stilbene oxide (OESCH and SCHMASSMANN, 1979) induces both liver and kidney activities whilst 1,2-dibromo-3-chloropropane (SUZUKI and LEE, 1981) induces liver, prostate, testis and kidney activities. In mouse, clofibrate and nafenopin induced cis-stilbene oxide and styrene oxide activity in both liver and kidney, but BPO activity was only increased in the liver (LOURY et al., 1985; WAECHTER et al., 1988). Finally, VOGEL-BINDEL et al. (1982) observed differential ontogeny for mEH$_b$ in rat testis, using BPO, OE and AE as substrates, as compared to liver and ovary. However, despite these various results, it remains unclear if they are indications of organ-specific forms of mEH$_b$ or if they constitute artefacts resulting from different membrane environments.

The immunological characteristics of the enzyme in various organs is also unclear: HASANI et al. (1979) have claimed that an immunological identity exists between rat liver, kidney, and lung mEH$_b$, whilst GUENGERICH, WANG, MASON and MITCHELL (1979a) obtained no reaction between liver, testis and lung in an Ouchterlony immunodiffusion experiment, whilst observing two

**Table 10b.** Species differentiation in organ distribution of cEH activity

| Species | Organ distribution | Substrate | Reference |
|---|---|---|---|
| Rabbit: | | | |
| New Zealand White | liver>kidney>duodenum >lung | Ethyl epoxide | GILL and HAMMOCK (1980) |
| Rat: | | | |
| Sprague-Dawley | kidney>heart>liver>brain >lung>testis>spleen | TSO | SCHLADT et al. (1986) |
| | kidney>liver>testis | TSO | MOODY et al. (1986) |
| Mouse: | | | |
| Swiss-Webster | liver>kidney>lung>testis | Ethyl epoxide | GILL and HAMMOCK (1980) |
| | liver>kidney>lung>testis | TSO | LOURY et al. (1985) |
| C57 BL | liver>kidney>testis>lung | TSO | DE PIERRE et al. (1983) |
| DBA/2 | liver > kidney > heart > testis > lung | TSO | WAECHTER et al. (1988) |

lines of reaction with kidney microsomes. This result is complicated, however, by the additional observation that 90% of the STO activity could be immuno-precipitated from solubilized microsomes prepared from all four organs using the same antibodies as above. Unfortunately, the authors provide no explanation for this phenomenon.

In addition to the normal organ distribution of epoxide hydrolase activities, $mEH_b$ has also been detected in various cell populations: for example, in human lymphocytes (GLATT et al., 1980; GLATT et al., 1983), human blood resting mononuclear leucocytes (SEIDEGÅRD et al., 1984), adult human hepato-cytes (RATANASAVANH et al., 1986), human and rodent established cell lines (BROWN and CHALMERS, 1986; CHALMERS and BROWN, 1987), rat hepatocytes and hepatoma cell lines (RAZZOUK et al., 1985) and monolayer cultures of human fetal hepatocytes (PENG et al., 1984). The last two studies have present-ed results showing differential stability and induction of BPO and STO acti-vities in the corresponding cell cultures.

cEH activity has been found in all extrahepatic tissues investigated until now (Table 10b). Within the species differences in organ distribution of prati-cular interest is the high activity in rat kidney and heart, as compared to liver (SCHLADT et al., 1986). Also, the reduced relative testicular activity, as compared to $mEH_b$, may be of significance, and will be discussed later.

Whilst $mEH_{ch}$ activity has been measured in rat liver, kidney, lung, testis, spleen, brain and intestinal epithelium, liver microsomal activity was at least 5-fold greater than that of the other organs (ÅSTRÖM et al., 1986b).

## 5.4. Species and strain differences

Epoxide hydrolase activity has been found in all organisms tested, including fungi (KOLATTUKUDY and BROWN, 1975), fruits and vegetable (CROTEAU and KOLATTUKUDY, 1975), insects (BROOKS et al., 1970; MULLIN and CROFT, 1984; JANSEN et al., 1986), fish (JAMES et al., 1979; BALK et al., 1980), birds (WALKER et al., 1987), primates (PACIFICI et al., 1983a) and humans (GUENGERICH et al., 1979b; LU et al., 1979) as well as a series of established human and rodent cell lines (CHALMERS and BROWN, 1987). However, the activity can only be classified as $mEH_b$ in the higher forms of life where the liver microsomal specific activity follows the sequence: larger mammals > rodents > birds > amphibia > fish. Between the species the ratios of the specific activities towards BPO, STO and HEOM do vary, possibly indicating the existence of multiple forms of $mEH_b$, although the evidence is by no means conclusive (WALKER et al., 1978). With CSO as substrate the livers of experimental animals show the following levels of specific activity of $mEH_b$ compared to man (MEIJER et al., 1987)

monkey > guinea pig > human ∼ rabbit > hamster > rat > mouse

With the advent of ELISA (enzyme-linked immunosorbent assay), the immunological relationships between $mEH_b$ from various species could be accurately characterized and the following trend observed: increasing immunological differentiation from rat $mEH_b$ is observed in the order rat ∼ mouse ∼ syrian hamster < guinea-pig < rabbit < monkey ∼ human (THOMAS et al., 1982; WOLF et al., 1983; LEVIN et al., 1978b).

However, exceptions to the above sequence have been reported; for example, a strong reaction of identity between anti-rat $mEH_b$ and purified human $mEH_b$, (TELAKOWSKI-HOPKINS et al., 1983), and a double immunoprecipitation line produced in immunodiffusion, when anti-rat $mEH_b$ raised in goat was incubated with rat microsomes (LEVIN et al., 1978b). These results are difficult to explain, but contamination of the antibodies preparations is a possible explanation. The immunological characterization of $mEH_b$ has also shown that the antigenic determinants in the enzyme molecule are not essential for catalytic activity, as evidenced by the absence of inhibition of this activity after immunoprecipitation (LEVIN et al., 1978b; WOLF et al., 1983).

Species-dependent differential chromatography characteristics of $mEH_b$, have also been reported during purification: HASANI et al., (1979) stated that human and guinea-pig liver microsomal epoxide hydrolase behaved differently from that of rat and mouse during DEAE-cellulose chromatography, which they explained by differences either in the primary sequence or tertiary conformation. GUENGERICH et al. (1979b) have also noted significant differences in substrate-specificity and amino acid composition between purified multiple forms of $mEH_b$ from humans and rat. Furthermore, WALZ et al. (1983) have reported the presence of two forms of $mEH_b$ in hamster liver microsomes, with identical peptide maps and approximately 90% homology with rat $mEH_b$.

A systematic study of 22 rat strains revealed an autosomal and codominant inheritance of $mEH_b$ activity, with only quantitative differences being observed (OESCH et al., 1983b) as was also the case in six strains of mice (WALKER et al., 1978). However, LYMAN et al. (1980) have presented evidence for a genetic polymorphism of $mEH_b$ in the mouse, which is inherited as an autosomal trait, with co-dominant expression in heterozygotes: on the basis of pH optima and heat-sensitivity they separated 26 inbred strains into two phenotypic classes, with C57 BL/6J and DBA/2J mice being representative for each of the two groups. Coincidentally, these are the same two strains from which LEVIN et al. (1978b) showed the $mEH_b$ to be immunologically identical. Furthermore, whilst WAECHTER et al. (1984) observed $mEH_b$ induction by nafenopin in these two strains and Balb c, no corresponding induction could be recorded in C3H mice. Therefore, both inter-species and inter-strain differences in $mEH_b$ are significant, and should not be ignored when comparing the enzyme from various sources.

Although the species distribution of cEH has not been as thoroughly investigated, it has been found to be present in most experimental animals including fish (BALK et al., 1980), rodents (SCHLADT et al., 1986; GILL, 1983), rabbits (WAECHTER et al., 1982b) and humans (WANG et al., 1982). The following ranking of the specific activity of liver cEH, using TESO, has been observed:  mouse > hamster > rabbit > human > guinea-pig > rat  (GILL and HAMMOCK, 1980; MULLIN and HAMMOCK. 1980; WANG et al., 1982; MEIJER et al., 1987).

In general, only one form of cEH has been discovered in most species, although WANG et al. (1982) as well as SCHLADT et al. (1988a) have purified two forms from human liver. However, interspecies differences have been observed; for example, phenobarbital induced rat liver, but not mouse liver cEH (HAMMOCK and OTA, 1983). However, induction of rat liver cEH by phenobarbital was not observed in a recent study (SCHLADT et al., 1986). The study of HAMMOCK and OTA (1983) also reported quantitative inter-strain differences in mice, although WAECHTER et al. (1984) observed no differences in induction with nafenopin. Quantitative differences in specific activity have also been found between individuals of outbred Sprague-Dawley rats. Interindividual differences varied by a factor of 38.5 whereas cEH activity of inbred Fischer 344 rats varied only by a factor of two. The mean specific activity of both rat strains was nearly the same (SCHLADT et al., 1986).

$mEH_{ch}$ activity has been measured in the mouse (WATABE et al., 1983), rat (LEVIN et al., 1983; ASTRÖM et al., 1986), rabbit (OESCH et al., 1984) and cow (WATABE et al., 1981); there is however, at the moment, no evidence for a multiplicity of $mEH_{ch}$, either within, or between species.

## 5.5. Sex differences and inter-/intra-individual variation

In both mouse (GILL and HAMMOCK, 1980; HAMMOCK and OTA, 1983) and rat (BINDEL et al., 1979; OESCH et al., 1983 b) liver as well as the livers of several other species including hamster, guinea pig, rabbit, monkey and man (MEIJER et al., 1987) the level of $mEH_b$ activity is approximately 50% higher in males than in females; similarly, cEH activity is $50-200\%$ higher in males (HAMMOCK and OTA, 1983). In extrahepatic organs, lung and adrenal $mEH_b$ activity is the same in both sexes, and kidney activity is even slightly greater in the females than in the males. A substrate-specificity might also exist, as CRAVEN (1977) has reported higher female rat liver $mEH_b$ activity towards HEOM than in the male. Induction studies however have not produced different results between the sexes using various inducers (OESCH and SCHMASSMANN, 1979; GONTOVNICK and BELLWARD, 1980).

In contrast to the situation in rat and mouse, no significant sex-differences could be observed in $mEH_b$ activity in human lymphocytes (GLATT et al., 1980) and liver (CRAVEN et al., 1982; MERTES et al., 1985). However, the intra- and inter-individual differences must be taken into account, particularly with human samples, and they can be considerable. Both GLATT et al. (1980) and SEIDEGÅRD et al. (1984) have reported variation in $mEH_b$ activity from native lymphocytes and the mononuclear blood fraction from the same donors, although in the latter case the variation was significantly less, possibly as a result of the greatly reduced experimental time period. In comparison to these variations however, the inter-individual differences are far more significant; for example, GLATT et al. (1980) have reported statistically significant differences in $mEH_b$ activity of various lymphocyte samples, OESCH et al. (1980b) observed a 5-fold difference between the maximal and minimal specific activities of $mEH_b$ in human lung samples, SEIDEGÅRD et al. (1984) obtained differences by a factor 2.5 in the mononuclear blood fraction and MERTES et al. (1985) a factor of 63 in human liver microsomes. Interestingly, the interindividual variation in lymphocytes could be greatly decreased by activation and mitofen-stimulation of the cells (GLATT et al., 1980).

In contrast, variation by a factor of five has been reported for cEH activity in the mononuclear blood fraction (SEIDEGÅRD et al., 1984) and a factor of 539 in human liver (MERTES et al., 1985). Therefore, it appears that the inter-individual variation is much larger for cEH than $mEH_b$, and this is also reflected in the situation in rat liver, where the variation is almost 3-fold greater for cEH than $mEH_b$ (SCHLADT et al., 1986).

## 5.6. Ontogeny

$mEH_b$ has been shown to be present in fetal rat liver, adrenals, skin, ovaries and testis, using various substrates (STOMING and BRESNICK, 1974; MUKHTAR and BICKERS, 1981; MUKHTAR et al., 1978a, b). Activity has also been observed in the human fetus (PACIFICI and RANE, 1982), and fetal liver cell cultures, this latter activity being inducible, using classical in vivo inducers (GOUJON et al., 1980; PENG et al., 1984). Also, neonatal rat liver activities were induced in vivo by application of Aroclor 1254 to skin (MUKHTAR and BICKERS, 1981). After birth, $mEH_b$ activity develops rapidly in rat ovary and skin, reaching a peak at $20-25$ days (MUKHTAR et al., 1978b; MUKHTAR and BICKERS, 1983) compared to liver, testis, adrenals and mammary epithelial cells where a maximal level is reached after $40-45$ days: slight decreases to the adult level have also been noted after this time (OESCH et al., 1971b; MUKHTAR et al., 1978a, b; GREINER et al., 1980). VOGEL-BINDEL et al. (1982) measured the development of activity in the livers of Sprague-Dawley rats using several substrates and reported maximal activity only after 65 days, after which it remained stable. However, evidence for fetal or post-natal development of multiple forms of $mEH_b$ has not been presented.

In contrast, cEH activity has been followed between 20 and 70 days of age in Swiss-Webster mice, during which time a steady ascending development of activity was observed, with a more dramatic increase in males than in females (GILL and HAMMOCK, 1980). Even after 70 days an activity plateau had not been reached for either sex and further increases were expected. Interestingly, PACIFICI et al. (1983) found that human fetal liver cEH activity followed Michaelis-Menten kinetics but that adult samples showed biphasicity. This could indicate the postnatal development of a second form of cEH which would support the purification of multiple forms by WANG et al. (1982).

# 6.  Purification

Using classical methods of chromatography, liver $mEH_b$ has been purified to apparent homogeneity, and only in one form, from the following species: rat (LU et al., 1975; BENTLEY and OESCH, 1975; KNOWLES and BURCHELL, 1977b), mouse (KNOWLES and BURCHELL, 1977a), rabbit (HALPERT et al., 1979) and human (LU et al., 1979; BEAUNE et al., 1988). Table 11 shows specific activity of purified rat liver $mEH_b$ with different substrates. From some of these schedules, indications for multiple forms of $mEH_b$ have been reported including differential purification factors, (LU et al., 1975; BENTLEY et al., 1976; GUENGE-RICH and MASON, 1980) and differential inhibition and induction characteristics OESCH et al., 1971a; DANSETTE et al., 1974; LEVIN et al., 1978b). However, further work showed that most of these differences may be due to interfering factors in the assays, e.g. detergents, lipids, variable kinetic factors, rather than to a multiplicity of $mEH_b$ (OESCH et al., 1980a; VOGEL-BINDEL et al., 1982; BULLEID et al., 1986). Indeed, when $mEH_b$ in rat liver microsomes has been immunoprecipitated, at least 90% of the activity towards STO, BPO (OESCH and BENTLEY, 1976), a range of polycyclic aromatic hydrocarbons (LEVIN

---

**Table 11.** Specific activities and yields of activities in purified mEH$_b$ fractions from rat liver

| Substrate | | Specific activity (nmol product/ mg protein/min) | Purification factor[a] | Reference |
|---|---|---|---|---|
| STO | | 516 | 57 | Bentley and Oesch (1975) |
| STO | | 479 | 461 | Knowles and Burchell (1977a) |
| STO | | 500—685 | 35—53 | Lu et al. (1977) |
| BPO | | 445 | 23 | |
| STO | | 576 | 77 | Vogel-Bindel et al. (1982) |
| OE | | 538 | 45 | |
| AE | | 520 | 10 | |
| STO | | | | Guengerich et al. (1979b) |
| Fraction: | A$_1$ | 87 | 8 | |
| | A$_2$ | 959 | 93 | |
| | B | 1,103 | 107 | |
| BPO | | | | |
| Fraction: | A$_1$ | 271 | —[b] | |
| | A$_2$ | 507 | — | |
| | B | 499 | — | |
| STO | | | | Bulleid et al. (1986) |
| Fraction: | CM A$_2$ | 256 | 15 | |
| | CM B$_1$ | 255 | 15 | |
| | CM B$_2$ | 736 | 43 | |

[a] Calculated from microsomal activity as 1
[b] No microsomal activities given

et al., 1978b), OE (Bindel et al., 1979) and AE (Vogel-Bindel et al., 1982), was precipitated. This would indicate that if multiple forms are responsible for the metabolism of these substrates, they are immunologically similar. Furthermore, immunoquantitation has shown that the concentration of mEH$_b$ can be directly correlated to the catalytic activity towards octene oxide in human liver microsomes (Thomas et al., 1982; Kapitulnik et al., 1977) and in control and induced rat liver microsomes (Thomas et al., 1982; Pickett et al., 1981). Guengerich et al. (1979b) reported no correlation between STO activity and mEH$_b$ concentration in human liver microsomes, but it may be significant that the antibodies used reacted only very weakly with the human liver mEH$_b$, and so quantification would have been difficult. Therefore, the large interindividual differences observed in mEH$_b$ activity in human samples

H. Thomas; C. W. Timms; F. Oesch

correlate well with the amount of enzyme present and are probably not caused by endogenous modifiers or multiple forms. However, GUENGERICH et al. (1979b), using a modified purification schedule of KNOWLES and BURCHELL (1977b), succeeded in separating and purifying three forms of $mEH_b$ from control and induced rat liver ($A_1$, $A_2$, B) and six forms from human liver (DE, $HA_a$, $HA_b$, $HA_c$, $HA_I$ and $HA_{II}$). Although all isolated forms had apparent molecular weights of approximately 50,000 daltons, differences in the amino acid composition and substrate-specificity were reported, as well as differences in complement fixation experiments (GUENGERICH et al., 1979a). As always with human samples, however, caution must be exercised in the interpretation of the results due to lability of the enzyme activities in the autopsy samples, although proteolysis was reduced by the addition of protease inhibitors, which had no effect on the enzymes' characteristics. A possible solution to this problem would be to decrease the time needed to purify the various isozymes. Of possible interest in this respect, has been the development of an immuno-affinity column, consisting of monoclonal antibodies raised against $mEH_b$, bound to a CNBr-activated Sepharose-4B column (KENNEDY and BURCHELL, 1983). Although this method is not without problems, e.g. the large quantity of antibodies needed, it does provide a fast and specific method of purifying $mEH_b$. Recently MOODY and HAMMOCK (1987a) succeeded in separating microsomal cis- and trans stilbene oxide hydrolases by a sequence of DEAE-Sephacel, CM-cellulose and hydroxylapatite chromatography from liver of rhesus monkey. Additionally inhibition and immunoprecipitation studies confirmed the structural diversity of both isoenzymes and thus proofed for the first time the indirectly deduced existence of multiple microsomal epoxide hydrolases.

cEH has been purified to apparent homogeneity from rabbit (WAECHTER et al., 1982b) and mouse liver (GILL, 1983; MEIJER and DE PIERRE, 1985a) in only a single form. In contrast, WANG et al. (1982) as well as SCHLADT et al. (1988a) have separated at least two forms of cEH from human liver: TESO is the model substrate for one form (TESO-hydrolase) which exhibits different substrate-specificity, peptide maps and immunological characteristics from $mEH_b$. The model substrates for the additional forms are p-nitrophenylstyrene oxide (PNSO-hydrolase; WANG et al., 1982) and cis-stilbene oxide (SCHLADT et al., 1988a), the former being a good substrate for $mEH_b$ itself. Indeed PNSO-hydrolase as well as cis-stilbene oxide hydrolase are immunologically similar to $mEH_b$, and possess similar substrate specificity. WANG et al. (1982) also saw indications for further forms of cEH, particularly of PNSO-hydrolase, but this speculation awaits confirmation. Recently, elegant affinity chromatography techniques using 7-methoxycitronellyl thiol (PRESTWICH and HAMMOCK, 1985) and different alkyl and aryl thiols coupled to epoxy activated Sepharose have opened rapid and specific access to cEH from a variety of species comprising rhesus monkey, baboon, rabbit, rat and mouse as well as human

**Table 12.** Summary of kinetic data for mEH$_b$

| Substrate | Species | Vmax (nmol/ mg pro- tein/min) | K$_m$ (μM) | K$_m$ (μM/mg protein) | Reference |
|---|---|---|---|---|---|
| STO | rat microsomes | 6.3 | 180 | 298 | GANU and ALWORTH (1978) |
| | | 10 | 125 | 525 | OESCH, et al. (1983) |
| | guinea-pig (partially purified) | 3.8 | 230 | — | OESCH et al. (1971 c) |
| BPO (Non-linear Lineweaver- Burk plots) | rat microsomes rabbit microsomes (PB-treated) rat microsomes | 7.1 | 6 | 97 | LU et al. (1977) HALPERT et al. (1979) SCHMASSMANN et al. (1976) |
| | | 17.5 | 0.25 | 313 | LU et al. (1977) |
| | pure mEH$_b$ | 300 | 0.50 | — | IBRAHIM et al. (1985) |
| OE | rat microsomes | 12.8 | 10 | 952 | BINDEL et al. (1979) |
| AE | rat microsomes | 55 | 100 | 6 192 | OESCH et al. (1981) |
| HEOM | rat: partially purified | 59 | 10.4 | — | CRAVEN (1977) |
| | rabbit: partially purified | 714 | 12.5 | — | |
| BP 7,8-oxide | rat microsomes | 17.4 | 15.9 | 636 | LU et al. (1977) |
| BP 9,10-oxide | | 16.2 | 17.1 | 684 | |
| BP 11,12-oxide | | 2.4 | 119 | 1 190 | |
| Benzene oxide | | 24.5 | 32 500 | 6 500 | OESCH et al. (1980 a) |
| Octene 1,2-oxide | pure rat mEH$_b$ | 1 000 | 4.2 | 83 333 | LEVIN et al. (1978 b) |
| BP 11,12-oxide Naphthalene 1,2-oxide BA 5,6-oxide | | 4.000 | 10 | 667 | |
| (Other non-linear Lineweaver- Burk plots) | | | | | LU et al. (1977) |

liver (SILVA and HAMMOCK, 1987; WIXTROM et al., 1988). Among others, cEH from control and clofibrate induced mice has been purified and shown to be the same form in both cases (PRESTWICH and HAMMOCK, 1985). In addition, the identity of both forms was confirmed by HPLC peptide mapping and partial N-terminal peptide sequencing (MEIJER et al., 1987 b). Table 12 gives a summary of kinetic parameters for the cEH and mEH$_b$.

Although mEH$_{ch}$ has not been purified to homogeneity, WATABE et al. (1986 b) have succeeded in separating it from mEH$_b$ in mouse liver microsomes by chromatography on a Sepharose 4B column.

Leukotriene A$_4$ hydrolase has been purified from human and rat leukocytes (RÅDMARK et al., 1984; EVANS et al., 1985) as well as from human erythrocytes (McGEE and FITZPATRICK, 1985). But, although hepatic location of this enzyme has been demonstrated for human and mouse liver (HAEGGSTRÖM et al., 1985; HAEGSTRÖM et al., 1986) and an ubiquitous organ distribution of LTA$_4$ hydrolase has been shown in guinea pig (IZUMI et al., 1986) there has been no report on a purification from either of these sources.

## 7. Topology

Most of the polypeptide chain of mEH$_b$ is localized within the endoplasmic reticulum, with only a small part being exposed at the cytoplasmic surface (SEIDEGÅRD et al., 1982). Also, HEINEMANN and OZOLS (1984 b) have proposed that three segments of the polypeptide chain are responsible for membrane binding, including the amino-terminal segment which could act as a "signalpeptide" for membrane insertion. A similar idea has been proposed by OKADA et al. (1982), who compared the amino-terminal peptide sequence to the transient insertion signal found in presecretory proteins. Recently, PORTER et al. (1986) have delineated six membrane-spanning segments, and have further claimed that much of the epoxide hydrolase is constrained to maintain its hydrophobic character, consistent with its intramembranous location. More practically, SEIDEGÅRD et al. (1982), have warned that results, obtained from incorporation of mEH$_b$ into liposomes with differential constitution, must be treated with caution; they observed that the redistribution of the enzyme within the membrane is strongly effected by the constituent lipids.

## 8. Molecular cloning, gene structure and expression in cultured cells

First attempts to elucidate the cDNA-structure for rat liver mEH$_b$ date back to 1981, when GONZALEZ and KASPER (1981 b) succeeded in the isolating and restriction mapping of a 1,310 base pairs spanning cDNA-segment, coding for approxi-

---

mately 95% of the $mEH_b$ message sequence. But it was not until 1986 that the complete cDNA and deduced amino acid sequence of Sprague Dawley rat liver $mEH_b$ were published (PORTER et al., 1986). According to these data the cDNA sequence comprises 1365 nucleotides coding for a 455 amino acid polypeptide with a molecular weight of 52,581. The deduced amino acid composition agrees well with those obtained from direct amino acid analysis of the rat protein (LU et al., 1975; BEUTLEY et al., 1975; DUBOIS et al., 1978), and the amino acid sequence is 81% identical to that of rabbit $mEH_b$ (HEINEMANN and OZOLS, 1984 b). A comparison of the codon usage for rat liver $mEH_b$ with the codon usage for cytochromes P-450b (P 450 II B1), d (P450 I A) and PCN (P 450 I) as well as NADPH-cytochrome P-450 reductase revealed that $mEH_b$ is more conserved than the cytochromes P-450b and P-450 PCN.

Characterization of the gene for $mEH_b$ from Wistar Furth rats (FALANY et al., 1987) indicated a length of approximately 16 kilobases with nine exons ranging in size from 109 to 420 base pairs, eight intervening sequences, the largest of which is 3.2 kilobases, and regulatory sequences in the 5'-flanking region of the gene.

Although the cDNA and gene for $mEH_b$ were isolated from different rat strains, they showed sequence identity. Only a single functional $mEH_b$ gene was found in this study and there was no evidence of hybridization to the genes for $mEH_{ch}$ or cEH. However, a nonfunctional pseudogene for mEH was characterized lacking substantial DNA sequences corresponding to exon 1 and other exons. This pseudogen is suggested to have arisen by gene duplication and subsequent deletion of one or more exons.

Identical to the rat, only one gene copy per haploid genome and a single mRNA of 1.8 kilobases has been identified by Southern and Northern blotting following cDNA cloning of human liver $mEH_b$ (SKODA et al., 1988). The corresponding cDNA codes for a protein of 455 amino acids with a calculated molecular weight of 52,956 (JACKSON et al., 1987; SKODA et al., 1988). The nucleotide sequence was found to be 77% homologous to rat liver $mEH_b$, while the deduced amino acid sequence was reported 84% similar to the deduced rat (PORTER et al., 1986) and 80% similar to the previously reported rabbit protein sequence. The gene for human $mEH_b$ was assigned to the long arm of chromosome 1 in contrast to the proposed location on chromosome 6 which had been tentatively assigned by BROWN and CHALMERS (1986) based on indirect immunological evidence for lack of expression of the human enzyme in a series of human-mouse hybrid cell lines deficient in chromosome 6.

Also, human leukotriene $A_4$ hydrolase has recently been cloned (FUNK et al., 1987), and the coding sequence was shown to determine a 610 amino acid protein with a calculated molecular weight of 69,140. No apparent homologies with $mEH_b$ have been detected and there was only one discrete mRNA of 2,250 nucleotides found in lung and leukocytes.

No reports have appeared so far on the cloning of cytosolic epoxide hydrolase or $mEH_{ch}$ and the study of regulatory phenomena such as xenobiotic epoxide hydrolase induction on the molecular level.

However, cDNA-directed expression of human $mEH_b$ has been reported in cultured monkey kidney cells (COS-1) (SKODA et al., 1988) and the yeast Saccharomyces pombe (JACKSON and BURCHELL, 1988).

COS-1 cells, which originally lack immunodetectable epoxide hydrolase protein and measurable enzyme activity were transfected with the SV 40 expression vector p91023(B) containing the coding insert, resulting in immunologically detectable appearance of $mEH_b$ protein. Subcellular fractionation suggested an incorporation of the enzyme into its physiological compartment, i.e. the endoplasmic reticulum, which could be confirmed by activity determination using styrene oxide.

Correspondingly, human liver $mEH_b$ cDNA was introduced into Saccharomyces pombe by the use of the yeast expression vector pEVP11. Epoxide hydrolase protein and enzyme activity were expressed and identified in the 105,000 × g pellet following subcellular fractionation.

Both expression systems are prone to providing useful sources for the large scale purification of human liver $mEH_b$ and the study of xenobiotic metabolism, enzyme induction and catalytic function following modification of catalytic and hydrophobic sites.

# 9. Physiological significance

## 9.1. Preneoplastic antigen (PNA)

After several studies had characterized the microsomal protein (PNA) which increased in concentration in correlation with preneoplastic and neoplastic states (OKITA and FARBER, 1975; LIN et al., 1977; GRIFFIN and KIZER, 1978), its identity as $mEH_b$ was confirmed by LEVIN et al. (1978a) and GRIFFIN and NODA (1980). of, A significant induction of $mEH_b$ activity has been reported in rat liver nodules, induced by 2-acetylaminofluorene (NOVIKOFF et al., 1979; LEVIN et al., 1978a). Intensive characterization of $mEH_b$ activity in these nodules however showed the activity towards STO, BPO and OE to be qualitatively and quantitatively similarly induced (OESCH et al., 1983a). During the later stages of the carcinogenic process, however, the status of $mEH_b$ is unclear: whilst some authors claim that the activity stays high in hepatomas (NOVIKOFF et al., 1979; LEVIN et al., 1978a), others have reported no induced activity (KUHLMANN et al., 1981). Similarly, in human liver tumour (CRAVEN et al., 1982) and bronchiogenic carcinoma tissue (OESCH et al., 1980b), significantly reduced activity towards HEOM and BPO respectively was measured in comparison to normal tissue. Therefore, it would appear that

$mEH_b$ is significantly induced in nodules, but as the carcinogenic process proceeds the activity decreases until in carcinomas and tumours the activity is actually lower than in normal tissue.

In addition to these changes in $mEH_b$ activity, GRIFFIN and NODA (1980) observed that whilst hyperplastic nodule and hepatoma cytosolic fractions contained significant amounts of $mEH_b$, control cytosol contained none. As a result of this observation, various workers have investigated the possibility of using this phenomenon as a marker for carcinogenesis. For example, GRIFFIN and GENGOZIAN, (1984) detected circulating $mEH_b$ in the sera of rats with either large hyperplastic nodules or hepatomas, but not in control sera. Correspondingly, MOODY et al. (1987) found no increased serum mEH-preneoplastic antigen levels only with mEH induction in rat liver upon clofibrate or phenobarbital treatment. From these results the authors conclude that increases in serum preneoplastic antigen levels concur with neoplastic or necrotic events and that serum mEH levels may indeed be a suitable marker for hepatocellular carcinoma. In contrast, however, HAMMOCK et al. (1984) could see no statistically significant difference between the amount of $mEH_b$ in sera from apparently healthy patients and that from patients with diagnosed liver cancer. In fact, the authors themselves admit that a time-course study of $mEH_b$ release into serum during carcinogenesis and cell-injury is an essential prerequisite to be able to determine if serum levels are specifically due to the carcinogenic process or simply a more generalized hepatic necrosis. This induction and release of $mEH_b$ during the development of hyperplastic nodules has been interpreted as an adaptive response to a hazardous environment, which increases the resistance to many toxic xenobiotics. It is paralleled by a decreased metabolism of carcinogens by the microsomal mixed-function oxidase system as a consequence of differential inhibition of cytochrome P-450 isozymes (BUCHMANN et al., 1985), an increased conjugation of reactive metabolites with glutathione and glucuronic acid by an increase in glutathione concentration and glutathione-S-transferases, particularly GST-A and GST-P (SATO et al., 1984) and GST-B and GST-C (BUCHMANN et al., 1985), and increased UDP-glucuronosyl transferase (SATO et al., 1984), a decrease in sulpho-transferases and the accelerated reduction of quinones by DT-diaphorase (for reviews see: FARBER, 1984; HICKS, 1983).

The only observation which has been made with respect to the behaviour of the other forms of epoxide hydrolase during preneoplasia, is that $mEH_{ch}$ activity is not increased in rat liver nodules, indicating a differential regulation from $mEH_b$ (BATT et al., 1984).

H. THOMAS; C. W. TIMMS; F. OESCH

## 9.2. Endogenous roles

The first indications that $mEH_b$ might be involved in steroid metabolism were provided by the high specific activity, observed in the testis and ovary of mouse and rat (OESCH et al., 1977; DePIERRE et al., 1983). Indeed, the biosynthesis of $16\alpha,17\alpha$-epoxy-estratetraenol (OE) (WATABE et al., 1979; SIEKMANN et al., 1980) and $16\alpha,17\alpha$-epoxy-4-androsten-3-one (AE) (DISSE et al., 1980) in the microsomal fraction of rat liver confirmed these suspicions, especially as they were reported to be good substrates for $mEH_b$ (WATABE et al., 1979; BINDEL et al., 1979; OESCH et al., 1980a). Similarly, the epoxidation of androsta-5,16-dien-3$\beta$-ol by rat liver microsomal lipid hydroperoxides, with the subsequent conversion of the 5,6-epoxides to the $5\alpha,6\beta$-glycol by $mEH_b$, indicates further involvement of this enzyme in steroid metabolism (WATABE et al., 1986a). However, this evidence implicates only a few particular substrates, and it is also possible that the situation in vivo is somewhat different to in vitro: therefore, the overall endogenous role of $mEH_b$ requires further extensive investigation but it appears at the moment that this is only a minor role of $mEH_b$.

In contrast, the sole substrates found until now for $mEH_{ch}$, are steroids (Table 5). Since most of these epoxides arise as a result of lipid peroxidation, it is credible that the enzyme has a protective role, to dispose of such epoxides (SEVANIAN et al., 1979; WATABE et al., 1981; WATABE et al., 1984). It is also possible that the cholesterol epoxides are obligatory precursors to cholestanetriol in the regulation of the tissue cholesterol level (WATABE et al., 1980): the observation that the 5,6-epoxides inhibit cholesterol $7\alpha$-hydrolase might also be of relevance here (ARINGER and ENEROTH, 1974). Cholesterol $5,6\alpha$-epoxide itself, is probably weakly mutagenic (SEVANIAN and PETERSEN, 1984; BLACK, 1980) and can covalently attach to DNA (BLACKBURN et al., 1979): therefore, it is also important for the organism to keep control over this potentially toxic, endogenous substance.

A possible role in lipid control has also been proposed for cEH, as a result of the fact that oxygenated sterols, and their precursors, such as squalene 2,3-24,25-dioxide and lanosterol 24,25-epoxide, are excellent substrates for this enzyme in mouse liver (HAMMOCK et al., 1980b; HAMMOCK and OTA, 1983). Also of interest in this respect are the observations that mouse liver cEH metabolizes cis- and trans-epoxymethyl stearates more rapidly than $mEH_b$ (GILL and HAMMOCK, 1979), and that the epoxy-fatty acid content is significantly increased in rat lung tissue on exposure to nitrogen, probably being due to lipid autoxidation (SEVANIAN et al., 1979). The increased epoxy-fatty acids were primarily the 18-carbon species, including epoxystearic and epoxyoctadecanoic acids derived from oleic and linoleic acids; in addition, epoxides of palmitoleic and arachidonic acids were also detected (CAPDEVILA et al., 1988). Exclusive metabolism of the latter by cEH has been claimed (CHACOS et al.,

1983), although other workers report the ability of mEH$_b$ to metabolize this substrate (OLIW et al., 1982). CHACOS et al. (1983) regarded the lower reactivity of 5,6-epoxyeicosatrienoic acids to enzymatic hydration with special interest, particularly with respect to its possible involvement as a selective mediator for the release of somatostatin and luteinizing hormone from in vitro incubations of median eminence fragments and anterior pituitary cells respectively. Therefore, cEH could have regulatory roles as a sequester of fatty acid epoxides formed by autoxidation and lipid peroxidation, and in prostaglandin synthesis.

For leukotriene A$_4$ hydrolase, which catalyzes the hydrolysis of the allylic epoxide, leukotriene A$_4$, to the dihydroxy acid leukotriene B$_4$ in rodent and human leukocytes, erythrocytes and several other tissues a possible regulatory role in the leucotriene pathway can be envisaged.

## 10. Conclusion

At least six forms of epoxide hydrolase, mEH$_t$, mEH$_{ch}$, cEH, a microsomal form with substrate specificity for trans-stilbene oxide as well as a cytosolic form with substrate specificity for cis-stilbene oxide and leukotriene A$_4$ hydrolase have been characterized in different species and organs by means of their subcellular localization, substrate specificity, inducibility as well as physical, biochemical and immunological proteins.

At least two different cytosolic epoxide hydrolases and two forms of microsomal epoxide hydrolase have been purified to apparent homogeneity and characterized. Although enzyme purification and induction studies have provided evidence for a multiplicity of microsomal and cytosolic isoenzymes, the molecular biology data remain conflicting. Analysis of the genes for rat and human liver mEH$_t$ revealed only a single functional copy for this enzyme with an additional, but nonfunctional, pseudogene being present in the rat liver genome. This supports in part the idea that some of the proposed multiple forms may in fact represent artifacts due to different purification conditions including varying lipid contents and detergent interaction.

In conclusion, the epoxide hydrolase family of isoenzymes reflects a much simpler structure of organization compared to the multigene family of cytochrome P-450 dependent monooxygenases (NEBERT et al., 1987) to which they may be closely related regarding their metabolic destination and evolutionary development.

## Acknowledgements

The authors thank Mrs. I. BÖHM for typing the manuscript.

H. THOMAS; C. W. TIMMS; F. OESCH

# 11. References

A. AITIO, M. AHOTUPA, and M. G. PARKKI (1978), Biochem. Biophys. Res. Commun. **83**, 850—856

L. ARINGER and P. ENEROTH (1974), J. Lipid Res. **15**, 389—398.

R. N. ARMSTRONG, W. LEVIN, and D. M. JERINA (1980), J. Biol. Chem. **255**, 4698—4705.

R. N. ARMSTRONG, B. KEDZIERSKI, W. LEVIN, and D. M. JERINA (1981), J. Biol. Chem. **256**, 4726—4733.

A. ÅSTRÖM and J. W. DE PIERRE (1981), Biochim. Biophys. Acta **673**, 225—233.

A. ÅSTRÖM, W. BIRBERG, A. PILOTTI, and J. W. DE PIERRE (1986a), Eur. J. Biochem. **154**, 125—134.

A. ÅSTRÖM, M. ERIKSSON, L. C. ERIKSSON, W. BIRBERG, A. PILOTTI, and J. W. DE PIERRE (1986b), Biochim. Biophys. Acta **882**, 359—366.

A. ÅSTRÖM, S. MÅNER, and J. W. DE PIERRE (1987), Xenobiotica **17**, 155—163.

L. BALK, J. MEIJER, J. SEIDEGÅRD, R. MORGENSTERN, and J. W. DE PIERRE (1980), Drug Metab. Disp. **8**, 98—103.

J. BARON, J. A. REDICK, and F. P. GUENGERICH (1980), Life Sci. **26**, 489—493.

A. M. BATT, G. SIEST, and F. OESCH (1984), Carcinogenesis **5**, 1205—1206.

P. H. BEAUNE, T. CRESTEIL, J.-P. FLINOIS, and J.-P. LEROUX (1988), J. Chromatogr. **426**, 169—176.

G. BELLUCCI, G. BERTI, M. FERRETTI, F. MARIONI, and F. RE (1981), Biochem. Biophys. Res. Commun. **102**, 838—844.

A. M. BENSON, Y.-N. CHA, E. BUEDING, H. S. HEINE, and P. TALALAY (1979), Cancer Res. **39**, 2971—2977.

P. BENTLEY, F. OESCH, and A. TSUGITA (1975), FEBS Lett. **59**, 296—299.

P. BENTLEY and F. OESCH (1975), FEBS Lett. **59**, 291—295.

P. BENTLEY, H. SCHMASSMANN, P. SIMS, and F. OESCH (1976), Eur. J. Biochem. **69**, 97—103.

P. BENTLEY, F. WAECHTER, F. OESCH, and W. STÄUBLI (1979), Biochem. Biophys. Res. Commun. **91**, 1101—1108.

U. BINDEL, A. SPARROW, H. SCHMASSMANN, M. GOLAN, P. BENTLEY, and F. OESCH (1979), Eur. J. Biochem. **97**, 275—281.

H. S. BLACK and W. A. LENGER (1979), Anal. Biochem. **94**, 383—385.

H. S. BLACK (1980), Lipids **15**, 705—709.

G. M. BLACKBURN, A. RASHID, and M. H. THOMPSON (1979), J.C.S. Chem. Commun., 420—421.

B. M. BLOOM and G. M. SHULL (1955), J. Am. Chem. Soc. **77**, 5767—5768.

W. A. BORNSTEIN, W. LEVIN, P. E. THOMAS, O. E. RYAN, and E. BRESNICK (1979), Arch. Biochem. Biophys. **197**, 436—444.

E. BOYLAND (1950), Biochem. Soc. Symp. **5**, 40—54.

E. BRESNICK, H. MUKHTAR, T. A. STOMING, P. M. DANSETTE, and D. M. JERINA (1977a), Biochem. Pharmacol. **26**, 891—892.

E. BRESNICK, J. B. VAUGHT, A. Y. L. CHUANG, T. A. STOMING, D. BOCKMAN, and H. MUKHTAR (1977b), Arch. Biochem. Biophys. **181**, 257—269.

H. BREUER and R. KNUPPEN (1961), Biochim. Biophys. Acta **49**, 620—621.

G. T. BROOKS (1966), World Rev. Pest Control **5**, 62—84.

G. T. BROOKS (1973), Nature **245**, 382—384.

G. T. BROOKS, A. HARRISON, and S. E. LEWIS (1970), Biochem. Pharmacol. **19**, 255—273.

S. BROWN and D. E. CHALMERS (1986), Biochem. Biophys. Res. Commun. **137**, 775—780.

A. BUCHMANN, W. KUHLMANN, M. SCHWARZ, W. KUNZ, C. R. WOLF, E. MOLL, T. FRIEDBERG, and F. OESCH (1985), Carcinogenesis **6**, 513—521.

---

N. J. Bulleid and J. A. Craft (1984), Biochem. Pharmacol. **33**, 1451—1457.

N. J. Bulleid, A. B. Graham, and J. A. Craft (1986), Biochem. J. **233**, 607—611.

B. Burchell, P. Bentley, and F. Oesch (1976), Biochim. Biophys. Acta **444**, 531 to 538.

J. H. Capdevila, P. Mosset, P. Yadagiri, S. Lumin, and J. R. Falck (1988), Arch. Biochem. Biophys. 261, 122—133.

Y.-N. Cha, F. Martz, and E. Bueding (1978), Cancer Res. **38**, 4496—4498.

N. Chacos, J. Capdevilla, J. R. Falck, S. Manna, C. Martin-Wixtrom, S. S. Gill, B. D. Hammock, and R. W. Estabrook (1983), Arch. Biochem. Biophys. **223**, 639 to 648.

D. Chalmers and S. Brown (1987), Xenobiotica **17**, 71—77.

J. A. Craft, N. J. Bulleid, M. R. Jackson, and B. Burchell (1988), Biochem. Pharmacol. **37**, 297—302.

A. C. C. Craven (1977), Ph. D. thesis, Department of Physiology and Biochemistry, University of Reading, England.

A. C. C. Craven, C. H. Walker, and I. M. Murray-Lyon (1982), Biochem. Pharmacol. **31**, 1321—1324.

T. Cresteil, J.-L. Mahu, P. M. Dansette, and J.-P. Leroux (1980), Biochem. Pharmacol. **29**, 1127—1133.

R. Croteau and P. E. Kolattukudy (1975), Arch. Biochem. Biophys. **170**, 73—81.

P. M. Dansette, H. Yagi, D. Jerina, J. W. Daly, W. Levin, A. Y. H. Lu, R. Kuntzmann, and A. H. Conney (1974), Arch. Biochem. Biophys. **164**, 511—517.

P. M. Dansette, V. B. Makedonska, and D. M. Jerina (1978), Arch. Biochem. Biophys. **187**, 290—298.

P. M. Dansette, G. C. Du Bois, and D. M. Jerina (1979), Anal. Biochem. **97**, 340—345.

B. Davidow and J. L. Radomski (1953), J. Pharmacol. Expt. Therap. **107**, 259—265.

J. G. Dent and M. E. Graichen (1982), Carcinogenesis **3**, 733—738.

J. W. De Pierre, J. Meijer, W. Birberg, A. Pilotti, L. Balk, and J. Seidegård (1983), In: Extrahepatic Drug Metabolism and Chemical Carcinogenesis (J. Rydström, J. Montelius, and M. Bengtsson, eds.), pp. 95—103, Elsevier Science Publishers, Stockholm.

K. W. Di Biasio, M. H. Silva, B. D. Hammock, and L. R. Shull (1989), Fundam. Appl. Toxicol. **12**, 449—459.

B. Disse, L. Siekmann, and H. Breuer (1980), Acta Endocr. **95**, 58—66.

G. Doerjer, M. A. Bedell, and F. Oesch (1984), In: Mutations in Man (G. Obe, ed.), pp. 20—34, Springer Verlag, Berlin.

G. C. Du Bois, E. Appella, W. Levin, A. Y. H. Lu, and D. M. Jerina (1978), J. Biol. Chem. **253**, 2932—2939.

G. C. Du Bois, E. Appella, R. Armstrong, W. Levin, A. Y. H. Lu, and D. M. Jerina (1979), J. Biol. Chem. **254**, 6240—6243.

G. C. Du Bois, E. Appella, D. E. Ryan, D. M. Jerina, and W. Levin (1982), J. Biol. Chem. **257**, 2708—2712.

J. F. Evans, P. Dupuis, and A. W. Ford-Hutelmison (1985), Biochim. Biophys. Acta **840**, 43—50.

W. E. Fahl, C. R. Jefcoate, and C. B. Kasper (1978), J. Biol. Chem. **253**, 3106—3113.

C. N. Falany, P. McQuiddy, and C. B. Kasper (1987), J. Biol. Chem. **262**, 5924—5930.

E. Farber (1984), Carcinogenesis **5**, 1—5.

C. D. Funk, D. Rådmark, J. Y. Fu, T. Matsumoto, H. Jörvall, T. Shimizu, and B. Samuelsson (1987), Proc. Natl. Acad. Sci. USA **84**, 6677—6681.

M.-M. Galteau, B. Antoine, and H. Reggio (1985), EMBO J. **4**, 2793—2800.

V. S. Ganu, S. O. Nelson, L. Verlander, and W. L. Alworth (1977), Anal. Biochem. **78**, 451—458.

V. S. Ganu and W. L. Alworth (1978), Biochemistry 17, 2876—2881.

G. Gazotti, E. Garattini, and M. Salmona (1981), Chem.-Biol. Interactions 35, 311—318.

S. S. Gill, B. D. Hammock, and J. E. Casida (1974), J. Agric. Food Chem. 22, 386—395.

S. S. Gill and B. D. Hammock (1979), Biochem. Biophys. Res. Commun. 89, 965—971.

S. S. Gill and B. D. Hammock (1980), Biochem. Pharmacol. 29, 389—395.

S. S. Gill and B. D. Hammock (1981), Nature 291, 167—168.

S. S. Gill, S. I. Wie, T. M. Guenthner, F. Oesch, and B. D. Hammock (1982), Carcinogenesis 3, 1307—1310.

S. S. Gill (1983), Biochem. Biophys. Res. Commun. 112, 763—769.

S. S. Gill, K. Ota, and B. D. Hammock (1983a), Anal. Biochem. 131, 273—282.

S. S. Gill, K. Ota, B. Ruebner, and B. D. Hammock (1983b), Life Sci. 32, 2693—2700.

S. S. Gill (1984), Biochem. Biophys. Res. Commun. 122, 1434—1440.

S. S. Gill and S. Kaur (1987), Biochem. Pharmacol. 36, 4221—4227.

K. A. Giuliano, E. P. Lau, and R. R. Fall (1980), J. Chromatography 202, 447—452.

H. R. Glatt, E. Kaltenbach, and F. Oesch (1980), Caucer Res. 40, 2552—2556.

H. R. Glatt, T. Wölfel, and F. Oesch (1983), Biochem. Biophys. Res. Commun. 110, 525—529.

L. S. Gontovnick and G. D. Bellward (1980), Biochem. Pharmacol. 29, 3245—3251.

L. S. Gontovnick and G. D. Bellward (1981), Drug Metab. Disp. 9, 265—269.

F. J. Gonzalez and C. B. Kasper (1980), Biochem. Biophys. Res. Commun. 93, 1254—1258.

F. J. Gonzalez and C. B. Kasper (1981a), Mol. Pharmacol. 21, 511—516.

F. J. Gonzalez and C. B. Kasper (1981b), J. Biol. Chem. 256, 4697—4700.

F. J. Gonzalez, M. Samore, P. McQuiddy, and C. B. Kasper (1982), J. Biol. Chem. 257, 11032—11036.

F. M. Goujon, J. Van Cantfort, and J. E. Gielen (1980), Chem.-Biol. Interactions 32, 361—375.

M. E. Graichen and J. G. Dent (1984), Carcinogenesis 5, 23—28.

J. W. Greiner, A. H. Bryan, L. B. Malan-Shibley, and D. H. Janss (1980), J. Natl. Cancer Inst. 64, 1127—1132.

M. J. Griffin and D. E. Kizer (1978), Cancer Res. 38, 1136—1141.

M. J. Griffin and K. Noda (1980), Cancer Res. 40, 2768—2773.

M. J. Griffin and N. Gengozian (1984), Ann. Clin. Lab. Science 14, 27—31.

P. L. Grover, A. Hewer, and P. Sims (1971), FEBS Lett. 18, 76—80.

F. P. Guengerich, P. Wang, P. S. Mason, and M. B. Mitchell (1979a), J. Biol. Chem. 254, 12255—12259.

F. P. Guengerich, P. Wang, M. B. Mitchell, and P. S. Mason (1979b), J. Biol. Chem. 254, 12248—12254.

F. P. Guengerich and P. S. Mason (1980), Anal. Biochem. 104, 445—451.

F. P. Guengerich (1982), in: Reviews in Biochemical Toxicology, (E. Hodgson, J. R. Bend, and R. M. Philpott, eds.), Vol. 4, pp. 5—30, Elsevier Science Publishing Company, New York.

F. P. Guengerich and N. K. Davidson (1982), Arch. Biochem. Biophys. 215, 462 to 477.

T. M. Guenthner, B. D. Hammock, V. Vogel, and F. Oesch (1981), J. Biol. Chem. 256, 3163—3166.

T. M. Guenthner and F. Oesch (1983), J. Biol. Chem. 258, 15054—15061.

T. M. Guenthner and T. A. Karnezis (1986a), Drug. Metab. Disp. 14, 208—213.

T. M. Guenthner (1986a), Biochem. Pharmacol. 35, 839—845.

T. M. Guenthner (1986b), Biochem. Pharmacol. 35, 3261—3266.

T. M. Guenthner and T. A. Karnezis (1986b), J. Biochem. Toxicol. 1, 67—81.

---

Epoxide Hydrolases

W. H. Habig, M. J. Pabst, and W. B. Jakoby (1974), J. Biol. Chem. **249**, 7130—7139.

J. Haeggström, O. Rådmark, and F. A. Fitzpatrick (1985), Biochim. Biophys. Acta **835**, 378—384.

J. Haeggström, J. Meijer, and O. Rådmark (1986), J. Biol. Chem. **261**, 6332—6337.

P. P. Halarnka, R. N. Wixtrom, M. H. Silva, and B. D. Hammock (1989), Arch. Biochem. Biophys. **272**, 226—236.

J. Halpert, H. Glaumann, and M. Ingelmann-Sundberg (1979), J. Biol. Chem. **254**, 7434—7441.

B. D. Hammock, M. El Tantawy, S. S. Gill, L. Hasegawa, C. A. Mullin, and K. Ota (1980a), in: Microsomes, Drug Oxidations and Chemical Carcinogenesis (M. J. Coon, A. Y. H. Conney, R. W. Estabrook, H. V. Gelboin, J. R. Gillette, and P. J. O'Brien, eds.), pp. 655—658, Academic Press, New York.

B. D. Hammock, S. S. Gill, S. M. Mumby, and K. Ota (1980b), in: Molecular Basis of Toxicology (R. S. Bhatnagar, ed.), pp. 229—272, Ann. Arbor Science Publishers, Ann. Arbor.

B. D. Hammock and L. S. Hasegawa (1983), Biochem. Pharmacol. **32**, 1155—1164.

B. D. Hammock and K. Ota (1983), Toxicol. Appl. Pharmacol. **71**, 254—265.

B. D. Hammock, D. N. Loury, D. E. Moody, B. Ruebner, R. Baselt, K. M. Milam, P. Volberding, A. Kettermann, and R. Talcott (1984), Carcinogenesis **5**, 1467 to 1473.

R. P. Hanzlik, M. Edelman, W. J. Michaely, and G. Scott (1976), J. Am. Chem. Soc. **98**, 1952—1955.

R. P. Hanzlik, S. Heideman, and D. Smith (1978), Biochem. Biophys. Res. Commun. **82**, 310—315.

S. Hasani and B. Burchell (1979), Biochem. Pharmacol. **28**, 2175—2179.

S. Hasani, R. G. Knowles, and B. Burchell (1979), Eur. J. Biochem. **10**, 589—594.

F. S. Heinemann and J. Ozols (1984a), J. Biol. Chem. **259**, 791—796.

F. S. Heinemann and J. Ozols (1984b), J. Biol. Chem. **259**, 797—804.

A. Helenius and K. Simons (1975), Biochim. Biophys. Acta **415**, 29—79.

R. M. Hicks (1983), Carcinogenesis **4**, 1209—1214.

R. M. Hodgson, P. D. Cary, P. L. Grover, and P. Sims (1983), Carcinogenesis **4**, 1153—1158.

R. M. Hodgson, A. Seidel, W. Bochnitschek, H. R. Glatt, F. Oesch, and P. L. Grover (1985), Carcinogenesis **6**, 135—139.

M. W. A. C. Hukkelhoven, E. W. M. Vromans, A. J. M. Vermorken, and H. Bloemendal (1982), FEBS Lett. **144**, 104—108.

J. M. Hunt, C.-J. Guo, and P. A. Desai (1987), Cancer Lett. **37**, 285—291.

M. Ibrahim, P. Hubert, E. Dellacherie, J. Magdalou, J. Müller, and G. Siest (1985), Enzyme Microb. Technol. **7**, 66—72.

T. Izumi, T. Shimizu, Y. Seyama, N. Olishi, and F. Takaku (1986), Biochem. Biophys. Res. Commun. **135**, 139—145.

M. R. Jackson, J. A. Craft, and B. Burchell (1987), Nucl. Acid Res. **15**, 7188.

M. R. Jackson and B. Burchell (1988), Biochem. J. **251**, 931—933.

M. O. James, E. R. Bowen, P. M. Dansette, and J. R. Bend (1979), Chem.-Biol. Interactions **25**, 321—344.

M. Jansen, A. J. Baars, and D. D. Breimer (1986), Biochem. Pharmacol. **35**, 2229 to 2232.

D. M. Jerina, J. W. Daly, B. Witkop, P. Zaltzman-Nirenberg, and S. Udenfriend (1970a), Biochemistry **9**, 147—156.

D. M. Jerina, H. Ziffer, and J. W. Daly (1970b), J. Am. Chem. Soc. **92**, 1056—1061.

D. M. Jerina and J. W. Daly (1974), Science **185**, 573—582.

H. Thomas; C. W. Timms; F. Oesch

D. M. Jerina, P. M. Dansette, A. Y. H. Lu, and W. Levin (1976), Mol. Pharmocal. 13, 342—351.

D. M. Jerina, R. Lehr, M. Schaeffer-Ridder, H. Yagi, J. M. Karle, and D. R. Thakker (1977), in: Mechanism of Carcinogenesis (H. H. Hiatt, J. D. Watson, and J. A. Winster, eds.), pp. 639—658, Cold Spring Harbor Laboratory.

D. M. Jerina, H. Yagi, R. E. Lehr, D. R. Thakker, M. Schaeffer-Ridder, J. M. Karle, W. Levin, A. W. Wood, R. L. Chang, and A. H. Conney (1978), in: Polycyclic Hydrocarbons and Cancer (H. V. Gelboin and P. O. Ts'o, eds.), Vol. 1, pp. 173—188, Academic Press Inc, London.

B. Jernström, H. Vardi, and S. Orrenius (1976), Cancer Res. 36, 4107—4113.

V. Joste and J. Meijer (1989), FEBS Lett. 249, 83—88.

R. Kahl (1980), Biochem. Biophys. Res. Commun. 95, 163—169.

J. Kapitulnik, W. Levin, A. Y. H. Lu, R. Morecki, P. M. Dansette, D. M. Jerina, and A. H. Conney (1977), Clin. Pharmacol. Ther. 21, 158—165.

J. Kapitulnik, P. G. Wislocki, W. Levin, H. Yagi, D. M. Jerina, and A. H. Conney (1978), Cancer Res. 38, 354—358.

S. Kaur and S. S. Gill (1986), Biochem. Pharmacol. 35, 1299—1308.

T. T. Kawabata, F. P. Guengerich, and J. Baron (1981), Mol. Pharmacol. 20, 709 to 714.

S. M. E. Kennedy and B. Burchell (1983), Biochem. Pharmacol. 32, 2029—2032.

D. E. Kizer, J. A. Clouse, D. P. Ringer, O. Hanson-Painton, A. D. Vaz, R. M. Palokedety, and M. J. Griffin (1985), Biochem. Pharmacol. 34, 1795—1800.

R. G. Knowles and B. Burchell (1977a), Biochem. Soc. Trans. 5, 731—732.

R. G. Knowles and B. Burchell (1977b), Biochem. J. 163, 381—383.

K. K. Kohli, H. Mukhtar, J. R. Bend, P. W. Albro, and J. D. McKinney (1979), Biochem. Pharmacol. 28, 1444—1446.

P. E. Kolattukudy and L. Brown (1975), Arch. Biochem. Biophys. 166, 599—607.

D. J. Kroll, M. E. Graichen, and T. B. Leonard (1988), Carcinogenesis 9, 193—198.

W. D. Kuhlmann, R. Krischan, W. Kunz, T. M. Guenthner, and F. Oesch (1981), Biochem. Biophys. Res. Commun. 98, 417—423.

W. M. Lafranconi, H. Glatt, and F. Oesch (1986), Toxicol. Appl. Pharmacol. 84, 500—511.

K. C. Leibmann and E. Ortiz (1968), Fed. Proc. 27, 302—310.

W. Levin, A. W. Wood, P. G. Wislocki, R. L. Chang, J. Kapitulnik, H. D. Mah, H. Yagi, D. M. Jerina, and A. H. Conney (1978), in: Polycyclic hydrocarbons and cancer, (H. V. Gelboin and P. O. P. Ts, o, eds.), Academic Press, New York, p. 189.

W. Levin, A. Y. H. Lu, P. E. Thomas, D. Ryan, D. E. Kizer, and M. J. Griffin (1978a), Proc. Natl. Acad. Sci., U.S.A. 75, 3240—3243.

W. Levin, P. E. Thomas, D. Korzeniowski, H. Seifried, D. M. Jerina, and A. Y. H. Lu (1978b), Mol. Pharmacol. 14, 1107—1120.

W. Levin, D. P. Michaud, P. E. Thomas, and D. M. Jerina (1983), Arch. Biochem. Biophys. 220, 485—494.

D. Lichtenberg, R. J. Robson, and E. A. Dennis (1983), Biochim. Biophys. Acta 737, 285—304.

J.-C. Lin, Y. Hiasa, and E. Farber (1977), Cancer Res. 37, 1972—1981.

D. N. Loury, D. E. Moody, B. W. Kim, and B. D. Hammock (1985), Biochem. Pharmacol. 34, 1827—1833.

A. Y. H. Lu, D. Ryan, D. M. Jerina, J. W. Daly, and W. Levin (1975), J. Biol. Chem. 250, 8283—8288.

A. Y. H. Lu, D. M. Jerina, and W. Levin (1977), J. Biol. Chem. 252, 3715—3723.

A. Y. H. Lu, P. E. Thomas, D. Ryan, D. M. Jerina, and W. L. vin (1979), J. Biol. Chem. 254, 5878—5881.

A. Y. H. Lu and G. T. Miwa (1980), Ann. Rev. Pharmacol. Toxicol. **20**, 513—531.

B. Lundgren, J. Meijer, and J. W. De Pierre (1987), Drug Metab. Dispos. **15**, 114—121.

S. D. Lyman, A. Poland, and B. A. Taylor (1980), J. Biol. Chem. **255**, 8650—8654.

J. Magdalou and B. D. Hammock (1988), Biochem. Pharmacol. **37**, 2717—2722.

B. Mannervik (1985), Adv. Enzymol. Rel. Areas Mol. Biol. **57**, 357—417.

M. McGee and F. Fitzpatrick ( 985), J. Biol. Chem. **260**, 12832—12837.

J. Meijer and J. W. De Pierre (1985a), Eur. J. Biochem. **148**, 421—430.

J. Meijer and J. W. De Pierre (1985b), Eur. J. Biochem. **150**, 7—16.

J. Meijer and J. W. De Pierre (1987), Chem.-Biol. Interactions **62**, 249—269.

J. Meijer, G. Ludqvist, and J. W. De Pierre (1987), Eur. J. Biochem. **167**, 269—279.

J. Meijer, J. W. De Pierre, and H. Jörnvall (1987b), Bioscience Reports **7**, 891—89ℓ.

I. Meijer and J. W. De Pierre (1988), Chem.-Biol. Interactions **64**, 207—249.

J. Meijer and J. W. De Pierre (1988a), Arch. Toxicol. Suppl. **12**, 283—287.

I. Mertes, R. Fleischmann, H. R. Glatt, and F. Oesch (1985), Carcinogenesis **6**, 219—223.

I. Miki, T. Shimizu, Y. Seyama, S. Kitamura, K. Yamaguchi, H. Sano, H. Ueno, A. Hiratsuka, and T. Watabe (1989) J. Biol. Chem. **264**, 5799—5805.

U. Milbert, W. Wörner, and F. Oesch (1986), In: Primary Changes and Control Factors in Carcinogenesis (T. Friedberg and F. Oesch, eds.), pp. 14—21, Deutscher Fachschriften-Verlag, Wiesbaden, FRG.

T. Miyamoto, M. Silva, and B. D. Hammock (1987), Arch. Biochem. Biophys. **254**, 203—213.

D. E. Moody, M. H. Silva, and B. D. Hammock (1986), Biochem. Pharmacol. **35**, 2073—2080.

D. E. Moody and B. D. Hammock (1987), Toxicol. Appl. Pharmacol. **89**, 37—48.

D. E. Moody and B. D. Hammock (1987a), Arch. Biochem. Biophys. **258**, 156—166.

D. E. Moody, D. N. Loury, and B. D. Hammock (1987), Biochem. Pharmacol. **36**, 570—572.

D. E. Moody, G. A. Clawson, D. A. Geller, L. A. Taylor, J. Button, D. N. Loury, B. D. Hammock, and E. A. Smuckler (1988) Biochem. Pharmocol. 37, 1331—1341.

H. Mukhtar, I. P. Lee, G. L. Foureman, and J. R. Bend (1978a), Chem.-Biol. Interactions **22**, 153—165.

H. Mukhtar, R. M. Philpott, and J. R. Bend (1978b), Drug Metab. Disp. **6**, 577—581.

H. Mukhtar, T. H. Elmambouk, and J. R. Bend (1980), Arch. Biochem. Biophys. **192**, 10—21.

H. Mukhtar and D. R. Bickers (1981), Drug Metab. Disp. **9**, 311—314.

H. Mukhtar and D. R. Bickers (1983), Drug Metab. Disp. **11**, 562—567.

C. A. Mullin and B. D. Hammock (1980), Anal. Biochem. **106**, 476—485.

C. A. Mullin and B. D. Hammock (1982), Arch. Biochem. Biophys. **216**, 423—439.

C. A. Mullin and B. A. Croft (1984), Experientia **40**, 176—178.

S. M. Mumby and B. D. Hammock (1979), Pestic. Biochem. Physiol. **11**, 275—284.

S. M. Mumby and B. D. Hammock (1979a), Anal. Biochem. **92**, 16—21.

M. Mushtag, H. B. Weems, and S. K. Yang (1986), Arch. Biochem. Biophys. **246**, 478—487.

N. T. Nashed, D. P. Michaud, W. Levin, and D. M. Jerina (1985), Arch. Biochem. Biophys. **241**, 149—162.

D. W. Nebert, W. F. Benedict, J. E. Gielen, F. Oesch, and J. W. Daly (1972), Mol. Pharmacol. 8, 374—379.

D. W. Nebert, M. Adesnik, M. J. Coon, R. W. Estabrook, F. J. Gonzalez, F. P. Guengerich, I. C. Gunsalus, E. F. Johnson, B. Kemper, W. Levin, I. R. Phillips, R. Sato, and M. R. Waterman (1987), DNA 6, 1—11.

A. B. NOVIKOFF, P. M. NOVIKOFF, R. J. STOCKERT, F. F. BECKER, A. YAM, M. S. PORU-
CHYNSKY, W. LEVIN, and P. E. THOMAS (1979), Cell Biol. **76**, 5207—5211.
F. OESCH and J. W. DALY (1971a), Biochim. Biophys. Acta **227**, 692—697.
F. OESCH, D. M. JERINA, and J. W. DALY (1971b), Arch. Biochem. Biophys. **144**,
253—261.
F. OESCH, N. KAUBISCH, D. M. JERINA, and J. W. DALY (1971c), Biochemistry **10**,
4858—4866.
F. OESCH (1973), Xenobiotica **3**, 305—340.
F. OESCH, D. M. JERINA, J. W. DALY, and J. M. RICE (1973), Chem.-Biol. Interactions
**6**, 189—202.
F. OESCH (1974), Biochem. J. **139**, 77—88.
F. OESCH and P. BENTLEY (1976), Nature **259**, 53—55.
F. OESCH, H. R. GLATT, and H. SCHMASSMANN (1977), Biochem. Pharmacol. **26**, 603 to
607.
F. OESCH (1979), In: Progress in Drug Metabolism (J. W. BRIDGES and L. F. CHAS-
SEAUD, eds.), Vol. 3, pp. 253—301, John Wiley, Chichester, England.
F. OESCH and H. U. SCHMASSMANN (1979), Biochem. Pharmacol. **28**, 171—176.
F. OESCH and M. GOLAN (1980), Cancer Lett. **9**, 169—175.
F. OESCH, P. BENTLEY, K. L. PLATT, and M. D. GOLAN (1980a), Arch. Biochem. Bio-
phys. **199**, 538—544.
F. OESCH, H. SCHMASSMANN, E. OHNHAUS, V. ALTHAUS, and J. LORENZ (1980b),
Carcinogenesis **1**, 827—835.
F. OESCH, D. BEERMAN, A. J. SPARROW, P. BENTLEY, and U. VOGEL-BINDEL (1981),
Anal. Biochem. **117**, 223—230.
F. OESCH (1982), In: Chemical Carcinogenesis (C. NICOLINI, ed.), p. 1—22, Plenum
Publishing Corporation, London.
F. OESCH, U. VOGEL-BINDEL, T. M. GUENTHNER, R. CAMERON, and E. FARBER (1983a),
Cancer Res. **43**, 313—319.
F. OESCH, A. ZIMMER, and H. R. GLATT (1983b), Biochem. Pharmacol. **32**, 1783—1788.
F. OESCH, C. W. TIMMS, C. H. W. WALKER, T. M. GUENTHNER, A. SPARROW, T. WA-
TABE, and C. R. WOLF (1984), Carcinogenesis **5**, 7—9.
F. OESCH, L. SCHLADT, P. STEINBERG, and H. THOMAS (1988), Arch. Toxicol. Suppl.
**12**, 248—255.
Y. OKADA, A. B. FREY, T. M. GUENTHNER, F. OESCH, D. D. SABATINI, and G. KREI-
BICH (1982), Eur. J. Biochem. **122**, 393—402.
K. OKITA and E. FARBER (1975), Gann **17**, 282—299.
E. H. OLIW, F. P. GUENGERICH, and J. A. OATES (1982), J. Biol. Chem. **257**, 3771—3781.
C. R. PACE-ASCIAK, J. KLEIN, and S. P. SPEILBERG (1986), Biochim. Biophys. Acta
**875**, 406—409.
G. M. PACIFICI and A. RANE (1982), Drug Metab. Disp. **10**, 302—305.
G. M. PACIFICI, C. COLIZZI, L. GIULIANI, and A. RANE (1983), Arch. Toxicol. **54**,
313—334.
G. M. PACIFICI, B. LINDBERG, H. GLAUMANN, and A. RANE (1983a), J. Pharmacol. Exp.
Therap. **226**, 869—875.
G. M. PACIFICI, M. FRANCHI, C. BENCINI, and A. RANE (1986), Br. J. Clin. Pharmac.
**22**, 269—274.
A. PARKINSON, S. H. SAFE, L. W. ROBERTSON, P. E. THOMAS, D. E. RYAN, L. M. REIK,
and W. LEVIN (1983), J. Biol. Chem. **258**, 5967—5976.
M. G. PARKKI and A. AITIO (1978), Arch. Toxicol. Suppl. **1**, 261—265.
M. G. PARKKI (1980), Xenobiotica **10**, 307—310.
M. G. PARKKI (1982), Ph. D. thesis, Department of Physiology and Biochemistry,
University of Turku.

---

B. N. Patel, M. I. Mackness, V. Nwosu, and M. J. Connock (1986), Biochem. Pharmacol. **35**, 231—235.

D. I. Peng, G. M. Pacifici, and A. Rane (1984), Biochem. Pharmacol. **33**, 71—77.

H. Peter, R. Jung, H. M. Bolt, and F. Oesch (1981), J. Steroid Biochem. **14**, 83—90.

C. B. Pickett, R. L. Jeter, J. Morin, and A. Y. H. Lu (1981), J. Biol. Chem. **256**, 8815—8820.

T. D. Porter, T. W. Beck, and C. B. Kasper (1986), Arch. Biochem. Biophys. **248**, 121—129.

P. Pott (1963), Reprinted, Natl. Cancer Inst. Monogr. **10**, 7.

G. D. Prestwich and B. D. Hammock (1985), Proc. Natl. Acad. Sci. USA **82**, 1663 to 1667.

G. D. Prestwich, J.-W. Kuo, S. K. Park, D. N. Lowry, and B. D. Hammock (1985), Arch. Biochem. Biophys. **242**, 11—15.

G. D. Prestwich, I. Lucarelli, S.-K. Park, D. N. Loury, D. E. Moody, and B. D. Hammock (1985a), Arch. Biochem. Biophys. **237**, 361—372.

O. Rådmark, T. Shimizu, H. Jörnvall, and B. Samuelsson (1984), J. Biol. Chem. **259**, 12339—12345.

D. Ratanasavanh, P. Beaune, G. Baffet, M. Rissel, P. Kremers, F. P. Guengerich, and A. Guillouzo (1986), J. Histochem. Cytochem. **34**, 527—533.

C. Razzouk, M. E. McManus, S. Hayashi, D. Schwartz, and S. S. Thorgeirsson (1985), Biochem. Pharmacol. **34**, 1537—1542.

D. K. Robbins, P. J. Wedlund, S. Elsberg, F. Oesch, and H. Thomas (1990), Biochem. Pharmacol., in press.

J. A. Redick, T. T. Kawabata, F. P. Guengerich, P. A. Krieter, T. K. Shires and J. Baron (1980), Lite. Sci. 27, 2465—2470.

M. Romano, G. Gazotti, V. Clos, B. M. Assael, R. M. Facino, and M. Salmona (1983), Chem.-Biol. Interactions 47, 213—222.

K. Sato, A. Kitahara, K. Satoh, T. Ishikawa, M. Tatematsu, and N. Ito (1984), Gann **75**, 199—202.

J. M. Sayer, H. Yagi, P. J. Van Bladeren, W. Levin, and D. M. Jerina (1985), J. Biol. Chem. **260**, 1630—1640.

L. Schladt, W. Wörner, F. Setiabudi, and F. Oesch (1986), Biochem. Pharmacol. **35**, 3309—3316.

L. Schladt, R. Hartmann, C. Timms, M. Strolin-Benedetti, P. Dostert, W. Wörner, and F. Oesch (1987), Biochem. Pharmacol. **36**, 345—351.

L. Schladt, H. Thomas, R. Hartmann, and F. Oesch (1988a), Eur. J. Biochem., 176, 715—723.

L. Schladt, R. Hartmann, W. Wörner, H. Thomas, and F. Oesch (1988b), Eur. J. Biochem., 176, 31—37.

H. U. Schmassmann, H. R. Glatt, and F. Oesch (1976), Anal. Biochem. **74**, 94—104.

H. Schmassmann, A. Sparrow, K. Platt, and F. Oesch (1978), Biochem. Pharmacol. **27**, 2237—2245.

B. Schoket and I. Vincze (1986), Acta Pharmacol. Toxicol. **58**, 156—158.

J. Seidegård, J. W. De Pierre, M. S. Moron, K. A. M. Johannesen, and L. Ernster (1977), Cancer Res. **3**, 1075—1082.

J. Seidegård and J. W. De Pierre (1980), Eur. J. Biochem. **112**, 643—648.

J. Seidegård, J. W. De Pierre, R. Morgenstern, Å. Pilotti, and L. Ernster (1981), Biochim. Biophys. Acta **672**, 65—78.

J. Seidegård, J. W. De Pierre, T. M. Guenthner, and F. Oesch (1982), Acta Chem. Scand. **B36**, 555—557.

J. Seidegård and J. W. De Pierre (1983), Biochim. Biophys. Acta **695**, 251—270.

J. Seidegård, J. W. De Pierre, and R. W. Pero (1984), Cancer Res. **44**, 3654—3660.

J. Seidegård, J. W. De Pierre, T. M. Guenthner, and F. Oesch (1986), Eur. J. Biochem. **159**, 415—423.

A. Sevanian, J. F. Mead, and R. A. Stein (1979), Lipids **14**, 634—643.

A. Sevanian, R. A. Stein, and J. F. Mead (1980), Biochim. Biophys. Acta **614**, 489 to 500.

A. Sevanian, R. A. Stein, and J. F. Mead (1981), Lipids 16, 781—798.

A. Sevanian and A. R. Peterson (1984), Proc. Natl. Acad. Sci. USA **81**, 4198—4202.

A. Sevanian and L. McLeod (1985), Toxicologist **5**, 80.

A. Sevanian and L. L. McLeod (1986), J. Biol. Chem. **261**, 54—59.

M. R. Share and R. M. Roe (1988), Anal. Biochem. **169**, 81—88.

L. Siekmann, P. Thull, and H. Breuer (1980), Acta Endocr. **95**, 49—57.

M. H. Silva and B. D. Hammock (1987), Comp. Biochem. Physiol. **87B**, 95—102.

P. Sims (1966), Biochem. J. **98**, 215—228.

J. E. Sinsheimer, E. van den Eeckhout, B. H. Hooberman, and V. G. Geylin (1987), Chem.-Biol. Interactions **63**, 75—90.

R. C. Skoda, A. Demierre, O. W. McBride, F. J. Gonzales, and U. A. Meyer (1988), J. Biol. Chem. **263**, 1549—1554.

F. Snyder (1964), Anal. Biochem. **9**, 183—196.

R. Snyder, D. V. Parke, J. J. Koscis, D. J. Jollow, C. G. Gibson, and C. M. Witmer (1982), Advances in Experimental Medicine and Biology — Biological Reactive Intermediates-II, Chemical Mechanisms and Biological Effects, Plenum Press, New York.

P. Stasiecki, F. Waechter, P. Bentley, and F. Oesch (1979), Biochim. Biophys. Acta **568**, 446—453.

P. Stasiecki, F. Oesch, G. Bruder, E.-D. Jarasch, and W. W. Franke (1980), Eur. J. Cell Biol. **21**, 79—82.

B. Steinberg W. M. Lafranconi, and F. Oesch (1987), Arch. Toxicol. Suppl. **10**, 148—156.

P. Steinberg, L. Schladt, H. P. Dienes, C. Timms, and F. Oesch (1988), Eur. J. Biochem. **176**, 39—45.

T. M. Stoming and E. Bresnick (1974), Cancer Res. **34**, 2810—2813.

M. Sugimura, Y. Yamazoe, T. Kamataki, and R. Kato (1980), Cancer Res. **40**, 2910—2914.

K. Suzuki and I. P. Lee (1981), Toxicol. Appl. Pharmacol. **58**, 151—155.

C. A. Telakowski-Hopkins, A. Y. H. Lu, and C. B. Pickett (1983), Arch. Biochem. Biophys. **221**, 79—88.

D. R. Thakker, H. Yagi, H. Akagi, M. Koreeda, A. Y. H. Lu, W. Levin, A. W. Wood, A. H. Conney, and D. M. Jerina (1977a), Chem.-Biol. Interactions **16**, 281—300.

D. R. Thakker, H. Yagi, W. Levin, A. Y. H. Lu, A. H. Conney, and D. M. Jerina (1977b), J. Biol. Chem. **252**, 6328—6334.

H. Thomas, L. Schladt, R. Hartmann, and F. Oesch (1988) In: Abstracts of the IInd International ISSX Meeting — "Xenobiotic Metabolism and Disposition", Kobe, Japan, p. 73.

P. E. Thomas, D. Korzeniowski, E. Bresnick, W. A. Bornstein, C. B. Kasper, W. E. Fahl, C. R. Jefcoate, and W. Levin (1979), Arch. Biochem. Biophys. **192**, 22—26.

P. E. Thomas, L. M. Reik, D. E. Ryan, and W. Levin (1981), J. Biol. Chem. **256**, 1044—1052.

P. E. Thomas, D. E. Ryan, C. von Bahr, H. Glaumann, and W. Levin (1982), Mol. Pharmacol. **22**, 190—195.

C. W. Timms, L. Schladt, H. Schramm, U. Milbert, M. Arand, F. Setiabudi, M. Mackness, H. Thomas, and F. Oesch (1988), Arch. Biochem. Biophys., submitted.

---

P. J. van Bladeren, J. M. Sayer, D. E. Ryan, P. E. Thomas, W. Levin, and D. M. Jerina (1985), J. Biol. Chem. **260**, 10226–10235.

E. van den Eeckhout, P. de Moerloose, and J. E. Sinsheimer (1985), J. Chromatogr. **318**, 343–349.

A. D. Vaz, V. M. Fiorica, and M. J. Griffin (1981), Biochem. Pharmacol. **30**, 651–656.

U. Vogel-Bindel, P. Bentley, and F. Oesch (1982), Eur. J. Biochem. **126**, 425–431.

F. Waechter, P. Bentley, M. Germann, F. Oesch, and W. Stäubli (1982a), Biochem. J. **202**, 677–686.

F. Waechter, M. Merdes, F. Bieri, W. Stäubli, and P. Bentley (1982b), Eur. J. Biochem. **125**, 457–461.

F. Waechter, P. Bentley, F. Bieri, W. Stäubli, K. Völkl, and H. D. Fahimi (1983). FEBS Lett. **158**, 225–228.

F. Waechter, F. Bieri, W. Stäubli, and P. Bentley (1984), Biochem. Pharmacol. **33**, 31–34.

F.Waechter, P. Beutley, F. Bieri, S. Muakkassah-Kelly, W. Stäubli, and M. Villermain (1988) Biochem. Pharmacol. **37**, 3897–3903.

C. H. Walker, P. Bentley, and F. Oesch (1978), Biochim. Biophys. Acta **539**, 427 to 434.

C. H. Walker, C. W. Timms, C. R. Wolf, and F. Oesch (1986), Biochem. Pharmacol. **35**, 499–503.

C. H. Walker, I. Newton, S. D. Hallam, and M. J. J. Ronis (1987), Comp. Biochem. Physiol. **86C**, 379–382.

F. G. Walz, G. P. Vlasuk, and A. W. Steggles (1983), Biochemistry **22**, 1547–1556.

P. Wang, J. Meijer, and F. P. Guengerich (1982), Biochem. **21**, 5769–5776.

T. Watabe and E. W. Maynert (1968), Pharmacologist **10**, 203.

T. Watabe and S. Kanehira (1970), Chem. Pharm. Bull. (Tokyo) **18**, 1295–1296.

T. Watabe, K. Akamatsu, and K. Kiyonaga (1971a), Biochem. Biophys. Res. Commun. **44**, 199–204.

T. Watabe, K. Kiyonaga, and S. Hara (1971b), Biochem. Pharmacol. **20**, 1700–1702.

T. Watabe and K. Akamatsu (1972), Biochim. Biophys. Acta **279**, 297–305.

T. Watabe and S. Suzuki (1972), Biochem. Biophys. Res. Commun. **46**, 1120–1127.

T. Watabe and K. Akamatsu (1974), Biochem. Pharmacol. **23**, 1845–1851.

T. Watabe, S. Ichihara, and T. Sawahata (1979), J. Biol. Chem. **254**, 10720–10727.

T. Watabe and T. Sawahata (1979), J. Biol. Chem. **254**, 3854–3860.

T. Watabe, M. Isobe, and M. Kanai (1980), J. Pharm. Dyn. **3**, 553–560.

T. Watabe, M. Kanai, M. Isobe, and N. Ozawa (1981), J. Biol. Chem. **256**, 2900–2907.

T. Watabe, T. Komatsu, M. Isobe, and A. Tsubaki (1983), Chem.-Biol. Interactions **44**, 143–154.

T. Watabe, A. Tsubaki, M. Isobe, N. Ozawa, and A. Hiratsuka (1984), Biochim. Biophys. Acta **795**, 60–66.

T. Watabe, K. Kobayashi, Y. Saitoh, T. Komatsu, N. Ozawa, A. Tsubaki, K. Endoh, and A. Hiratsuka (1986a), J. Biol. Chem. **261**, 3200–3207.

T. Watabe, N. Ozawa, H. Ishii, K. Chiba, and A. Hiratsuka (1986b), Biochem. Biophys. Res. Commun. **140**, 632–637.

W. B. Westkaemper and R. P. Hanzlik (1981), Arch. Biochem. Biophys. **208**, 195 to 204.

J. N. H. White (1980), Chem.-Biol. Interactions **29**, 103–115.

R. N. Wixtrom and B. D. Hammock (1985), in: Biochemical Pharmacology and Toxicology (D. Zakim and D. A. Vessey, eds.), Vol. 1, pp. 1–94, Methodological aspects of drug-metabolising enzymes, John Wiley and Sons, New York.

R. N. Wixtrom, M. H. Silva, and B. D. Hammock (1988), Anal. Biochem. **169**, 71–80.

C. R. Wolf, F. Oesch, C. W. Timms, T. M. Guenthner, R. Hartmann, M. Maruhn, and R. Burger (1983), FEBS Lett. **157**, 271—276.

C. R. Wolf, E. Moll, T. Friedberg, F. Oesch, A. Buchmann, W. D. Kuhlmann, and H. W. Kunz (1984), Carcinogenesis **5**, 993—1001.

S. K. Yang, D. W. McCourt, J. C. Leutz, and H. V. Gelboin (1977a), Science **196**, 1199—1201.

S. K. Yang, P. R. Roller, and H. V. Gelboin (1977b), Biochemistry **16**, 3680—3687.

S. K. Yang, M. Mushtag, and H. B. Weems (1987a), Arch. Biochem. Biophys. **255**, 48—63.

S. K. Yang, M. Mushtag, H. B. Weems, D. W. Miller, and P. P. Fu (1987b), Biochem. J. **245**, 191—204.

S. K. Yang (1988), Biochem. Pharmacol. **37**, 61—70.

A. Zhiri, J. Muller, S. Fournel, J. Magdalou, M. Willman-Bednawska, and G. Siest (1987), Anal. Biochem. **163**, 298—302.

# Supplement

The unified nomenclature of the P450 supergene family as presented in the present volume by NEBERT and GONZALEZ (Chapter 2) requires to assign the cytochrome P-450 isozymes to this nomenclature which have been used by GUENGERICH and other authors in volume 1 of this series.

The editors are grateful to Dr. GUENGERICH for solving that difficult task by attributing the diverse trivial names of cytochrome P-450 isozymes to the unified nomenclature which is listed in the following supplement.

## Classification of Cytochromes P-450 into Gene Families

F. P. GUENGERICH

| Cytochrome P-450 (trivial name)[a] | Systematic protein name | Gene locus symbol |
|---|---|---|
| **rat liver P-450s** (Table 1)[b] | | |
| 1. $P\text{-}450_{UT-A}$ | IIC11 | *CYP2C11* |
| 2. $P\text{-}450_{\beta NF-B}$ | IA1 | *CYP1A1* |
| 3. $P\text{-}450_{PB-B}$ | IIB1 | *CYP2B1* |
| 4. $P\text{-}450_{PB-C}$ | IIC6 | *CYP2C6* |
| 5. $P\text{-}450_{PB-D}$ | IIB2 | *CYP2B2* |
| 6. $P\text{-}450_{PCN-E}$ | IIIA1 | *CYP3A1* |
| 7. $P\text{-}450_{PCN-2}$ | IIIA2 | *CYP3A2* |
| 8. $P\text{-}450_{UT-F}$ | IIA1 | *CYP2A1* |
| 9. $P\text{-}450_{ISF-G}$ | IA2 | *CYP1A2* |
| 10. $P\text{-}450_{UT-H}$ | IID1 | *CYP2D1* |
| 11. $P\text{-}450_{UT-I}$ | IIC12 | *CYP2C12* |
| 12. P-450j | IIE1 | *CYP2E1* |
| 13. P-452 | IVA1 | *CYP4A1* |
| | IVA2 | *CYP4A2* |
| | IVA3 | *CYP4A3* |
| 14. P-450f | IIC7 | *CYP2C7* |
| 15. P-450g | IIC13 | *CYP2C13* |
| 16. $P\text{-}450_{RLM2}$ | ? | ? |
| 17. $P\text{-}450_{RLM5a}$ | ? | ? |
| 18. P-450 PB-6 | ? | ? |
| 19. Taurodeoxycholate 7$\alpha$-hydroxylase | ? | ? |
| 20. P-450, P-451 | ? | ? |
| 21. P-450 7 | ? | ? |
| 22. $P\text{-}450_{14\alpha-DM}$ | ? | ? |
| **rabbit liver P-450s** (Table 2) | | |
| 1. P-450 1 | IIC5 | *CYP2C5* |
| 2, 3. P-450 2 | IIB4 | *CYP2B4* |
| 4. P-450 3a | IIE1 | *CYP2E1* |

Continued

| Cytochrome P-450 (trivial name)[a] | Systematic protein name | Gene locus symbol |
|---|---|---|

| | | | |
|---|---|---|---|
| 5, 6. | P-450 3b | IIC3 | *CYP2C3* |
| 7. | P-450 3c | IIIA6 | *CYP3A6* |
| 8, 9. | P-450 4 | IA2 | *CYP1A2* |
| 10. | P-450 5 | IVB1 | *CYP4B1* |
| 11. | P-450 6 | IA1 | *CYP1A1* |
| 12. | P-450 7 | ? | ? |
| 13. | P-450 CN | ? | ? |
| 14. | P-450$_2$ | ? | ? |
| 15. | P-450$_4$, P-450$_6$, P-450$_7$, P-450$_8$ | ? | ? |

**mouse liver P-450s** (Table 3)

| | | | |
|---|---|---|---|
| 1. | P$_1$-450 | IA1 | *CYP1A1* |
| 2. | P$_3$-450 | IA2 | *CYP1A2* |
| 3. | P$_2$-450 | IA2 | *CYP1A2* |
| 4. | P-450A$_2$ | ? | ? |
| 5. | P-450C$_2$ | ? | ? |
| 6. | P-450$_{15\alpha}$ | IIA3 | *CYP2A3* |
| 7. | P-450$_{16\alpha}$ | IID9 | *CYP2D9* |

**extrahepatic microsomal P-450s** (Table 4)

| | | | |
|---|---|---|---|
| 1. | P-450 2 | IIB4 | *CYP2B4* |
| 2. | P-450 5 | IVB1 | *CYP4B1* |
| 3. | P-450 6 | IA1 | *CYP1A1* |
| 4. | P-450$_{PG-\omega}$ | IVA1 | *CYP4A1* |
| | | IVA2 | *CYP4A2* |
| | | IVA3 | *CYP4A3* |
| 5. | P-450 3a | ? | ? |
| 6. | P-450$_{17\alpha}$ | XVIIA1 | *CYP17* |
| 7. | P-450$_{C-21}$ | XXIA1 | *CYP21A1* |
| 8. | Aromatase | XIXA1 | *CYP19* |
| 9. | P$_1$-450 | IA1 | *CYP1A1* |
| 10. | P-450$_{PB-B}$ | IIB1 | *CYP2B1* |
| 11. | P-450b | IVA1 | *CYP4A1* |
| | | IVA2 | *CYP4A2* |
| | | IVA3 | *CYP4A3* |
| 12. | (P-450) | ? | ? |
| 13, 14. | P-450a,b | ? | ? |
| | | ? | ? |
| 15, 16. | P-450ia,ca | IVA1 | *CYPA1* |
| | | IVA2 | *CYPA2* |
| | | IVA3 | *CYPA3* |
| 17. | P-450cb | ? | ? |
| 18. | P-448c | ? | ? |

Continued

| Cytochrome P-450 (trivial name)[a] | Systematic protein name | Gene locus symbol | |
|---|---|---|---|
| **Human P-450s** | | | chromosomal location |
| 1. | P-450$_{DB}$ | IID6 | CYP2D6 | 22q11.2-qter |
| 2. | P-450$_{PA}$ | IA2 | CYP1A2 | 15q22-qter |
| 3. | P$_1$-450 | IA1 | CYP1A1 | 15q22-qter |
| 4. | P-450$_{MP-1}$[c] | IIC? | CYP2C? | 10q24.1-24.3 |
| 5. | P-450$_{MP-2}$[c] | IIC? | CYP2C? | 10q24.1-24.3 |
| 6. | P-450$_{BufII}$ | ? | ? | ? |
| 7. | P-450$_{NF}$ | IIIA4 | CYP3A4 | 7q21.3-q22 |
| 8. | P-450 HLp | IIIA3 | CYP3A3 | 7q21.3-q22 |
| 9. | P-450 9 | ? | ? | ? |
| 10. | P-450HFLa | IIIA5 | CYP3A5 | 7q21.3-q22 |
| 11. | P-450j | IIE1 | CYP2E1 | 10 |
| 12. | P-450pHB$_1$ | IIA3 | CYP2A3 | 19q13.1-13.2 |
| 13. | P-450$_{c-21}$ | XXIA1 | CYP21A1 | 6p |

[a] In some cases definitive assignments to sequences cannot be made, and either question marks are left or a group of possibilities are presented.
Since the original article was prepared (January, 1987), other proteins have been isolated (or nucleic acid sequences determined), and the reader is referred to NEBERT et al. (1989) for further consideration. See also GONZALEZ (1989).

[b] The table numbers are related to the article of F. P. GUENGERICH (1989) in Vol. 1, 101—150 of the series Frontiers in Biotransformation.

[c] The specific sequences related to the proteins termed P-450$_{MP-1}$ and P-450$_{MP-2}$, the mephenytoin 4'-hydroxylases, are unknown. P-450$_{MP-3}$, a closely-related protein (GED et al., 1988), corresponds to IIC8 (CYP2C8). Also closely related are the proteins IIC9 (MP-4) and IIC10 (MP-8) (GED et al., 1988). The IIC10 protein is P-450$_{TB}$, shown by inhibition, amino acid sequencing, and vector expression experiments to be the tolbutamide hydroxylase (BRIAN et al., 1989).

# References

BRIAN, W. R., P. K. SRIVASTAVA, D. R. UMBENHAUER, R. S. LLOYD, and F.P. GUEN-GERICH, (1989), Biochemistry, **28**, 4993—4999.

DISTLERATH, L. M., and F. P. GUENGERICH, (1987), in: Mammalian Cytochromes P-450, Vol. 1, (F. P. GUENGERICH, ed.), CRC Press, Boca Raton, Florida, 133—198.

GED, C., D. R. UMBENHAUER, T. M. BELLEW, R. W. BORK, P. K. SRIVASTAVA, N. SHIN-RIKI, R. S. LLOYD, and F. P. GUENGERICH, (1988), Biochemistry **27**, 6929—6940.

GONZALEZ, F. J., (1989), Pharmacol. Reviews **40**, 243—288.

NEBERT, D. W., D. R. NELSON, M. ADESNIK, M. J. COON, R. W. ESTABROOK, F. J. GONZALEZ, F. P. GUENGERICH, I. C. GUNSALUS, E. F. JOHNSON, B. KEMPER, W. LE-VIN, I. R. PHILLIPS, R. SATO, and M. R. WATERMAN, (1989), DNA 8, 1—13.

# List of Authors

Dr. H.-H. BORCHERT
Humboldt-University
Section of Chemistry
Division of Pharmacy
Goethestr. 54

**1120 Berlin**
GDR


Prof. Dr. E. BRESNICK
Department of Pharmacology and
Toxicology
Dartmouth Medical School

**Hanover New Hampshire 03756**
USA


Dr. F. GONZALES
NICHHD, Department of Health
and Human Service
Laboratory of Developmental
Pharmacology
Room 6 C-101 Building 10

**Bethesda, Maryland 20205**
USA


Dr. W. H. HOUSER
Departments of Biochemistry
and Pharmacology and the
Epploy Institute for Research
in Cancer and Allied Diseases
University of Nebraska
Medical Center

**Omaha, Nebraska 68105**
USA


Prof. Dr. R. KATO
Department of Pharmacology
School of Medicine
Keio University
35 Shinanomachi, Shinjuku-ku

**Tokyo 160**
Japan

Prof. Dr. B. KETTERER
Courtauld Institute of
Biochemistry
Middlesex Hospital
Medical School

**London W1P 7 PN**
England


Prof. Dr. W. KLINGER
Friedrich-Schiller-University
Institute of Pharmacology
and Toxicology
Löbderstraße 1

**6900 Jena**
GDR


Dr. A. LANGNER
Humboldt-University
Section of Chemistry
Division of Pharmacy
Goethestr. 54

**1120 Berlin**
GDR


Prof. Dr. P. I. MACKENZIE
Department of Clinical
Pharmacology
Flinders Medical Centre
Bedford Park

**South Australia 5042**


Prof. Dr. D. W. NEBERT
Laboratory of Molecular Toxicology
Institute of Environmental Health
University of Cincinnati
Medical Center

**Cincinnati, Ohio 45267-0056**
USA


Prof. Dr. F. OESCH
University Mainz
Institute of Toxicology
Obere Zahlbacher Straße 67

**D-6500 Mainz**
FRG

Prof. Dr. D. V. Parke
Department of Biochemistry
University of Surrey

**Guildford Surrey GU2 5 XH**
England


Prof. Dr. S. Pfeifer
Humboldt-University
Section of Chemistry
Division of Pharmacy
Goethestraße 54

**1120 Berlin**
GDR


Dr. J. B. Taylor
Courtauld Institute of
Biochemistry
Middlesex Hospital
Medical School

**London W1P 7 PN**
England


Dr. H. Thomas
University Mainz
Institute of Toxicology
Obere Zahlbacher Straße 67

**D-6500 Mainz**
FRG


Dr. C. W. Timms
Procter & Gamble,
European Technical Center
Temselaan 100

**B-1820 Strombeek-Bever**
Belgium


Prof. Dr. Y. Yamazoe
Department of Pharmacology
School of Medicine
Keio University
35 Shinanomachi, Shinjuku-ku

**Tokyo 160**
Japan

# Subject Index